Lecture Notes in Computer Science 13963

FoLLI Publications on Logic, Language and Information

Subline of Lectures Notes in Computer Science

More information about this series at https://link.springer.com/bookseries/558

Mohua Banerjee · A. V. Sreejith
Editors

Logic and Its Applications

10th Indian Conference, ICLA 2023
Indore, India, March 3–5, 2023
Proceedings

Springer

Editors
Mohua Banerjee
Indian Institute of Technology Kanpur
Kanpur, India

A. V. Sreejith
Indian Institute of Technology Goa
Ponda, India

ISSN 0302-9743 ISSN 1611-3349 (electronic)
Lecture Notes in Computer Science
ISBN 978-3-031-26688-1 ISBN 978-3-031-26689-8 (eBook)
https://doi.org/10.1007/978-3-031-26689-8

This Springer imprint is published by the registered company Springer Nature Switzerland AG
The registered company address is: Gewerbestrasse 11, 6330 Cham, Switzerland

Preface

The Indian Conference on Logic and Its Applications (ICLA) is a biennial conference organized under the aegis of the Association for Logic in India. The tenth edition of the conference was held during March 3–5, 2023, at the Indian Institute of Technology (IIT) Indore. This volume contains papers presented at the 10th ICLA.

A variety of themes are covered by the papers published in the volume. These are related to modal and temporal logics, intuitionistic connexive and imperative logics, systems for reasoning with vagueness and rough concepts, topological quasi-Boolean logic and quasi-Boolean based rough set models, and first-order definability of path functions of graphs. Three single blind reviews for each submission were ensured. Aside from reviews by the Program Committee (PC) members, there were reviews by external experts. In some cases, in order to reach a final decision on acceptance, there were further reviews by PC members or external experts. The Easy Chair system was used for submission and reviews; it proved to be quite convenient. We would like to express our deep appreciation to all the PC members for their efforts and support. We also thank all the external reviewers for their invaluable help.

ICLA 2023 included 8 invited talks, and 6 of these appear in the volume as full papers. We are immensely grateful to Mihir K. Chakraborty, Supratik Chakraborty, Marie Fortin, Giuseppe Greco, Kamal Lodaya, Sandra Müller, R. Ramanujam and Yde Venema for kindly accepting our invitations.

Special thanks are due to IIT Indore, the organizing committee steered by Md. Aquil Khan and all the volunteers, for making this edition of ICLA possible.

We are grateful to Springer, for agreeing to publish this volume in the LNCS series.

February 2023

Mohua Banerjee
A. V. Sreejith

Organization

Program Chairs

Mohua Banerjee	Indian Institute of Technology Kanpur, India
A. V. Sreejith	Indian Institute of Technology Goa, India

Program Committee

C. Aiswarya	Chennai Mathematical Institute, India
Michael Benedikt	University of Oxford, UK
Thomas Colcombet	IRIF, France
Laure David	City, University of London, UK
Soma Dutta	University of Warmia and Mazury, Poland
Deepak D'Souza	IISc Bangalore, India
John Horty	University of Maryland, USA
Juliette Kennedy	University of Helsinki, Finland
Minghui Ma	Sun Yat-Sen University, China
Arnaldev Manuel	Indian Institute of Technology Goa, India
Luca Motto Ros	University of Turin, Italy
Alessandra Palmigiano	Vrije Universiteit Amsterdam, The Netherlands
M. Praveen	Chennai Mathematical Institute, India
Prakash Saivasan	IMSc Chennai, India
H. P. Sankappanavar	SUNY at New Paltz, USA
Manidipa Sanyal	University of Calcutta, India
Andrzej Szałas	University of Warsaw, Poland, and Linkoping University, Sweden
Giorgio Venturi	Universidade Estadual de Campinas, Brazil
Zach Weber	University of Otago, New Zealand

Local Organization

Organizing Chair

Md. Aquil Khan	Indian Institute of Technology Indore, India

Organizing Committee

Mohd. Arshad	Indian Institute of Technology Indore, India
Somnath Dey	Indian Institute of Technology Indore, India
Vijay Kumar Sohani	Indian Institute of Technology Indore, India
M. Tanveer	Indian Institute of Technology Indore, India

Additional Reviewers

Claudio Agostini
Alessandro Andretta
A. Baskar
Mihir K. Chakraborty
Amita Chatterjee
Koduri Siddharth Choudary
Andrea De Domenico
David Gabelaia
Sujata Ghosh
Md. Aquil Khan
Khushraj Madnani
Krishna Balajirao Manoorkar
Anup Basil Mathew
Heinrich Wansing

Contents

A Note on the Ontology of Mathematics

Mihir Kumar Chakraborty[1,2](✉)

[1] School of Cognitive Science, Jadavpur University, Kolkata 700032, India
mihirc4@gmail.com
[2] Indraprastha Institute of Information Technology, Delhi, Delhi 110020, India

Abstract. Provocation behind writing this paper has come from celebrated French philosopher Alain Badiou's slogan "Mathematics is ontology" and subsequent reading of his book [2]. However, this is not a critique of the book or a response to his philosophy. Some philosophico-mathematical issues have been raised by the author of the book in order to clarify and establish the slogan. In this paper, responses to some such issues have been presented such as the issues of continuum, Continuum Hypothesis, constructible sets and Axiom of Foundation. Remarks on these issues are made, though in brief. Finally, it is remarked that in the present era ontology of mathematics has to be pluralistic and inconsistency-tolerant.

Keywords: Ontology · Continuum hypothesis · Constructible sets · Paraconsistency

1 Introduction

To me, the main problem of mathematics lies in that it fails to establish its own consistency. What is meant by this? A huge corpus of mathematical entities has piled up over the centuries, from almost the beginning of human civilization. Primordial mathematical objects are positive whole numbers and geometric figures. Then other entities came into existence such as rationals, irrationals, reals and complex numbers. At one point of history appeared infinitesimals and the notion of limit point. The ontic status of mathematical entities changed over time. Modern era is predominantly Cantorian. From the Cantorian standpoint, each mathematical object is ultimately a set – a pure collection, 'Pure multiple' in Badiou's terminology. Only multiples remain. The natural numbers 1, 2, 3, ... are all multiples. 'Every multiple is a multiple of multiples' [2]. Every set is a collection of sets.

What is understood by the statement that this corpus of objects is inconsistent? Are there mathematical objects contradicting each other or itself in some sense? Yes, there are, at least as a first answer – there are Euclidean and non-Euclidean triangles, of the first kind the angle sum is 180°, of the second it is not so. There are Cantorian and non-Cantorian sets. According to the former, there does not exist any multiplicity between the naturals and the reals (the Continuum Hypothesis, or CH in short), on the other hand, according to the later there

M. Banerjee and A. V. Sreejith (Eds.): ICLA 2023, LNCS 13963, pp. 1–10, 2023.
https://doi.org/10.1007/978-3-031-26689-8_1

does exist. And so on. Badiou has aired the slogan 'Mathematics is ontology.' He means being, though he is not a Platonist. I quote from the translator's preface of his book 'Being and Event' [2]:

In Badiou's terms, the proposition 'mathematics is ontology' is a philosophical idea conditioned by an event and its consequent truth procedure in the domain of science. The event was Cantor's invention of set theory and the truth procedure its subsequent axiomatization by Zermelo and Fraenkel.

Badiou does not subscribe to the phrase "Philosophy of mathematics" as that may mean philosophy objectifies mathematics – as if philosophy has its own categories such as , realism, antirealism, nominalism conceptualism etc. and mathematics is to fall in one of these. The task of a researcher of the above field designed by a 'specialised bureaucracy in the academic authority' is to investigate which of these already existing categories applies to this or that mathematics. In [1] he says:

It is only through preliminary reduction to logical and linguistic problems that mathematics is ***forcibly*** *incorporated into a specialised objective area of philosophical* ***interrogation***. (emphasis by present author)

Though my position in this regard is by and large different, I refer to Badiou since his observation on Cantor's invention which he calls an 'event' because of 'rupturing with the order' is precisely my own attitude about which I shall say something later. My position is different in that I consider mathematical objects as artefacts and a mathematician as a mathematical artist [5]. However, after being created by some math-artist or artists, a mathematical object starts an existence of its own. Public, including the artist, looks at it with awe and wonder and discovers its unseen properties. Its ontology or being begins. And simultaneously with the creation of set, the pure multiple, begins its doom. Quoting Badiou [2]:

I showed how ontology, the doctrine of pure multiple prohibits the belonging of a multiple to itself, and consequently posits that the event is not. This is the function of the Axiom of Foundation.

Cantor led us to the paradise that he himself had created. (Recall Hilbert's declaration, "From the paradise, that Cantor created for us, no-one shall be able to expel us" [13]). We now know that the paradise is lost.

And what does the Axiom of Foundation say? For all sets A there is a member a of it such that a and A do not share a common member, i.e., $a \cap A = \emptyset$. Or equivalently, any chain

$$\cdots \in a_3 \in a_2 \in a_1 \in A$$

should terminate. From this, it follows that $A \notin A$.

Axiomatization of Cantorian set theory was a necessity, and there are several of them, ZFC, NBG and others. Badiou used ZFC in his philosophy. To quote Badiou,

> *Of course, there are other characterizations of set theory such as W. V. O. Quine's, but this multiplicity simply reveals the contingency of philosophy's conditioning: a conditioning that can only be contrasted by developing another metaontology on the basis of another axiomatization of set theory.*

As mentioned earlier, Badiou names the Zermelo-Fraenkel axiomatization of set theory as a truth procedure that follows upon the 'Cantor-event'. He transforms it into a 'condition' for his philosophy.

Our objective, however, it not to make a critique of Badiou's philosophy, but rather to point out some significant items of Cantorian set theory and its axiomatization ZFC in shaping his ideas. More specifically speaking, set theory and its subsequent axiomatization made an impact on his system of thought. The issue of Continuum Hypothesis plays a key role in it.

2 Continuum and Continuum Hypothesis

In the body of mathematics taken in its totality, there do exist as we have seen earlier, contradictory objects. But such problems may be negotiated by the process of segregation – let there be two domains non-overlapping, one Cantorian domain, the other non-Cantorian. This type of resolution may be somewhat satisfactory for mathematics as epistemology (a tool for solving problems of other fields, e.g. physics, Euclidean for Newton, non-Euclidean for Einstein), not as ontology. In both the cases, the natural linguistic word 'set' is used and hence an obligation remains as to what it really means if it is not considered simply a symbol. As ontology the fundamental threat looms via Gödel's second incompleteness theorem. Arithmetic (Number theory), the fundamental mathematical entity, if assumed to be consistent, cannot prove its own consistency. Speaking a bit formally, if the number theory $\mathbb{N}$ is consistent then there is a formal sentence A such that it can not be proved to be a theorem of $\mathbb{N}$ and A is the formal form of the statement "$\mathbb{N}$ is consistent."

One may question: what is the obligation on the part of mathematics (or $\mathbb{N}$) to establish its consistency? Does physics prove its own consistency? As an answer, it may be said that the nature of physics does not need that. The objective of physics is to discover answers to 'How'-questions/'Why'-questions and predict future/newly observed phenomena. Physics is definitely concerned about consistency, but that is local, temporal, and it is ready to drop/reshuffle old beliefs or hypotheses and adapt an opposing view. Mathematics cannot do that. Whenever genuinely contradictory entities appear, e.g., non-Euclidean geometry or non-Cantorian set theory, the mathematician accepts both but places them in two compartments and says, that makes a different mathematics. A new domain emerges. But the problem with the totality of mathematics as one entity remains.

The totality with all its branches rests at present upon Cantorian set theory that has been formalized, one of these formalizations being ZF and the consistency of ZF-axioms cannot be established by itself. If one assumed that the consistency of ZF may be proved from outside, that is by another theory, the same problem will be shifted to this later theory. Thus an infinite regress would occur.

This problem had persisted in all great minds, from Hilbert to Gödel. Hilbert in a meeting declared, "We must know, we shall know." – there cannot be unanswered mathematical queries; there cannot be inconsistency in mathematics. Gödel similarly cherished the hope that there is reality in the basic mathematical objects, it was not simply a linguistic game. At the same timc, he suffered for not being able to construct a 'rational proof' of that realism. He established that Continuum Hypothesis (CH) is consistent with the ZF-axioms. There was a relief in the mathematics community. But that was temporary. Cohen established that ZF was consistent with the negation of CH too. The two findings together show the independence of CH with ZF-axioms. On the other hand Gödel had a kind of belief that the Continuum Hypothesis should not be true. In Badiou's terms [2]:

> *These hypotheses are in reality pure decisions. Nothing in fact, allows them to be verified or refuted.*

There are at least two issues related with the Continuum Hypothesis: one, cognition of continuity (geometric line) and two, its measure namely the enigmatic phenomenon of the one-one correspondence between the power set of naturals and the points on the geometric line. In my view, in spite of the spectacular success of calculus in predicting the positions of a projectile the first issue has remained unsolved. Computer science seems to be satisfied with the "drawing" of a line joining two points but what is actually done is to place before our eyes, the visual organ, a discrete sequence of pixels. This, perhaps, is okay for all the 'practical purposes' but certainly is not ontologically the same as drawing a line on a piece of paper by ruler and pencil. Fortunately (or unfortunately) a postulate is thrown in viz. so called Cantor-Dedekind Axiom presuming a bijection between the real numbers and the 'real' line. (For the very exciting period of this development in the realm of mathematics I will refer to the correspondences between Cantor and Dedekind [3]). This, in turn, gives rise to the second problem namely non-denumerability (un-measure) of the continuum. As Badiou puts it [2],

> *The impasse of ontology – the quantitive un-measure of the set of parts of a set – tormented Cantor: it threatened his very desire for foundation.*

In fact, Cantor proposed CH in the context of seeking an answer to the question, "What is the identifying nature of continuity?" Results by Gödel and Cohen put together establish the independence of CH relative to ZF axioms. These independence results however, do not provide an answer to the original query as regards the identity of continuity, just knowing the cardinality should not be enough.

Of course, conceptualizing the geometric continuum in terms of discrete points and eventually as real numbers faced criticism almost from the very inception of the concept – both from the philosophical as well as the cognitive standpoints. I invite readers of the present article to have a look into the wonderful treatise by George Lakoff viz. "Where mathematics comes from" [15], especially Chaps. 12 and 13. The author explains why the geometric line is not the number line and "Why Continuum Hypothesis is not about the continuum". Philosopher Charles Sanders Peirce also differs from Cantor-Dedekind characterization and attributes to the continuum the following chracteristic properties: inextensibility, supermultitude, reflexivity, potentiality and genereticity. For the details see [18] from which I quote,

> *It is important to highlight the enormous distance that separates the mature Peircean ideas on the continuum (particularly around the turn of the century) from today's dominant conception which identifies it with the real numbers.*

Peirce coined the term 'synechism' in 1893 meaning thereby "the doctrine that continuity rules the whole domain of experience." (MS 946, p. 5) and thus leading towards the withdrawal of the atomistic belief in ultimate constituent components.

As a solution, Gödel proposed the notion of constructible universe, a decision known in the literature as the Axiom of Constructibility:

> *For every multiple γ, there exists a level of constructible hierarchy to which it belongs* [2].

Within von Neumann universe V of all sets Gödel proposed to focus on those sets which are constructible in the following sense. First, from a set A the collection denoted by **Definable**(A) is generated by,

$$\mathbf{Definable}(A) := \{y \mid y \in A \text{ and } (A, \in) \vdash \varphi(y, z_1, \ldots, z_n) \text{ for some first order formula } \varphi \text{ and } z_1, \ldots, z_n \in A\}.$$

The constructible universe L is next defined by,

$$L_0 = \emptyset \text{ (null set)},$$
$$L_{\alpha+1} = \mathbf{Definable}(L_\alpha) \text{ for a non-limit ordinal } \alpha$$
$$\text{and } L_\lambda = \bigcup_{\alpha<\lambda} \mathbf{Definable}(L_\alpha) \text{ when } \lambda \text{ is a limit ordinal.}$$

Finally, $L = \bigcup_{\alpha \in \mathsf{Ord}} L_\alpha$, Ord being the class of all ordinals in V.

Elements of L are called constructible sets. Axiom of Constructibility says that every set is constructible i.e., $V = L$. In other words, the excess is trimmed. Gödel proved that if ZF is consistent, so is ZF + $(V = L)$ and that ZF + $(V = L)$ implies CH (in fact, GCH i.e., Generalized Continuum Hypothesis) and also Axiom of Choice (C). L is called the standard inner model of ZFC.

The axiom $V = L$ is another imposition. Within the framework of ontology, one could consider that there are constructible sets as well as non-constructible. The second category is defined negatively, that is multiples without belonging to any level of the above mentioned hierarchy. The role of language (formal language) is crucial. A quote from Badiou [2]:

The only multiples which are admitted into existence are those extracted from the inferior level by means of constructions which can be articulated in the formal language, and not 'all' the parts, including the undifferentiated, the unnamable and the indeterminate.

But whenever one adds a new axiom or meta axiom, it again becomes a matter of decision, not necessarily a fact. And it has been established that if ZF is consistent then enhancement of it with the Axiom of Constructibility still remains consistent. Thus the ontology is restricted to constructible sets only and the Generalized Continuum Hypothesis is true in the constructible universe.

Similar is the case with the Axiom of Choice (C). It was Gödel again who proved that if ZF is consistent so is $\mathsf{ZF} + \mathsf{C}$. And again it was Cohen who established that if ZF is consistent, so is $\mathsf{ZF} + \neg\mathsf{C}$. Thus the mathematician is placed before the following picture of the mathematical world (Fig. 1):

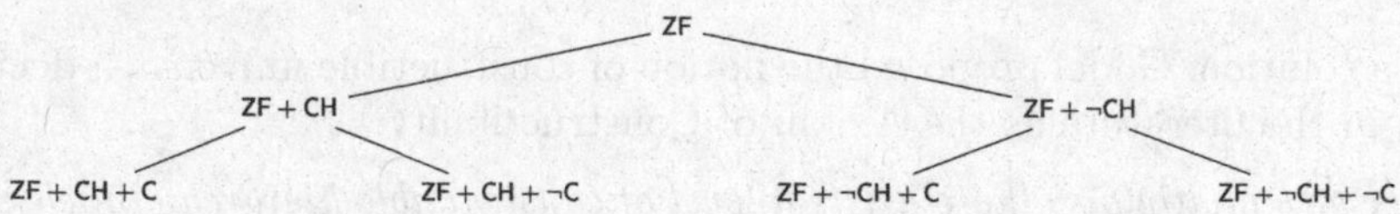

Fig. 1. Mathematical world

A mathematician can situate himself/herself at any of the nodes of the above tree – a matter of pure decision. The standard mathematics occupies the left-most node, namely $\mathsf{ZF} + \mathsf{CH} + \mathsf{C}$. It is not allowed that someone will respect both CH and $\neg\mathsf{CH}$ or both C and $\neg\mathsf{C}$ since then the system would collapse, any statement will be derivable. This happens due to the explosive nature of classical two-valued logic on which rests the deductive methodology of existing mathematics: from a contradiction any statement is derivable. Thus admission of contradictory statements in a mathematical system trivializes the system. A more diversified picture of mathematical world is perceived by Friend (see [12]).

3 Inconsistency-Tolerance

My programme is that of inconsistency-tolerance, the project of democracy. By this I would like to mean that the general scenario of existence demands incorporation of the case 'both, is and is not'. Let me narrate a story, a true experience of mine. It was an arranged chat-session between me and a group of school students from 7th to 10th standards (of Indian schools). They were children of age group 13 to 16. They were naturally shaky. To make them speak, I presented Russell's Paradox and finally ended up with: "You see, if the set of all ordinary sets is assumed to be ordinary then it turns out to be extraordinary and if assumed extraordinary then it turns out ordinary – an obvious contradiction, a paradox."

I was quite happy to see that I could communicate the message to the young minds and they were amazed. But one girl, the youngest of the lot, of the 7th standard stood up and said, "Sir, I do not understand what is the problem in it, the set is both ordinary and extraordinary." Accepting the contradiction was not problematic to her and she was a normal child. Russell's Paradox is cited as an example of dialethea (actual situation of contradiction). Graham Priest in [17] posed the question "What is so bad about contradictions?", discussed possible objections about believing in **some** contradictions and arrived at the answer "Maybe nothing." I felt extremely gratified at the discovery of a footnote in Latta's translation of Leibniz's "Monadology" in which his celebrated Law of Continuity is presented as:

> *Everything is continually changing, and in every part of this change there is both a permanent and varying element. That is to say, at any moment everything is both 'is' and 'is not', everything is becoming something else – something which is nevertheless, not entirely 'other'* [16].

In fact, understanding of any change has to adopt this kind of ontology; Zeno's arrow at any instant of time is static as well as moving. We find acceptance of simultaneous existence of opposites in the classical Eastern thoughts, for example Catuskoti (tetralemma) in the Buddhist system where four possible states are envisaged viz. X is P, X is non-P, X is neither P nor non-P, X **is both** P **and non-**P. The last (4th) one is the inconsistent state of existence. In ancient Chinese culture the Yin-Yang symbol (Tai Chi) represents two aspects of nature which are opposing and complementing together. The symbol is especially significant because of the curvy (non-straight) dividing line between black and white and the black dot in white as well as the white dot in black. The first indicates that the two halves are intertwined and the second shows each carries the seed of the other. This means that the universe consists of co-existent, contradictory forces, in other words 'yes' and 'no' together, reflecting the fourth koti (corner) of Catuskoti in the Indian thought (Figs. 2 and 3).

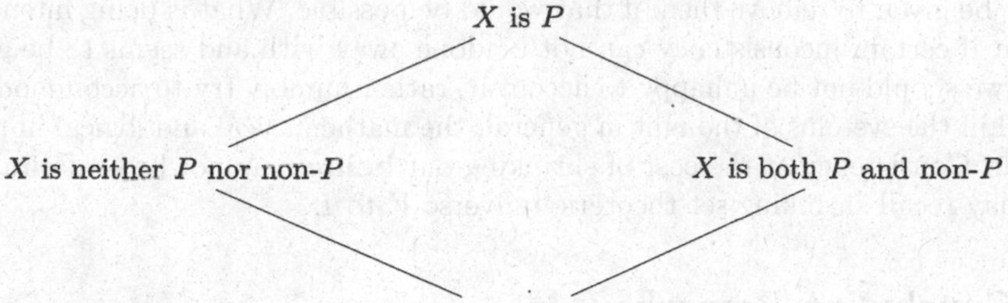

Fig. 2. Catuskoti (Tetralemma)

Fig. 3. Tai Chi (Yin and Yang)

In the field of mathematics, at least two ways are in sight towards accommodation of contradiction. First, to localize inconsistency. A set S of sentences may be inconsistent with respect to a sentence α i.e. α and $\neg\alpha$ both follow from S but S is non-trivial in the sense that not every sentence follows form S. As a consequence there shall be a sentence β with respect to which S is not inconsistent. This situation arises in paraconsistent logics [9,11]. Second is the project of graded inconsistency or consistency: α and $\neg\alpha$ are both true to some degree and false to some degree. This happens in many-valued logics, fuzzy logics and the theory of graded consequence [4,6]. In other words, banishing universal bi-valence. Both the approaches may occur together. In fact, most of the paraconsistent logic systems are at least trivalent. Trivalence appeared around 1920 through the Polish logician-mathematician Jan Łukasiewicz. Almost simultaneously Kleene, Post, Bochvar and others proposed trivalent systems, the third value being interpreted in various ways giving rise to various kinds of truth tables. It took no time to pass on to n (finite)-valued models. But an infinite valued truth set viz., the interval $[0, 1]$ was proposed only as a mathematical generalization. It was in 1965, after the advent of fuzzy set theory due to Lotfi Zadeh, the truth set $[0, 1]$ gained a real significance [19].

The above is not to advocate that inconsistencies are welcome. Utmost efforts are to be given to remove them if that would be possible. What is being intended is that if certain inconsistency can not be done away with and seems to be genuine, we should not be unhappy to accept it, rather humbly try to accommodate it within the systems of thought in general, the mathematical and logical in particular. Consistency at the cost of shrinking our **being** may not be so desirable. We may recall shrinking set-theoretic universe V to L.

4 Concluding Remarks

Pluralism in mathematics is a reality of the present era. This arises from various angles that include at least:

- pluralism in methodology or procedure,
- pluralism in theories (individuated by sets of axioms or rules or methods),

- pluralism in underlying logic,
- pluralism in truth,
- pluralism in ontology and,
- pluralism in foundations.

These categories are taken from the preface of [7] edited by Chakraborty and Friend. Friend's book "Pluralism in Mathematics: A New Position in the Philosophy of Mathematics" [12] is a significant addition in this direction. From this book I quote below what Friend has to say about pluralism in the foundation of mathematics.

> *The* pluralist in foundations *believes that there is insufficient evidence to think that there is a unique foundation for mathematics. Moreover, the pluralist in foundations works under the assumption that there is no reason to think that there will be a convergence to a unique theory in the future. He takes seriously the possibility that there are several,* ***together inconsistent****, foundations for mathematics. (emphasis mine)*

Besides this, pluralism in mathematics from the cultural angle is also in the agenda (see [8,10,14]). These studies have revealed the presence of cognitive differences in terms of

- categorization and abstraction of the world that includes units of counting process,
- perception of geometric shape,
- deduction and inference and
- reasoning and problem-solving.

For some details see [8].

As discussed in Sect. 3, it is a necessity of today that inconsistency (opposing view) is to be accommodated within the body of the system be it social, political, scientific and philosophical – in particular, it is true of mathematics. So, as it appears to me, the ontology of mathematics has to be pluralistic and inconsistency tolerant in multiple reasonable ways and forms.

Acknowledgement. I thank my student Mr. Sayantan Roy for extending technical help in writing this paper and making some important remarks.

References

1. Badiou, A.: Mathematics and philosophy. In: Duffy, S. (ed.) Virtual Mathematics: The Logic of Difference, pp. 16–30. Clinamen Press (2006)
2. Badiou, A.: Being and Event. Continuum International Publishing Group (2012). Translated from the 2015 French original by Oliver Feltman
3. Bell, J.L.: The Continuous, the Discrete and the Infinitesimal in Philosophy and Mathematics. TWOSPS, vol. 82. Springer, Cham (2019). https://doi.org/10.1007/978-3-030-18707-1

4. Bolc, L., Borowik, P.: Many-Valued Logics. Theoretical Foundations, vol. 1. Springer, Berlin (1992)
5. Chakraborty, M.K.: Mathematical RUPAs and their artists. History Sci. Philos. Cult. Indian Civilization **13**(2), 527–533 (2012). Edited by P. K. Sengupta
6. Chakraborty, M.K., Dutta, S.: Theory of Graded Consequence. LASLL, Springer, Singapore (2019). https://doi.org/10.1007/978-981-13-8896-5
7. Chakraborty, M.K., Friend, M.: Preface. J. Indian Council Philos. Res. Spec. Issue: Pluralism Math. **34**(2), 205–207 (2017). Edited by M. K. Chakraborty and M. Friend
8. Chakraborty, M.K., Sirkar, S.: Aspects of mathematical pluralism. J. Math. Cult. **10**(1), 21–52 (2016)
9. da Costa, N.C.A.: On the theory of inconsistent formal systems. Notre Dame J. Formal Logic **15**, 497–510 (1974)
10. D'Ambrosio, U.: Ethnomathematics, the nature of mathematics and mathematics education. In: Mathematics, Education and Philosophy. Stud. Math. Ed. Series, vol. 3, pp. 230–242. Falmer, London (1994)
11. Dutta, S., Chakraborty, M.K.: Consequence–inconsistency interrelation: in the framework of paraconsistent logics. In: Beziau, J.-Y., Chakraborty, M., Dutta, S. (eds.) New Directions in Paraconsistent Logic. SPMS, vol. 152, pp. 269–283. Springer, New Delhi (2015). https://doi.org/10.1007/978-81-322-2719-9_12
12. Friend, M.: Pluralism in Mathematics: A New Position in Philosophy of Mathematics. LEUS, vol. 32. Springer, Dordrecht (2014). https://doi.org/10.1007/978-94-007-7058-4
13. Hilbert, D.: Über das Unendliche. Mathematische Annalen **95**(1), 161–190 (1926)
14. Joseph, G.G.: Different ways of knowing: contrasting styles of argument in Indian and Greek mathematical traditions. In: Mathematics, Education and Philosophy. Stud. Math. Ed. Series, vol. 3, pp. 194–204. Falmer, London (1994)
15. Lakoff, G., Núñez, R.E.: Where Mathematics Comes From. Basic Books, New York (2000). How the embodied mind brings mathematics into being
16. Leibniz, G.W.: The Monadology and Other Philosophical Writings. Oxford University Press (1925). Second impression, Translated, with an Introduction and notes by R. Latta (the first edition was published in 1898)
17. Priest, G.: What's so bad about contradictions? In: Priest, G., Beall, J., Armour-Garb, B. (eds.) The Law of Non-contradiction, pp. 23–38. Oxford University Press, New York (2004)
18. Vargas, F.: A full model for Peirce's continuum. In: Bellucci, F., Pietarinen, A.V. (eds.) Advances in Peircean Mathematics, Peirceana, vol. 7, pp. 230–242. De Gruyter, Berlin/Boston (2022)
19. Zadeh, L.A.: Fuzzy sets. Inf. Control **8**, 338–353 (1965)

Boolean Functional Synthesis: From Under the Hood of Solvers

Supratik Chakraborty(✉)

Indian Institute of Technology Bombay, Mumbai, India
supratik@cse.iitb.ac.in

Abstract. Boolean functional synthesis concerns the automatic generation of Boolean functions satisfying given logical specifications. This problem has numerous applications, and has attracted significant attention from researchers over the past decade. Complexity-theoretic arguments indicate that it is extremely unlikely that the problem has any polynomial-time algorithm. Yet, state-of-the-art tools for this problem routinely handle problems with several thousands of variables. What makes these algorithms tick? In this paper, we provide an overview of some of the techniques that underlie the practical efficiency of these solvers.

1 Introduction

Boolean functional synthesis concerns the algorithmic construction of Boolean functions as circuits (or programs) from Boolean specifications. At first sight, the problem appears deceptively easy: if a specification has spelt out what a function must do, shouldn't it be easy to construct the function as a circuit? On closer inspection, however, subtler intricacies of the problem begin to emerge. Logical specifications are often stated as relations between inputs and outputs of a system, rather than explicitly specifying the outputs as functions of the inputs. In such cases, how does one find efficiently (in practice) which functions of inputs will work for the outputs, such that the specification is satisfied?

To appreciate the difficulty of solving this problem, consider a simple specification represented by a Boolean formula $\varphi_{\text{fact}}(\mathbf{X}, \mathbf{Y}_1, \mathbf{Y}_2)$, where $\mathbf{X}$ is sequence of $2n$ Boolean variables, and each of $\mathbf{Y}_1, \mathbf{Y}_2$ is a sequence of n Boolean variables. We define $\varphi_{\text{fact}}(\mathbf{X}, \mathbf{Y}_1, \mathbf{Y}_2)$, with some abuse of notation, to be $(\mathbf{X} = \mathbf{Y}_1 \times_{[n]} \mathbf{Y}_2) \wedge (\mathbf{Y}_1 \neq \mathbf{1}) \wedge (\mathbf{Y}_2 \neq \mathbf{1})$, where $\times_{[n]}$ is short-hand for n-bit unsigned integer multiplication, $\mathbf{1}$ denotes the n-bit representation of the unsigned integer 1, and $=$ denotes component-wise equivalence. It is easy to see that φ_{fact} can be written as a Boolean formula on the $4n$ Boolean variables represented by $\mathbf{X}, \mathbf{Y}_1$ and $\mathbf{Y}_2$, and the size of the formula is in $\mathcal{O}(n^2)$. Informally, this formula specifies that if we view $\mathbf{X}$, $\mathbf{Y}_1$ and $\mathbf{Y}_2$ as encodings of unsigned integers, then $\mathbf{Y}_1$ and $\mathbf{Y}_2$ are non-trivial factors of $\mathbf{X}$. If we now require $\mathbf{Y}_1$ and $\mathbf{Y}_2$ to be generated as functions of $\mathbf{X}$ such that $\varphi_{\text{fact}}(\mathbf{X}, \mathbf{Y}_1, \mathbf{Y}_2)$ is satisfied, we are effectively asking for a circuit that can factorize a $2n$-bit unsigned composite

M. Banerjee and A. V. Sreejith (Eds.): ICLA 2023, LNCS 13963, pp. 11–22, 2023.
https://doi.org/10.1007/978-3-031-26689-8_2

integer! While there certainly exists a circuit that does this job, it remains an open question whether there exists a circuit of size polynomial in n that achieves factorization of $2n$-bit integers. This immediately hints at the inherent hardness of Boolean functional synthesis. Indeed, if we had a polynomial-time algorithm for solving Boolean functional synthesis, we could feed the above specification to the algorithm, and obtain a circuit for factorizing a $2n$-bit product of two prime numbers. Since a circuit constructed in polynomial time can be at most polynomial sized (and hence can be evaluated in polynomial time), an efficient solution for Boolean functional synthesis would have serious ramifications for public-key cryptography.

The above example hints at fundamental roadblocks in designing efficient algorithms for Boolean functional synthesis. Indeed, as has been shown by Akshay et al [3,4], unless some long-standing complexity theoretic conjectures are falsified, it is impossible to have any worst-case polynomial-time algorithm that solves Boolean functional synthesis [3,4]. Does this mean the end of the road for practical synthesis tools? Fortunately not. Despite complexity-theoretic hurdles, practical applications have motivated researchers to continue to chip at the problem, and significant advances have been made in the past decade. This has resulted in multiple Boolean functional synthesis tools that are able to scale to benchmarks involving thousands of variables [1,3–5,8,13,15,16,21,22,27]. So, what makes these tools tick in practice? Is their success purely coincidental, or are there principled approaches at work? In this paper, we delve behind the scenes and look at some of the techniques that make modern Boolean functional synthesis tools scale to large problem instances in practice. Specifically, we consider the tools BFSS [3,4], Manthan [16], Manthan2 [15], CADET [21,22], BaFSyn [8] and C2Syn [1].

The remainder of the paper is organized as follows. We define the Boolean functional synthesis problem precisely in Sect. 2 and comment on how it differs from some other variants of synthesis problems in the literature. We present a very brief overview of some applications of Boolean functional synthesis in Sect. 3. In Sect. 4 we delve into six different techniques that work behind the scenes of various Boolean functional synthesis tools to make them scale to large problem instances. Finally, we conclude in Sect. 5.

2 Problem Statement

Let $\mathbf{X} = (x_1, \ldots x_m)$ be a sequence of Boolean variables representing inputs of a system to be designed, and $\mathbf{Y} = (y_1, \ldots y_n)$ be a sequence of Boolean variables representing the system outputs. A relational specification is a Boolean formula $\varphi(\mathbf{X}, \mathbf{Y})$ that specifies what assignments of outputs $\mathbf{Y}$ are "desirable" for each assignment of inputs $\mathbf{X}$. Note that multiple assignments of $\mathbf{Y}$ may satisfy the specification for the same assignment of $\mathbf{X}$ in general. Boolean functional synthesis requires us to algorithmically construct a sequence of Boolean function $\mathbf{F}(\mathbf{X}) = (F_1(\mathbf{X}), \ldots F_n(\mathbf{X}))$ such that $|\mathbf{Y}| = |\mathbf{F}(\mathbf{X})|$ and the following is a tautology

$$\forall \mathbf{X}\, (\exists \mathbf{Y} \varphi(\mathbf{X}, \mathbf{Y}) \leftrightarrow \varphi(\mathbf{X}, \mathbf{F}(\mathbf{X})).$$

The function $F_i(X)$ is also called a *Skolem function* for y_i in $\varphi(\mathbf{X}, \mathbf{Y})$. The problem of synthesizing these functions is therefore also called *Boolean Skolem function synthesis* in the literature.

A note about the problem definition is worth mentioning here. In the literature, a specification $\varphi(\mathbf{X}, \mathbf{Y})$ is said to be *realizable* if $\forall \mathbf{X} \exists \mathbf{Y}\, \varphi(\mathbf{X}, \mathbf{Y})$ is a tautology; otherwise it is called *unrealizable*. There is a general belief in the community that unless a specification is realizable, it is not meaningful for synthesis purposes. We submit that this is a restrictive view that is worth relaxing. Specifically, the problem definition above *does not* require $\forall \mathbf{X} \exists \mathbf{Y}\, \varphi(\mathbf{X}, \mathbf{Y})$ to be a tautology. It only requires that for every assignment of inputs $\mathbf{X}$, the formulas $\exists \mathbf{Y}\, \varphi(\mathbf{X}, \mathbf{Y})$ and $\varphi(\mathbf{X}, \mathbf{F}(\mathbf{X}))$ evaluate to the same truth value, *regardless of what that truth value is.* To see why synthesis from non-realizable specifications can be meaningful, let us re-visit the factorization example considered above. The specification $\forall \mathbf{X} \exists \mathbf{Y}_1 \exists \mathbf{Y}_2\, \varphi_{\mathsf{fact}}(\mathbf{X}, \mathbf{Y}_1, \mathbf{Y}_2)$ is clearly not a tautology since there is no assignment of $\mathbf{Y}_1, \mathbf{Y}_2$ that satisfies the specification if $\mathbf{X}$ encodes a prime number. Hence, this is an unrealizable specification. Notwithstanding this, suppose we go ahead and synthesize Skolem functions for all outputs in $\mathbf{Y}_1$ and $\mathbf{Y}_2$. Let $\mathbf{F}_1(\mathbf{X})$ and $\mathbf{F}_2(\mathbf{X})$ denote the corresponding sequences of synthesized Skolem functions. It is an easy exercise to show that $\varphi_{\mathsf{fact}}(\mathbf{X}, \mathbf{F}_1(\mathbf{X}), \mathbf{F}_2(\mathbf{X}))$ evaluates to false iff $\mathbf{X}$ encodes a prime number. Thus, synthesizing Skolem functions from the unrealizable specification $\varphi_{\mathsf{fact}}(\mathbf{X}, \mathbf{Y}_1, \mathbf{Y}_2)$ gives us a circuit to check primality of integers representable by $2n$-bits. Whenever this check indicates that $\mathbf{X}$ is composite, the Skolem functions $\mathbf{F}_1(\mathbf{X})$ and $\mathbf{F}_2(\mathbf{X})$ also give us non-trivial factors of $\mathbf{X}$.

3 Some Applications

In this section, we present a very brief of overview some applications of Boolean functional synthesis. It is clear from the definition of the problem that Skolem functions serve as witnesses of existentially quantified variables when deciding the satisfiability of quantified Boolean formulas (QBF). Hence, Boolean functional synthesis has direct applications in certified QBF satisfiability checking. In addition, QBFs have recently been used to model a wide variety of problems such as in non-monotonic reasoning, planning, games and the like (see [24] for an excellent survey). In each of these applications, certificates of existentially quantified variables serve a different purpose. For example, in planning, a certificate may correspond to a plan, while in games, it may correspond to a winning strategy of a player. Clearly, the ability to generate Skolem functions explicitly is of immense value in all such applications. In addition, variants of Boolean functional synthesis have also been used to automatically synthesize a restricted class of programs [14, 26] and also for circuit repair [17]. Finally, the synthesis of controllers from temporal logic specifications can also be posed as one of Boolean functional synthesis once a winning region has been identified [12].

4 Under the Hood of Modern Solvers

Given that Boolean functional synthesis is unlikely to have worst-case polynomial time algorithms [3,4], why do state-of-the-art tools work at all? While it is hard to pin-point any specific reason, certain techniques have been found to be extremely useful in state-of-the-art Boolean functional synthesis tools. We overview some of these below.

4.1 Unique Functional Dependencies

In a large class of specifications, several outputs may be uniquely determined by the inputs and remainder of the outputs. Identifying such functional dependencies has proved to be extremely helpful in practical Boolean functional synthesis. To illustrate how this helps, consider the specification $\varphi(x_1, \ldots, y_1, y_2, \ldots)$ given in conjunctive normal form as $(\neg x_1 \vee \neg y_2 \vee y_1) \wedge (y_2 \vee x_2) \wedge \cdots \wedge (y_2 \vee \neg y_1) \wedge \cdots \wedge (x_1 \vee \neg y_1) \wedge \cdots \wedge (\neg y_2 \vee \neg x_2)$, where x_1, x_2 are inputs, y_1, y_2 are outputs and $\cdots$ represents unspecified sub-formulas in the specification. Clearly, every assignment of $x_1, x_2, \ldots, y_1, y_2, \ldots$ that satisfies φ satisfies $(y_1 \leftrightarrow (x_1 \wedge y_2))$ as well as $(y_2 \leftrightarrow \neg x_2)$. This immediately gives us Skolem functions for y_1 and y_2, i.e. $(x_1 \wedge \neg x_2)$ for y_1 and $\neg x_2$ for y_2. Therefore, we need not worry any further about generating Skolem functions for y_1 and y_2, and can treat y_1 and y_2 as inputs (much like $x_1, x_2, \ldots$) while synthesizing Skolem functions for the other outputs.

How does one detect functional dependencies like the ones illustrated above? A syntactic way of doing this is to look for patterns of sub-formulas that encode such dependencies. This is what we did in the example above, and it is usually very efficient and effective when identifying dependencies in specifications that make use of Tseitin encoding [28]. However, such a syntactic approach can miss functional dependencies that are implicit in the semantics of the specification, but not necessarily apparent in any pattern of sub-formulas. A more rigorous way of identifying functional dependencies is to make use of Padoa's theorem [20]. Specifically, let $\varphi(\mathbf{X}, y, \mathbf{Y}, \mathbf{Y}_d)$ be a specification, where $\mathbf{X}$ is a sequence of input variables, y is an output variable that we wish to check for functional dependence, $\mathbf{Y}$ is a sequence of output variables that have not yet been identified as being functionally dependent, and $\mathbf{Y}_d$ is a sequence of output variables already identified as functionally dependent on $\mathbf{X}$, y and $\mathbf{Y}$. Let y' denote a fresh variable, and let $\mathbf{Y}'_d$ denote a sequence of fresh variables of the same dimension as $\mathbf{Y}_d$. Then, y is functionally dependent on $\mathbf{X}$ and $\mathbf{Y}$ in φ iff the following formula is unsatisfiable:

$$\varphi(\mathbf{X}, y, \mathbf{Y}, \mathbf{Y}_d) \wedge y \wedge \varphi(\mathbf{X}, y', \mathbf{Y}, \mathbf{Y}'_d) \wedge \neg y'.$$

Furthermore, if the above formula is unsatisfiable, the functional dependence of y on $\mathbf{X}$ and $\mathbf{Y}$ can be obtained by finding an interpolant between $\varphi(\mathbf{X}, y, \mathbf{Y}, \mathbf{Y}_d) \wedge y$ and $\neg(\varphi(\mathbf{X}, y', \mathbf{Y}, \mathbf{Y}'_d) \wedge \neg y')$. Fortunately, checking satisfiability and generating interpolants for Boolean formulas can be done reasonably efficiently, thanks to significant advances in the theory and practice of satisfiability solving over the past two decades. As a result, identifying functionally dependent variables can often be done efficiently in practice in modern Boolean functional synthesis tools [1,3,4,15,16].

4.2 Unate Variables

Although not immediately obvious, it turns out that a significant percentage of benchmarks for Boolean functional synthesis have output variables that admit a constant Skolem function. If we have a practically efficient way of identifying these variables and the corresponding (constant) Skolem functions, we can simplify the specification by substituting the constant functions for the respective variables and synthesizing Skolem functions for the remaining output variables from the simplified specification.

A simple way to identify if an output variable y in $\mathbf{Y}$ admits a constant Skolem function in the specification $\varphi(\mathbf{X}, \mathbf{Y})$ is to check implications of *cofactors* of $\varphi(\mathbf{X}, \mathbf{Y})$ with respect to y. Specifically, let $\varphi(\mathbf{X}, \mathbf{Y}) \mid_{y=1}$ denote the formula obtained by substituting 1 for y in $\varphi(\mathbf{X}, \mathbf{Y})$. This is also called the cofactor of φ with respect to $y = 1$. The cofactor of φ with respect to $y = 0$ is similarly defined and is denoted $\varphi(\mathbf{X}, \mathbf{Y}) \mid_{y=0}$. A sufficient condition for the output variable y to admit the constant Skolem function 1 (resp. 0) is to check if the implication $\varphi(\mathbf{X}, \mathbf{Y}) \mid_{y=0} \rightarrow \varphi(\mathbf{X}, \mathbf{Y}) \mid_{y=1}$ (resp. $\varphi(\mathbf{X}, \mathbf{Y}) \mid_{y=1} \rightarrow \varphi(\mathbf{X}, \mathbf{Y}) \mid_{y=0}$) holds. If so, we say that the specification φ is *positive (resp. negative) unate* in y. Interestingly, the above implication check can be encoded as a check for unsatisfiability of $\varphi(\mathbf{X}, \mathbf{Y}) \mid_{y=0} \wedge \neg\varphi(\mathbf{X}, \mathbf{Y}) \mid_{y=1}$ (resp. $\varphi(\mathbf{X}, \mathbf{Y}) \mid_{y=1} \wedge \neg\varphi(\mathbf{X}, \mathbf{Y}) \mid_{y=0}$). Thanks again to the capabilities of modern propositional SAT solvers, these checks can often be done efficiently in practice, making them powerful components of modern Boolean functional synthesis tools.

Interestingly, even if a specification $\varphi(\mathbf{X}, \mathbf{Y})$ is not unate w.r.t. an output variable y_i to begin with, it may become so after some other output y_j is set to a constant (say 1) in $\varphi(\mathbf{X}, \mathbf{Y})$. For example, consider the specification $\varphi(x, y_1, y_2)$ defined as $(x \wedge (y_1 \vee y_2)) \vee (\neg x \wedge \neg y_2)$. Here φ is positive unate in y_1, but it is neither positive nor negative unate in y_2. However, once we substitute 1 for y_1 in φ, we obtain $\varphi \mid_{y_1=1}$, i.e. $x \vee (\neg x \wedge \neg y_2)$, which is negative unate in y_2. This cascaded discovery of unateness is extremely useful in identifying constant Skolem functions and thereby in simplifying the specification. To exploit this maximally, modern tools execute the unate variable identification step iteratively until they can't find any output variable in which the resulting cofactored specification is unate.

A finer point about the use of unate variables for Skolem functions is worth mentioning here. Let $\varphi(\mathbf{X}, y, \mathbf{Y})$ be a Boolean relational specification on inputs $\mathbf{X}$ and outputs y and $\mathbf{Y}$, where y is a single output variable. It follows from the definition of Boolean functional synthesis that the Skolem function 1 can be used for output variable y iff $\forall \mathbf{X}\ (\exists \mathbf{Y} \varphi(\mathbf{X}, 0, \mathbf{Y}) \rightarrow \exists \mathbf{Y} \varphi(\mathbf{X}, 1, \mathbf{Y}))$ is a tautology. Note that this condition is significantly weaker than $\forall \mathbf{X} \forall \mathbf{Y}\ (\varphi(\mathbf{X}, 0, \mathbf{Y}) \rightarrow \varphi(\mathbf{X}, 1, \mathbf{Y}))$, which is what the implication check between cofactors described above does. Therefore, it is indeed possible that the check based on implication between cofactors, which requires the use of a propositional satisfiability (SAT) solver, misses out opportunities for using constants for Skolem functions of some output variables. However, the check for the weaker condition requires checking (un)satisfiability of a formula with at least one quantifier alternation, and hence requires a QBF satisfiability solver – at least one that can reason about one quantifier alternation. Since state-of-the-art QBF solvers do not scale as well as SAT solvers in practice, the stronger check based on implication between cofactors is prefered in most Boolean functional synthesis tools [3,4,8,15,16].

4.3 Guess, Check and Repair

While the above two techniques are extremely useful in identifying simple Skolem functions for some output variables, typically not all Skolem functions can be synthesized using these methods. An approach that has worked remarkably well in practice for synthesizing the remaining Skolem function is that of guess, check and repair. At a high level, this approach works as follows. In the "guess"ing step, we identify promising candidates for Skolem functions of all output variables for which Skolem functions have not yet been determined. In the "check"ing step, we check whether the candidate Skolem functions actually meet the formal requirements of Boolean functional synthesis. If so, we are done and can report the current candidate Skolem functions as the solution. Otherwise, we "repair" the candidate Skolem functions guided by hints from the checking process, and repeat the checking and repairing steps. The repairing of candidate Skolem functions is always done in a way that ensures that we always converge to correct Skolem functions within a finite number of iterations of repair and check.

There are at least two different schools of thought about how "good" candidate Skolem functions can be identified. In [3,4], the candidates are identified based on certain structural properties of the specification represented as a Boolean circuit. In [15,16], the candidates are identified using constrained sampling and decision-tree learning techniques. In general, if "good" candidate Skolem functions are identified early enough, we require fewer iterations of check and repair to eventually arrive at correct Skolem functions. Therefore, techniques for identifying "good" candidate Skolem functions are extremely important in this approach to solving Boolean functional synthesis.

Given a sequence of candidate Skolem functions, say $\mathbf{F}(\mathbf{X})$, how do we check if these are correct Skolem functions? The naive way to do this is to directly ask if $\forall \mathbf{X}\left(\exists \mathbf{Y} \varphi(\mathbf{X}, \mathbf{Y}) \;\leftrightarrow\; \varphi(\mathbf{X}, \mathbf{F}(\mathbf{X}))\right)$ holds. However, this requires the use of a QBF solver, and we have already seen that QBF solvers don't scale as well as SAT solvers do in practice. So can we re-cast the above check as a SAT solving problem? While not all QBF satisfiability problems can be decided by invoking a SAT solver on a Boolean formula with at most polynomial blow-up, it turns out that we can do this for the above check. It was shown by John et al [18] that the sequence of candidate Skolem functions $\mathbf{F}(\mathbf{X})$ is indeed a sequence of correct Skolem functions iff the following *(Boolean) error formula* is unsatisfiable, where

$\mathbf{Y}'$ is a sequence of fresh Boolean variables with the same dimension as that of $\mathbf{Y}$:

$$\varphi(\mathbf{X}, \mathbf{Y}') \wedge \bigwedge_{i=1}^{n} (y_i \leftrightarrow F_i(\mathbf{X})) \wedge \neg\varphi(\mathbf{X}, \mathbf{Y}).$$

Notice that the size of this formula is linear in the size of the QBF formula we started off with, and there is no need to reason about quantifier alternation now. This makes it possible to use state-of-the-art SAT solvers to implement the crucial checking step in modern Boolean functional synthesis tools [3–5, 15, 16, 18].

Since the checking step invokes a SAT solver, if the error formula turns out to be satisfiable, we can easily obtain a satisfying assignment of the error formula. This is also called a *counterexample* to the claim that the candidate Skolem functions are correct. Having such a counterexample is extremely useful, since it pinpoints at least one assignment of the inputs $\mathbf{X}$ for which the candidate Skolem functions didn't give the right values for the output variables. Repairing the candidate Skolem functions to fix the problem for this single assignment of inputs is relatively easy. However, this is unlikely to scale in practice since there can be exponentially many assignments of $\mathbf{X}$ for which the candidate Skolem functions don't evaluate to the right values of the outputs. Therefore, it is important to generalize a counterexample to multiple assignments of $\mathbf{X}$ where the candidate Skolem functions evaluate to incorrect values. The number of iterations of repair and check is often the limiting factor for scalability of guess-check-repair based synthesis tools. Hence, choosing good initial candidate Skolem functions and generalizing counterexamples aggressively are crucially important in tools based on the guess, check and repair paradigm.

4.4 Knowledge Compilation

A completely different approach to Boolean functional synthesis was presented by Akshay et al. in [1] and generalized further by Shah et al. in [23]. In this approach, a Boolean relational specification is compiled to a special representation from which Skolem functions for all output variables can be generated in polynomial time. This is called *knowledge compilation* for Boolean functional synthesis.

A new normal form for representing Boolean relational specifications, called *Synthesis Negation Normal Form (SynNNF)*, was proposed in [1]. It was further shown that for every relational specification represented in SynNNF, Boolean circuits representing Skolem functions for all outputs can be synthesized in time at most quadratic in the size of the SynNNF specification. It was also shown that several well-studied normal forms for Boolean formulas like disjunctive normal form (DNF), decomposable negation normal form (DNNF) [9], deterministic decomposable negation normal form (dDNNF) [10, 11], binary decision diagrams (BDD) [6], sentential decision diagrams (SDD) [19] either already satisfy the conditions for being in SynNNF or can be compiled efficiently to SynNNF. Therefore, a specification represented in any of these forms can be efficiently processed to yield Skolem functions as Boolean circuits. To show that SynNNF strictly generalizes these other representation forms, the authors of [1] also showed exponential gaps in the size of representation of some families of functions in SynNNF and in these other forms.

A couple of points about SynNNF are worth highlighting here. First, the worst-case complexity of synthesizing Skolem functions from Boolean relational specifications depends not only on the representation of the specification but also on the representation of the Skolem function. The polynomial-time synthesis result for SynNNF representations holds if we are allowed to represent Skolem functions as Boolean circuits. Second, given (conditional) exponential lower bounds for the worst-case time complexity of Boolean functional synthesis, it is inevitable that compilation to SynNNF must incur an exponential blow-up in some cases. Indeed, compiling from conjunctive normal form (CNF) or general Boolean circuits to SynNNF can blow up exponentially in the worst-case. However, from experimental results reported in [1], this doesn't seem to happen frequently in practice. A CNF to SynNNF compilation algorithm was proposed in [1] and experiments using a preliminary implementation of the algorithm showed that knowledge compilation based Boolean functional synthesis works reasonably well in practice.

In [23], the SynNNF form was further generalized to *Subset And-Unrealizable Negation Normal Form (or SAUNF)*. Moreover, it was shown that a specification admits polynomial-time or polynomial-sized synthesis of Skolem functions iff the specification is either represented in SAUNF or can be efficiently compiled to SAUNF. Despite their promise, knowledge compilation based Boolean functional synthesis tools have however not received as much attention as some of the other techniques discussed in this paper. We expect this to change in the near future, with further developments in knowledge compilation techniques.

4.5 Incremental Determinization

A generalization of the idea of identifying functionally dependent variables was presented by Rabe et al [21,22] in a technique called *incremental determinization*. In this approach, we start with a specification in CNF and try to identify patterns of clauses that establish the unique functional dependence of a variable on other variables. This is similar to what we have already discussed in Sect. 4.1. However, Rabe et al. go further and identify patterns of clauses that don't necessarily yield a unique dependence of an output variable on other variables. In such cases, their algorithm "determinizes" the dependence by adding additional constraints to the original specification in a carefully controlled manner. Specifically, only those additional constraints are added that don't restrict the values of inputs for which the specification can be satisfied. Furthermore, the added constraints are such that the original specification with the added constraints imply a unique functional dependence of an output variable on the inputs and on other output variables. As new constraints get added in this manner, we may detect that some added constraints are in conflict with each other. In such cases, the approach backtracks on the decisions made with respect to adding constraints and tries to undo the inconsistencies that may have been introduced. The overall approach follows a style of reasoning very similar to that of conflict-driven clause learning (CDCL) based SAT solvers [25]. In fact, the corresponding tool (called CADET) uses and extends several heuristics that have stood the test of time in CDCL SAT solvers, and performs extremely well in practice.

4.6 Input Output Separation

Yet another approach for Boolean functional synthesis from CNF specifications was presented in [8]. In this approach, given a CNF specification $\varphi(\mathbf{X}, \mathbf{Y})$, two separate formulas are first derived. One of these is obtained by ignoring all input literals (i.e. variables or their negations) in individual clauses of the CNF specification, and the other is obtained by ignoring all output literals in these clauses. This allows us to view the constraints imposed by the given specification on input and output variables separately.

Clearly, if there exists an assignment of the output variables that satisfies each clause of the original CNF specification even while ignoring the input literals, then we can assign these constant values to the output variables and successfully synthesize (constant) Skolem functions. However, we are unlikely to get this lucky in practice, and the specification restricted to output literals will almost always be unsatisfiable. In such cases, we can find an assignment of outputs that satisfies a maximal subset of clauses while ignoring the inputs, and then insist that the remainder of the clauses be satisfied by the assignment of inputs. This effectively gives us an assignment of outputs that works to satisfy the given specification for a set of assignments of inputs. In other words, the above step gives us a partial definition of Skolem functions for all outputs considered together. One can then repeat this process until all combinations of inputs are covered to obtain the complete Skolem functions for all outputs. This yields a complete synthesis algorithm by itself, but one that is driven by "best" possible assignments of outputs, for the given CNF specification.

In a similar manner, suppose we examine the clauses in the original CNF specification, but ignoring output literals this time. Every maximal subset of clauses, all of whose input literals can be falsified at the same time, yields one of the "worst" assignment of inputs for our specification. For every such input assignment, we must ensure that the outputs are assigned values that satisfy each clause in the maximal subsets considered above. This suggests a way to determine assignments of outputs that serve to satisfy the specification for the "worst" input assignments. Once again, this gives a partial definition of Skolem functions for all outputs considered together. By repeating the above process and by ensuring that all such "worst" input assignments are taken care of, we obtain complete Skolem functions for all outputs. This yields a complete synthesis algorithm as well, but one that is driven by the worst assignment of inputs for the given CNF specification.

In practice, the above two approaches are complementary and can be interleaved to yield an algorithm that tries to benefit from the best of both. Experimental results presented in [8] show that this approach has good performance for certain classes of benchmarks for which most other techniques do not perform well.

5 Conclusion

In this paper, we briefly looked under the hood of several state-of-the-art Boolean function synthesis tools to understand what makes them tick. For a more detailed account of these techniques (and beyond), the interested reader is referred to [2]. Among the techniques discussed here, there are some that stand out in their utility – these include identification of unate variables and identification of functionally defined variables. The other techniques have their pros and cons, and experimental evaluation does not allow us to conclude that any single technique dominates all other technique in terms of performance and actual set of problem instances solved. This suggests that the right approach to building practical Boolean function synthesis tools is to use a portfolio approach. The guess-check-repair approach and knowledge compilation approaches have just begun to be explored, and they already appear to hold a lot of promise.

Just as in propositional SAT solving, we don't yet fully understand why state-of-the-art tools are able to scale to large problem instances when we know that fundamental complexity hurdles lurk all along the way. Building a theoretical understanding of what makes these tools tick, and when they choke is an ambitious project that we are far from solving satisfactorily. The knowledge compilation approach provides some insights on the intricate dependency of representation forms with complexity of synthesis. Whether this insight can be used to improve our understanding of the performance of Boolean function synthesis tools beyond those based on knowledge compilation, remains an open question. Finally, synthesizing a set of Skolem functions for a given specification is just the bare minimum that one can ask from a synthesis tool. Interesting questions that deserve serious investigation include synthesis of "optimal" Skolem functions, where optimality can be defined with respect to several metrics. An equally interesting question is to symbolically represent the entire space of Skolem functions that serve to satisfy a given specification. The world of Boolean functional synthesis has just begun to be explored. It promises to remain an active area of research for some time to come. Skolem function synthesis from first order specifications is yet another promising area that has only begun to be explored [7].

Acknowledgements. The author is grateful to several colleagues, collaborators and students including S. Akshay, Jatin Arora, Aman Bansal, Dror Fried, Shubham Goel, Priyanka Golia, Ajith John, Sumith Kulal, Kuldeep S. Meel, Markus Rabe, Divya Raghunathan, Subhajit Roy, Preey Shah, Shetal Shah, Krishna S., Lucas M. Tabajara, Ashutosh Trivedi and Moshe Y. Vardi for discussions and sustained research collaborations, many of which have contributed to some of the techniques discussed in this paper.

References

1. Akshay, S., Arora, J., Chakraborty, S., Krishna, S., Raghunathan, D., Shah, S.: Knowledge compilation for Boolean functional synthesis. In: Proceedings of Formal Methods in Computer Aided Design (FMCAD), pp. 161–169 (2019)
2. Akshay, S., Chakraborty, S.: Synthesizing Skolem functions: a view from theory and practice. In: Sarukkai, S., Chakraborty, M. (eds.) Handbook of Logical Thought in India, pp. 1–36. Springer, New Delhi (2020)
3. Akshay, S., Chakraborty, S., Goel, S., Kulal, S., Shah, S.: Boolean functional synthesis: hardness and practical algorithms. Formal Methods Syst. Des. **57**(1), 53–86 (2020). https://doi.org/10.1007/s10703-020-00352-2
4. Akshay, S., Chakraborty, S., Goel, S., Kulal, S., Shah, S.: What's hard about Boolean functional synthesis? In: Proceedings of CAV 2018, pp. 251–269 (2018)
5. Akshay, S., Chakraborty, S., John, A.K., Shah, S.: Towards parallel Boolean functional synthesis. In: Legay, A., Margaria, T. (eds.) TACAS 2017. LNCS, vol. 10205, pp. 337–353. Springer, Heidelberg (2017). https://doi.org/10.1007/978-3-662-54577-5_19
6. Bryant, R.E.: Graph-based algorithms for Boolean function manipulation. IEEE Trans. Comput. **35**(8), 677–691 (1986). https://doi.org/10.1109/TC.1986.1676819
7. Chakraborty, S., Akshay, S.: On synthesizing computable Skolem functions for first order logic. In: Szeider, S., Ganian, R., Silva, A. (eds.) 47th International Symposium on Mathematical Foundations of Computer Science, MFCS 2022, August 22–26, 2022, Vienna, Austria. LIPIcs, vol. 241, pp. 30:1–30:15. Schloss Dagstuhl - Leibniz-Zentrum für Informatik (2022)
8. Chakraborty, S., Fried, D., Tabajara, L.M., Vardi, M.Y.: Functional synthesis via input-output separation. In: 2018 Formal Methods in Computer Aided Design, FMCAD 2018, Austin, TX, USA, 30 October–2 November 2018, pp. 1–9 (2018)
9. Darwiche, A.: Decomposable negation normal form. J. ACM **48**(4), 608–647 (2001)
10. Darwiche, A.: On the tractable counting of theory models and its application to truth maintenance and belief revision. J. Appl. Non-classical Log. **11**(1–2), 11–34 (2001)
11. Darwiche, A.: A compiler for deterministic, decomposable negation normal form. In: Proceedings of the Eighteenth National Conference on Artificial Intelligence (AAAI), Menlo Park, California, pp. 627–634. AAAI Press (2002)
12. Finkbeiner, B.: Synthesis of reactive systems. In: Esparza, J., Grumberg, O., Sickert, S. (eds.) Dependable Software Systems Engineering, NATO Science for Peace and Security Series - D: Information and Communication Security, vol. 45, pp. 72–98. IOS Press (2016). https://doi.org/10.3233/978-1-61499-627-9-72
13. Fried, D., Tabajara, L.M., Vardi, M.Y.: BDD-based Boolean functional synthesis. In: Chaudhuri, S., Farzan, A. (eds.) CAV 2016. LNCS, vol. 9780, pp. 402–421. Springer, Cham (2016). https://doi.org/10.1007/978-3-319-41540-6_22
14. Golia, P., Roy, S., Meel, K.S.: Program synthesis as dependency quantified formula modulo theory. In: Zhou, Z.H. (ed.) Proceedings of the Thirtieth International Joint Conference on Artificial Intelligence, IJCAI 2021, pp. 1894–1900. International Joint Conferences on Artificial Intelligence Organization (2021). https://doi.org/10.24963/ijcai.2021/261. Main Track
15. Golia, P., Slivovsky, F., Roy, S., Meel, K.S.: Engineering an efficient Boolean functional synthesis engine. In: IEEE/ACM International Conference on Computer Aided Design (ICCAD), pp. 1–9 (2021). https://doi.org/10.1109/ICCAD51958.2021.9643583

16. Golia, P., Roy, S., Meel, K.S.: Manthan: a data-driven approach for Boolean function synthesis. In: Lahiri, S.K., Wang, C. (eds.) CAV 2020. LNCS, vol. 12225, pp. 611–633. Springer, Cham (2020). https://doi.org/10.1007/978-3-030-53291-8_31
17. Jo, S., Matsumoto, T., Fujita, M.: SAT-based automatic rectification and debugging of combinational circuits with LUT insertions. In: Proceedings of the 2012 IEEE 21st Asian Test Symposium, ATS 2012, pp. 19–24. IEEE Computer Society (2012)
18. John, A., Shah, S., Chakraborty, S., Trivedi, A., Akshay, S.: Skolem functions for factored formulas. In: FMCAD, pp. 73–80 (2015)
19. Oztok, U., Darwiche, A.: A top-down compiler for sentential decision diagrams. In: Proceedings of the 24th International Joint Conference on Artificial Intelligence (IJCAI), pp. 3141–3148 (2015)
20. Padoa, A.: Essai d'une théorie algébrique des nombres entiers, précédé d'une introduction logique à une théorie déductive quelconque. Bibliothèque du Congrès International de Philosophie **3**, 309 (1901)
21. Rabe, M.N., Seshia, S.A.: Incremental determinization. In: Creignou, N., Le Berre, D. (eds.) SAT 2016. LNCS, vol. 9710, pp. 375–392. Springer, Cham (2016). https://doi.org/10.1007/978-3-319-40970-2_23
22. Rabe, M.N.: Incremental determinization for quantifier elimination and functional synthesis. In: Dillig, I., Tasiran, S. (eds.) CAV 2019. LNCS, vol. 11562, pp. 84–94. Springer, Cham (2019). https://doi.org/10.1007/978-3-030-25543-5_6
23. Shah, P., Bansal, A., Akshay, S., Chakraborty, S.: A normal form characterization for efficient Boolean Skolem function synthesis. In: 36th Annual ACM/IEEE Symposium on Logic in Computer Science (LICS), pp. 1–13 (2021). https://doi.org/10.1109/LICS52264.2021.9470741
24. Shukla, A., Bierre, A., Siedl, M., Pulina, L.: A survey on applications of quantified Boolean formula. In: Proceedings of the Thirty-First International Conference on Tools with Artificial Intelligence (ICTAI), pp. 78–84 (2019)
25. Silva, J.P.M., Lynce, I., Malik, S.: Conflict-driven clause learning sat solvers, chap. 4. In: Biere, A., Heule, M., van Maaren, H., Walsch, T. (eds.) Handbook of Satisfiability, pp. 131–153. IOS Press (2021)
26. Srivastava, S., Gulwani, S., Foster, J.S.: Template-based program verification and program synthesis. STTT **15**(5–6), 497–518 (2013)
27. Tabajara, L.M., Vardi, M.Y.: Factored Boolean functional synthesis. In: 2017 Formal Methods in Computer Aided Design, FMCAD 2017, Vienna, Austria, 2–6 October 2017, pp. 124–131 (2017)
28. Tseitin, G.S.: On the complexity of derivation in propositional calculus. In: Structures in Constructive Mathematics and Mathematical Logic, Part II, Seminars in Mathematics, pp. 115–125 (1968)

Labelled Calculi for Lattice-Based Modal Logics

Ineke van der Berg[1,3], Andrea De Domenico[1], Giuseppe Greco[1(✉)], Krishna B. Manoorkar[1], Alessandra Palmigiano[1,2], and Mattia Panettiere[1]

[1] School of Business and Economics, Vrije Universiteit Amsterdam, Amsterdam, The Netherlands
{i.van.der.berg,a.de.domenico,g.greco,k.b.manoorkar,alessandra.palmigiano, m.panettiere}@vu.nl

[2] Department of Mathematics and Applied Mathematics, University of Johannesburg, Johannesburg, South Africa

[3] Department of Mathematical Sciences, Stellenbosch University, Stellenbosch, South Africa

Abstract. We introduce labelled sequent calculi for the basic normal non-distributive modal logic **L** and 31 of its axiomatic extensions, where the labels are atomic formulas of a first order language which is interpreted on the canonical extensions of the algebras in the variety corresponding to the logic **L**. Modular proofs are presented that these calculi are all sound, complete and conservative w.r.t. **L**, and enjoy cut elimination and the subformula property. The introduction of these calculi showcases a general methodology for introducing labelled calculi for the class of LE-logics and their analytic axiomatic extensions in a principled and uniform way.

Keywords: Non-distributive modal logic · Algorithmic proof theory · Algorithmic correspondence theory · Labelled calculi

1 Introduction

The present paper pertains to a line of research in structural proof theory aimed at generating analytic calculi for wide classes of nonclassical logics in a principled and uniform way. Since the 1990s, semantic information about given logical frameworks has proven key to generate calculi with excellent properties [19]. The contribution of semantic information has been particularly perspicuous in the introduction of labelled calculi for e.g. classical normal modal logic [16] and intuitionistic logic [17], and their axiomatic extensions defined by axioms for which first-order correspondents exist of a certain syntactic shape [10]. Moreover, recently, the underlying link between the principled and algorithmic generation of analytic rules for capturing axiomatic extensions of given logics and the systematic access to, and use of, semantic information for this purpose has been established also in the context of other proof-theoretic formats, such as proper display calculi [1,15], and relative to classes of logics as wide as the normal (D)LE-logics, i.e. those logics canonically associated with varieties of normal (distributive) lattice expansions [7] (cf. Definition 1.1). In particular, in [15], the same algorithm ALBA which computes the first-order correspondents of (analytic) inductive axioms in

M. Banerjee and A. V. Sreejith (Eds.): ICLA 2023, LNCS 13963, pp. 23–47, 2023.
https://doi.org/10.1007/978-3-031-26689-8_3

any (D)LE-signature was used to generate the analytic rules in a suitable proper display calculus corresponding to those axioms.

The algorithm ALBA [6,7] is among the main tools in unified correspondence theory [5], and allows not only for the mechanization of well known correspondence arguments from modal logic, but also for the uniform generalization of these arguments to (D)LE-logics, thanks to the fact that the ALBA-computations are motivated by and interpreted in an algebraic environment in which the classic model-theoretic correspondence arguments can be rephrased in terms of the order-theoretic properties of the algebraic interpretations of the logical connectives. These properties guarantee the soundness of the rewriting rules applied in ALBA-computations, thanks to which, the first-order correspondent of a given input axiom (in any given LE-language $\mathcal{L}$) is generated in a language $\mathcal{L}^+$ expanding $\mathcal{L}$, which is interpreted in the canonical extensions of $\mathcal{L}$-algebras.

In the present paper, we showcase how the methodology adopted in [15] for introducing proper display calculi for (D)LE-logics and their analytic axiomatic extensions can be used also for endowing LE-logics with labelled calculi. Specifically, we focus on a particularly simple LE-logic, namely the basic normal non-distributive (i.e. lattice-based) modal logic **L** [3,4], for which we introduce a labelled calculus and show its basic properties, namely soundness, completeness, cut-elimination and subformula property. Moreover, we discuss, by way of examples, how ALBA can be used to generate analytic rules corresponding to (analytic inductive) axiomatic extensions of the basic logic **L**.

Structure of the Paper. Section 2 recalls preliminaries on basic normal non-distributive logic, canonical extensions and the algorithm ALBA, Sect. 3 presents a labelled calculus for normal non-distributive logic and its extensions. Section 4 proves soundness, completeness, cut elimination and subformula property for basic normal non-distributive logic and some of its axiomatic extensions. Section 5 shows that the all calculi introduced in the paper are proper labelled calculi. We conclude in Sect. 6. In Appendix A we provide the formal definition of proper labelled calculi and we show that any calculus in this class enjoys the canonical cut elimination *à la* Belnap.

2 Preliminaries

2.1 Basic Normal Non-distributive Modal Logic, Its Associated ALBA-language, and Some of Its Axiomatic Extensions

The basic normal non-distributive modal logic is a normal LE-logic (cf. [7,8]) which was used in [3,4] as the underlying environment for an epistemic logic of categories and formal concepts, and in [2] as the underlying environment of a logical theory unifying Formal Concept Analysis [13] and Rough Set Theory [18].

Let Prop be a (countable or finite) set of atomic propositions. The language $\mathcal{L}$ is defined as follows:

$$\varphi := \bot \mid \top \mid p \mid \varphi \wedge \varphi \mid \varphi \vee \varphi \mid \Box\varphi \mid \Diamond\varphi,$$

where $p \in \mathsf{Prop}$. The extended language $\mathcal{L}^+$, used in ALBA-computations taking inequalities of $\mathcal{L}$-terms in input, is defined as follows:

$$\psi ::= \mathbf{j} \mid \mathbf{m} \mid \varphi \mid \psi \wedge \psi \mid \psi \vee \psi \mid \Box\psi \mid \Diamond\psi \mid \blacksquare\psi \mid \blacklozenge\psi,$$

where $\varphi \in \mathcal{L}$, and the variables $\mathbf{j} \in \mathsf{NOM}$ (resp. $\mathbf{m} \in \mathsf{CNOM}$), referred to as *nominals* (resp. *co-nominals*), range over disjoint sets which are also disjoint from Prop. The *basic*, or *minimal normal $\mathcal{L}$-logic* is a set $\mathbf{L}$ of sequents $\varphi \vdash \psi$, with $\varphi, \psi \in \mathcal{L}$, containing the following axioms:

$$p \vdash p \quad \bot \vdash p \quad p \vdash p \vee q \quad p \wedge q \vdash p \quad \top \vdash \Box\top \quad \Box p \wedge \Box q \vdash \Box(p \wedge q)$$
$$p \vdash \top \quad q \vdash p \vee q \quad p \wedge q \vdash q \quad \Diamond\bot \vdash \bot \quad \Diamond(p \vee q) \vdash \Diamond p \vee \Diamond q$$

and closed under the following inference rules:

$$\frac{\varphi \vdash \chi \quad \chi \vdash \psi}{\varphi \vdash \psi} \quad \frac{\varphi \vdash \psi}{\varphi(\chi/p) \vdash \psi(\chi/p)} \quad \frac{\chi \vdash \varphi \quad \chi \vdash \psi}{\chi \vdash \varphi \wedge \psi} \quad \frac{\varphi \vdash \chi \quad \psi \vdash \chi}{\varphi \vee \psi \vdash \chi} \quad \frac{\varphi \vdash \psi}{\Box\varphi \vdash \Box\psi} \quad \frac{\varphi \vdash \psi}{\Diamond\varphi \vdash \Diamond\psi}$$

An *$\mathcal{L}$-logic* is any extension of $\mathbf{L}$ with $\mathcal{L}$-axioms $\varphi \vdash \psi$. In what follows, for any set Σ of $\mathcal{L}$-axioms, we let $\mathbf{L}.\Sigma$ denote the axiomatic extension of $\mathbf{L}$ generated by Σ. Throughout the paper, we will consider all subsets Σ of the set of axioms listed in the table below. These axioms are well known from classical modal logic, and have also cropped up in [2] in the context of the definition of relational structures simultaneously generalizing Formal Concept Analysis and Rough Set Theory.

(4)	$\Diamond\Diamond A \vdash \Diamond A$	transitivity	(D)	$\Box A \vdash \Diamond A$	seriality
(T)	$\Box A \vdash A$	reflexivity	(C)	$\Diamond\Box A \vdash \Box\Diamond A$	confluence
(B)	$A \vdash \Box\Diamond A$	symmetry			

2.2 $\mathcal{L}$-algebras, Their Canonical Extensions, and the Algebraic Interpretation of the Extended Language of ALBA

In the present section, we recall the definitions of the normal lattice expansions canonically associated with the basic logic $\mathbf{L}$, their canonical extensions, the existence of which can be shown both constructively and non-constructively, and the interpretation of the extended language $\mathcal{L}^+$ in the canonical extensions of $\mathcal{L}$-algebras.

An *$\mathcal{L}$-algebra* is a tuple $\mathbb{A} = (L, \Diamond^{\mathbb{A}}, \Box^{\mathbb{A}})$, where L is a bounded lattice, $\Diamond^{\mathbb{A}}$ (resp. $\Box^{\mathbb{A}}$) is a finitely join-preserving (resp. finitely meet-preserving) unary operation. That is, besides the usual identities defining general lattices, the following identities hold:

$$\Diamond(x \vee y) = \Diamond x \vee \Diamond y \qquad \Diamond\bot = \bot \qquad \Box(x \wedge y) = \Box x \wedge \Box y \qquad \Box\top = \top.$$

In what follows, we let $\mathsf{Alg}(\mathcal{L})$ denote the class of $\mathcal{L}$-algebras. Let L be a (bounded) sublattice of a complete lattice L'.

1. L is *dense* in L' if every element of L' can be expressed both as a join of meets and as a meet of joins of elements from L. We let $K(L')$ (resp. $O(L')$) denote the meet-closure (resp. join-closure) of L in L'. That is, $K(L') = \{k \in L' \mid k = \bigwedge S$ for some $S \subseteq L\}$, and $O(L') = \{o \in L' \mid o = \bigvee T$ for some $T \subseteq L\}$.

2. L is *compact* in L' if, for all $S, T \subseteq L$, if $\bigwedge S \leq \bigvee T$ then $\bigwedge S' \leq \bigvee T'$ for some finite $S' \subseteq S$ and $T' \subseteq T$.
3. The *canonical extension* of a lattice L is a complete lattice L^δ containing L as a dense and compact sublattice. Elements in $K(L^\delta)$ (resp. $O(L^\delta)$) are the *closed* (resp. *open*) *elements* of L^δ.

As is well known (cf. [14]), the canonical extension of a lattice L exists and is unique up to an isomorphism fixing L. The non-constructive proof of existence can be achieved via suitable dualities for lattices, while the constructive proof uses the MacNeille completion construction on a certain poset obtained from the families of proper lattice filters and ideals of the original lattice L (cf. [11, 14] for details). In the latter case, the ensuing complete lattice L^δ can be shown to be *perfect*, i.e., to be both completely join-generated by the set $J^\infty(L^\delta) \subseteq K(L^\delta)$ of the completely join-irreducible elements of L^δ, and completely meet-generated by the set $M^\infty(L^\delta) \subseteq O(L^\delta)$ of the completely meet-irreducible elements of L^δ.[1]

For every unary, order-preserving operation $f : L \to L$, the *σ-extension* of f is defined first on any $k \in K(L^\delta)$ and then on every $u \in L^\delta$ as follows:

$$f^\sigma(k) := \bigwedge\{f(a) \mid a \in L \text{ and } k \leq a\} \qquad f^\sigma(u) := \bigvee\{f^\sigma(k) \mid k \in K(L^\delta) \text{ and } k \leq u\}.$$

The *π-extension* of f is defined first on every $o \in O(L^\delta)$, and then on every $u \in L^\delta$ as follows:

$$f^\pi(o) := \bigvee\{f(a) \mid a \in L \text{ and } a \leq o\} \qquad f^\pi(u) := \bigwedge\{f^\pi(o) \mid o \in O(L^\delta) \text{ and } u \leq o\}.$$

Defined as above, the σ- and π-extensions maps are monotone, and coincide with f on the elements of $\mathbb{A}$. Moreover, the σ-extension (resp. (resp. π-extension) of a finitely join-preserving (resp. finitely meet-preserving) map is *completely* join-preserving (resp. *completely* meet-preserving). This justifies defining the *canonical extension of an $\mathcal{L}$-algebra* $\mathbb{A} = (L, \Box, \Diamond)$ as the $\mathcal{L}$-algebra $\mathbb{A}^\delta := (L^\delta, \Box^\pi, \Diamond^\sigma)$. By construction, $\mathbb{A}$ is a subalgebra of $\mathbb{A}^\delta$ for any $\mathbb{A} \in \mathsf{Alg}(\mathcal{L})$. In fact, compared to arbitrary $\mathcal{L}$-algebras, $\mathbb{A}^\delta$ enjoys additional properties that make it a suitable semantic environment for the extended language $\mathcal{L}^+$ of Sect. 2.1. Indeed, the lattice reduct of $\mathbb{A}^\delta$ is a *complete* lattice. Together with the fact that the operations $\Diamond^\sigma$ and $\Box^\pi$ do not preserve only *finite* joins and meets respectively, but *arbitrary* joins and meets, this implies, by well known order-theoretic facts (cf. [9, Proposition 7.34]), that the right and left adjoint of $\Diamond^\sigma$ and of $\Box^\pi$ are well defined on $\mathbb{A}^\delta$, which we denote $\blacksquare^{\mathbb{A}^\delta}$ and $\blacklozenge^{\mathbb{A}^\delta}$ respectively,[2] and provide the interpretations of the corresponding logical connectives in $\mathcal{L}^+$. Moreover, by denseness, $\mathbb{A}^\delta$ is both completely join-generated by the elements in $K(L^\delta)$ and completely meet-generated by the elements in $O(L^\delta)$, and when considering the non-constructive proof, these families of generators can be further restricted to $J^\infty(L^\delta)$ and $M^\infty(L^\delta)$,

[1] For any complete lattice L, any $j \in L$ is completely join-irreducible if $j \neq \bot$ and for any $S \subseteq L$, if $j = \bigvee S$ then $j \in S$. Dually, any $m \in L$ is completely meet-irreducible if $m \neq \top$ and for any $S \subseteq L$, if $m = \bigwedge S$ then $m \in S$.

[2] The unary operations $\blacksquare^{\mathbb{A}^\delta}$ and $\blacklozenge^{\mathbb{A}^\delta}$ on $\mathbb{A}^\delta$ are the unique maps satisfying the equivalences $\Diamond^\sigma u \leq v$ iff $u \leq \blacksquare^{\mathbb{A}^\delta} v$ and $\blacklozenge^{\mathbb{A}^\delta} u \leq v$ iff $u \leq \Box^\pi v$ for all $u, v \in \mathbb{A}^\delta$.

respectively. These generating subsets provide the interpretation of the variables in NOM and CNOM, respectively. As is well known, for any set Σ of $\mathcal{L}$-sequents, if $\mathsf{K}(\Sigma) = \{\mathbb{A} \in \mathsf{Alg}(\mathcal{L}) \mid \mathbb{A} \models \Sigma\}$ is closed under taking canonical extensions,[3] then the axiomatic extension $\mathbf{L}.\Sigma$ is complete w.r.t. the subclass $\mathsf{K}^\delta(\Sigma) = \{\mathbb{A}^\delta \mid \mathbb{A} \in \mathsf{K}(\Sigma)\}$, because any non-theorem $\xi \vdash \chi$ will be falsified in the Lindenbaum-Tarski algebra $\mathbb{A}$ of $\mathbf{L}.\Sigma$, which is an element of $\mathsf{K}(\Sigma)$ by construction, and hence $\xi \vdash \chi$ will be falsified under the same assignment in $\mathbb{A}^\delta$, given that $\mathbb{A}$ is a subalgebra of $\mathbb{A}^\delta$.

2.3 The Algorithm ALBA

The algorithm ALBA is guaranteed to succeed on a large class of formulas, called (analytic) inductive axioms, and it can be used to automatically generate labelled calculi with good properties equivalently capturing the LE-logics axiomatized by means of those axioms. We refer the reader to [7, Section 6,8] for the proof of correctness and success in the general setting of LE-logics. In the present section, we informally illustrate how the algorithm ALBA works by means of examples, namely, we run ALBA on the modal axioms in $\Sigma = \{\Box p \vdash p, p \vdash \Box\Diamond p, \Box p \vdash \Diamond p, \Diamond\Box p \vdash \Box\Diamond p\}$ computing their first-order correspondent, which, in turn, can be automatically transformed into an analytic structural rule of a labelled calculus equivalently capturing the axioms (see the table at the end of Sect. 3). In what follows, $\mathbb{A}$ denotes an $\mathcal{L}$-algebra, and $\mathbb{A}^\delta$ denotes its canonical extension. We abuse notation and use the same symbol for the algebra and its domain. We recall that variables $\mathbf{j}, \mathbf{h}$ and $\mathbf{k}$ (resp. $\mathbf{m}$) range in the set of the complete join-generators (resp. complete meet-generators) of $\mathbb{A}^\delta$.

The following chain of equivalences is sound on $\mathbb{A}^\delta$:

$\forall p(\Diamond\Diamond p \leq \Diamond p)$	
iff $\forall p \forall \mathbf{j} \forall \mathbf{m}((\mathbf{j} \leq p \;\&\; \Diamond p \leq \mathbf{m}) \Rightarrow \Diamond\Diamond\mathbf{j} \leq \mathbf{m})$	join- and meet-generation, $\Diamond$ c. join-preserving
iff $\forall \mathbf{j} \forall \mathbf{m}(\Diamond\mathbf{j} \leq \mathbf{m} \Rightarrow \Diamond\Diamond\mathbf{j} \leq \mathbf{m})$	Ackermann's lemma
iff $\forall \mathbf{j} \forall \mathbf{h} \forall \mathbf{m}(\Diamond\mathbf{j} \leq \mathbf{m} \Rightarrow (\mathbf{h} \leq \Diamond\mathbf{j} \Rightarrow \Diamond\mathbf{h} \leq \mathbf{m}))$	

Indeed, the first equivalence in the chain above is due to the fact that, since the variable $\mathbf{j}$ (resp. $\mathbf{m}$) ranges over a completely join-generating (resp. completely meet-generating) subset of $\mathbb{A}^\delta$, and $\Diamond$ is completely join-preserving, we can equivalently rewrite the initial inequality as follows: $\forall p(\bigvee\{\Diamond\Diamond\mathbf{j} \mid \mathbf{j} \leq p\} \leq \bigwedge\{\mathbf{m} \mid \Diamond p \leq \mathbf{m}\})$, which yields the required equivalence by the definition of the least upper bound and the greatest lower bound of subsets of a poset. The second equivalence is an instance of the core rule of ALBA, which allows to eliminate the quantification over proposition variables. As to the direction from bottom to top, by the monotonicity of $\Diamond$, the inequalities $\mathbf{j} \leq p$ and $\Diamond p \leq \mathbf{m}$ immediately imply $\Diamond\mathbf{j} \leq \Diamond p \leq \mathbf{m}$, from which the required inequality $\Diamond\Diamond\mathbf{j} \leq \mathbf{m}$ follows by assumption. For the converse direction, for a given interpretations of $\mathbf{j}$ and $\mathbf{m}$ such that $\Diamond\mathbf{j} \leq \mathbf{m}$, we let p have the same interpretation as $\mathbf{j}$. Then this interpretation satisfies both inequalities $\mathbf{j} \leq p$ and $\Diamond p = \Diamond\mathbf{j} \leq \mathbf{m}$, from which the required inequality $\Diamond\Diamond\mathbf{j} \leq \mathbf{m}$ follows by assumption. The third equivalence immediately follows from considerations similar to those made for justifying the first equivalence; namely, that the inequality $\Diamond\Diamond\mathbf{j} \leq \mathbf{m}$ can be equivalently rewritten as $\bigvee\{\Diamond\mathbf{h} \mid \mathbf{h} \leq \Diamond\mathbf{j}\} \leq \mathbf{m}$,

[3] By the general theory of unified correspondence (cf. [7]), this is the case of every subset Σ of the set of axioms listed at the end of Sect. 2.1.

which yields the required equivalence by the definition of a subset of a poset. Analogous arguments can be made to justify the following chains of equivalences:

$$
\begin{array}{lll}
& \forall p(\Box p \leq p) & \\
\text{iff} & \forall p \forall \mathbf{j} \forall \mathbf{m}\, ((\mathbf{j} \leq \Box p \;\&\; p \leq \mathbf{m}) \Rightarrow \mathbf{j} \leq \mathbf{m}) & \text{join- and meet-generation} \\
\text{iff} & \forall \mathbf{j} \forall \mathbf{m}\, (\mathbf{j} \leq \Box \mathbf{m} \Rightarrow \mathbf{j} \leq \mathbf{m}) & \text{Ackermann's lemma}
\end{array}
$$

$$
\begin{array}{lll}
& \forall p(p \leq \Box \Diamond p) & \\
\text{iff} & \forall p \forall \mathbf{j} \forall \mathbf{m}\, ((\mathbf{j} \leq p \;\&\; \Diamond p \leq \mathbf{m}) \Rightarrow \mathbf{j} \leq \Box \mathbf{m}) & \text{join- and meet-generation} \\
\text{iff} & \forall \mathbf{j} \forall \mathbf{m}\, (\Diamond \mathbf{j} \leq \mathbf{m} \Rightarrow \mathbf{j} \leq \Box \mathbf{m}) & \text{Ackermann's lemma}
\end{array}
$$

$$
\begin{array}{lll}
& \forall p(\Box p \leq \Diamond p) & \\
\text{iff} & \forall p \forall \mathbf{j} \forall \mathbf{m}\, ((\mathbf{j} \leq \Box p \;\&\; \Diamond p \leq \mathbf{m}) \Rightarrow \mathbf{j} \leq \mathbf{m}) & \text{join- and meet-generation} \\
\text{iff} & \forall p \forall \mathbf{j} \forall \mathbf{m}\, ((\blacklozenge \mathbf{j} \leq p \;\&\; \Diamond p \leq \mathbf{m}) \Rightarrow \mathbf{j} \leq \mathbf{m}) & \blacklozenge \dashv \Box \text{ adjunction} \\
\text{iff} & \forall \mathbf{j} \forall \mathbf{m}\, (\Diamond \blacklozenge \mathbf{j} \leq \mathbf{m} \Rightarrow \mathbf{j} \leq \mathbf{m}) & \text{Ackermann's lemma} \\
\text{iff} & \forall \mathbf{j} \forall \mathbf{m} \exists \mathbf{k}\, ((\mathbf{k} \leq \blacklozenge \mathbf{j} \Rightarrow \Diamond \mathbf{k} \leq \mathbf{m}) \Rightarrow \mathbf{j} \leq \mathbf{m}) & \text{join-generation}
\end{array}
$$

$$
\begin{array}{lll}
& \forall p(\Diamond \Box p \leq \Box \Diamond p) & \\
\text{iff} & \forall p \forall \mathbf{j} \forall \mathbf{m}\, ((\mathbf{j} \leq \Box p \;\&\; \Diamond p \leq \mathbf{m}) \Rightarrow \Diamond \mathbf{j} \leq \Box \mathbf{m}) & \text{join- and meet-generation} \\
\text{iff} & \forall p \forall \mathbf{j} \forall \mathbf{m}\, ((\blacklozenge \mathbf{j} \leq p \;\&\; \Diamond p \leq \mathbf{m}) \Rightarrow \blacklozenge \Diamond \mathbf{j} \leq \mathbf{m}) & \blacklozenge \dashv \Box \text{ adjunction} \\
\text{iff} & \forall \mathbf{j} \forall \mathbf{m}\, (\Diamond \blacklozenge \mathbf{j} \leq \mathbf{m} \Rightarrow \blacklozenge \Diamond \mathbf{j} \leq \mathbf{m}) & \text{Ackermann's lemma} \\
\text{iff} & \forall \mathbf{j} \forall \mathbf{m} \forall \mathbf{h} \exists \mathbf{k}\, ((\mathbf{k} \leq \blacklozenge \mathbf{j} \Rightarrow \Diamond \mathbf{k} \leq \mathbf{m}) \Rightarrow (\mathbf{h} \leq \Diamond \mathbf{j} \Rightarrow \blacklozenge \mathbf{h} \leq \mathbf{m})) & \text{join- and meet-generation}
\end{array}
$$

The second equivalence in the chain above is based on the existence of the adjoints of the maps interpreting the original connectives on canonical extensions of $\mathcal{L}$-algebras (cf. Sect. 2.2). Finally, we remark that carrying out the correspondence arguments above in the algebraic environment of the canonical extensions of $\mathcal{L}$-algebras allows us to clearly identify their pivotal properties, and, in particular, to verify that no property related with the setting of (perfect) distributive lattices (viz. the complete join-primeness of the elements interpreting nominal variables) is required.

3 The Labelled Calculus A.L and Some of Its Extensions

In what follows, we use $p, q, \ldots$ for proposition variables, $A, B, \ldots$ for *formulas* metavariables (in the original language of the logic), $\mathbf{j}, \mathbf{i}, \ldots$ for nominal variables, $\mathbf{m}, \mathbf{n}, \ldots$ for conominal variables, $\mathbf{J}, \mathbf{H}, \ldots$ (resp. $\mathbf{M}, \mathbf{N}, \ldots$) for *nominal terms* metavariables (resp. *conominal terms*), $\mathbf{T}$ for *terms* metavariables, and $\Gamma, \Delta, \ldots$ for *meta-structures* metavariables. Given $p \in \mathsf{Prop}$, $\mathbf{j} \in \mathsf{NOM}$, $\mathbf{m} \in \mathsf{CNOM}$, the language of (labelled) formulas, terms and structures is defined as follows:

$$
\begin{array}{rcl}
\text{formulas} \ni A & ::= & p \mid \top \mid \bot \mid A \wedge A \mid A \vee A \mid \Box A \mid \Diamond A \\
\text{nominal terms} \ni \mathbf{J} & ::= & \mathbf{j} \mid \Diamond \mathbf{j} \mid \blacklozenge \mathbf{j} \\
\text{conominal terms} \ni \mathbf{M} & ::= & \mathbf{m} \mid \Box \mathbf{m} \mid \blacksquare \mathbf{m} \\
\text{terms} \ni \mathbf{T} & ::= & \mathbf{J} \mid \mathbf{M} \\
\text{labelled formulas} \ni a & ::= & \mathbf{j} \leq A \mid A \leq \mathbf{m} \\
\text{pure structures} \ni t^{*} & ::= & \mathbf{j} \leq \mathbf{T} \mid \mathbf{T} \leq \mathbf{m} \\
\text{structures} \ni \sigma & ::= & a \mid t \\
\text{meta-structures} \ni \Gamma & ::= & \sigma \mid \Gamma, \Gamma
\end{array}
$$

*Side condition: $\mathbf{j}$ and $\mathbf{m}$ do not occur in $\mathbf{T}$.

Let us first recall some terminology (see e.g. [21, Section 4.1]) and notation. A (**A.L**-)*sequent* is a pair $\Gamma \vdash \Delta$ where Γ and Δ (the *antecedent* and the *consequent* of the sequent, respectively) are metavariables for meta-structures separated by commas. An *inference* r, also called an instance of a rule, is a pair (S, s) of a (possibly empty) set of sequents S (the premises) and a sequent s (the conclusion). We identify a *rule* R with the set of all instances that are instantiations of R. A *rule* R, also referred to as a *scheme* is usually presented schematically using metavariables for meta-structures (denoted by upper-case Greek letters $\Gamma, \Delta, \Pi, \Sigma, \ldots, \Gamma_1, \Gamma_2, \ldots$), or metavariables for structures (denoted by lower-case Latin letters: $a, b, c, \ldots, a_1, a_2, \ldots$ for labelled formulas and $t_1, t_2, t_3, \ldots$ for pure structures), or metavariables for formulas (denoted by $A, B, C, \ldots A_1, A_2, \ldots$, or metavariables for terms (denoted by $\mathbf{j}, \mathbf{i}, \mathbf{h}, \ldots, \mathbf{j}_1, \mathbf{j}_2, \ldots$ for nominal terms and $\mathbf{m}, \mathbf{n}, \mathbf{o}, \ldots, \mathbf{m}_1, \mathbf{m}_2, \ldots$ for conominal terms). A rule R with no premises, i.e. $S = \emptyset$, is called an *axiom scheme*, and an instantiation of such R is called an *axiom*. The immediate subformulas of a principal formula (see Definition 1) in the premise(s) of an operational inference are called *auxiliary formulas*. The formulas that are not preserved in an inference instantiating the cut rule are called *cut formulas*. If the cut formulas are principal in an inference instantiating the cut rule, then the inference is called *principal cut*. A cut that is not principal is called parametric. A *proof* of (the instantiation of) a sequent $\Gamma \vdash \Delta$ is a tree where (the instantiation of) $\Gamma \vdash \Delta$ occurs as the end-sequent, all the leaves are (instantiations of) axioms, and each node is introduced via an inference. Before providing the list of the primitive rules of **A.L**, we need two preliminary definitions.

Definition 1 (Analysis). *The* specifications *are instantiations of meta-structure metavariables in the statement of R. The* parameters *of $r \in R$ are substructures of instantiations of (meta-)structure metavariables in the statement of R. A formula instance is* principal *in an inference $r \in R$ if it is not a parameter in the conclusion of r (except for switch rules).*

(Meta-)Structure occurrences in an inference $r \in R$ are in the (symmetric) relation of local congruence *in r if they instantiate the same metavariable occurring in the same position in a premise and in the conclusion of R, or they instantiate nonparametric structures in the application of switch rules (namely, in the case of* **A.L**Σ*, occurrences of labelled formulas $\mathbf{j} \leq A$ and $A \leq \mathbf{m}$, or occurrences of pure structures $\mathbf{j} \leq \mathbf{T}$ and $\mathbf{T} \leq \mathbf{m}$). Therefore, the local congruence is a relation between specifications.*

Two occurrences instatiating a (meta-)structure are in the inference congruence relation *if they are locally congruent in an inference r occurring in a proof π. The* proof congruent relation *is the transitive closure of the* inference congruence relation *in a derivation π.*

Definition 2 (Position). *For any well-formed sequent $\Gamma \vdash \Delta$,*

- *The occurrence of a labelled formula $\mathbf{j} \leq A$ (resp. $A \leq \mathbf{m}$) is in* precedent *position if $\mathbf{j} \leq A \in \Gamma$ (resp. $A \leq \mathbf{m} \in \Delta$), and it is in* succedent *position if $\mathbf{j} \leq A \in \Delta$ (resp. $A \leq \mathbf{m} \in \Gamma$);*
- *any occurrence of a pure structure $\mathbf{j} \leq \mathbf{T}$ in Γ (resp. Δ) is in* precedent *(resp.* succedent*) position; any occurrence of a pure structure $\mathbf{T} \leq \mathbf{m}$ in Γ (resp. Δ) is in* succedent *(resp.* precedent*) position.*

We follow the notational conventions as stated in Definition 1, which provides the so-called *analysis* of the rules of any proper labelled calculus. In particular, according to Definition 1, notice that if an occurrence σ is a substructure of $\Pi \in \{\Gamma, \Gamma', \Delta, \Delta'\}$ occurring in an instantiation r of a rule $R \in \mathbf{A.L}$ (including axioms, namely rules with no premises), then σ is a *parameter* of r and every other σ' is *nonparametric* in r;[4] moreover, if σ occurs in a premise and in the conclusion of r in the same position (namely, in *precedent* versus in *succedent position*: see Definition 2), then these two occurrences of σ are *locally congruent* in r.[5] Notice that in the display calculi literature 'being locally congruent' usually presupposes 'being parametric', but in labelled calculi this is not anymore the case due to the presence of switch rules (see Remark 1).

Initial rules*

$$\mathrm{Id}_{\mathbf{j}\leq p}\ \frac{}{\mathbf{j} \leq p \vdash \mathbf{j} \leq p} \qquad \frac{}{p \leq \mathbf{m} \vdash p \leq \mathbf{m}}\ \mathrm{Id}_{p\leq\mathbf{m}} \qquad \mathrm{Id}_{\bot}\ \frac{}{\bot \leq \mathbf{m} \vdash \bot \leq \mathbf{m}} \qquad \frac{}{\mathbf{j} \leq \top \vdash \mathbf{j} \leq \top}\ \mathrm{Id}_{\top}$$

The initial rules above encode identities for atomic propositions and zeroary connectives, namely the fact that the derivability relation $\vdash$ is reflexive. Identity sequents of the form $\mathbf{j} \leq A \vdash \mathbf{j} \leq A$ (resp. $A \leq \mathbf{m} \vdash A \leq \mathbf{m}$) are derivable in the calculus.

Initial rules for ⊤ and ⊥*

$$\bot_{\mathbf{j}}\ \frac{}{\mathbf{j} \leq \bot \vdash \mathbf{j} \leq A} \qquad \bot_{\mathbf{m}}\ \frac{}{B \leq \mathbf{m} \vdash \bot \leq \mathbf{m}} \qquad \frac{}{\top \leq \mathbf{m} \vdash B \leq \mathbf{m}}\ \top_{\mathbf{m}} \qquad \frac{}{\mathbf{j} \leq A \vdash \mathbf{j} \leq \top}\ \top_{\mathbf{j}}$$

*Side condition: $A \in \{p, A_1 \wedge A_2, \Box A_1\}$ and $B \in \{p, B_1 \vee B_2, \Diamond B_1\}$

The initial rules for $\bot$ (resp. $\top$) above encodes the fact that $\bot$ is interpreted as the minimal element (resp. $\top$ as the maximal element) in the algebraic interpretation.

The cut rules below encode the fact that the derivability relation $\vdash$ is transitive. Notice that the notion of 'cut formula' in standard Gentzen sequent calculi corresponds to 'labelled cut formula' in the present setting. Before defining the cut rules, we need the following definition.

Definition 3. *A labelled formula a is a* **j***-labelled (resp.* **m***-labelled) formula in a derivation π if the uppermost labelled formulas congruent with a in π are introduced via* $Id_{\mathbf{j}\leq p}, \bot_{\mathbf{j}}, \top_{\mathbf{j}}, \wedge_P, \wedge_S, \Box_P, \Box_S$ *(resp.* $Id_{p\leq\mathbf{m}}, \bot_{\mathbf{m}}, \top_{\mathbf{m}}, \vee_P, \vee_S, \Diamond_P, \Diamond_S$*).*

Cut rules*

$$\mathrm{Cut}_{\mathbf{j}\leq A}\ \frac{\Gamma \vdash \mathbf{j} \leq A, \Delta \qquad \Gamma', \mathbf{j} \leq A \vdash \Delta'}{\Gamma, \Gamma' \vdash \Delta, \Delta'} \qquad \frac{\Gamma \vdash B \leq \mathbf{m}, \Delta \qquad \Gamma', B \leq \mathbf{m} \vdash \Delta'}{\Gamma, \Gamma' \vdash \Delta, \Delta'}\ \mathrm{Cut}_{B\leq\mathbf{m}}$$

*Side condition: $\mathbf{j} \leq A$ and $B \leq \mathbf{m}$ are in display,
$\mathbf{j} \leq A$ is a $\mathbf{j}$-labelled formula and $B \leq \mathbf{m}$ is an $\mathbf{m}$-labelled formula.

---

[4] Therefore, every instantiation of a labelled formula (resp. a pure structure) occurring in $R \in \mathbf{A.L}$ is nonparametric. For instance, $\mathbf{j} \leq p$ is nonparametric in $\mathrm{Id}_{\mathbf{j}\leq p}$, and $A \leq \mathbf{m}, \mathbf{j} \leq \Box A, \mathbf{j} \leq \Box\mathbf{m}$ are all nonparametric in $\Box_P$. Moreover, according to Definition 1, every instantiation of a structure (resp. a labelled formula) in the conclusion of initial rules (resp. logical rules) is *principal*. For instance, $\mathbf{j} \leq p$ is principal in $\mathrm{Id}_{\mathbf{j}\leq p}$ and $\mathbf{j} \leq \Box A$ is principal in $\Box_P$.

[5] For instance, given an instantiation r of the rule $\wedge_S$, assuming $\sigma \in \Gamma$ in the first (resp. second) premise of r, then it occurs in the same position in the conclusion of r, and these two occurrences of σ are *locally congruent* in r. Nonetheless, notice that the two occurrences of σ in the premises of r are not locally congruent in r.

The switch rules below encode elementary properties of pairs of inequalities with the same approximant (either a nominal **j** or a conominal **m**) occurring in the same sequent with opposite polarity (namely, the first in precedent position and the second in succedent position: see Definition 2). Notice that we might use S as a generic name denoting a specific switch rule in the following set. If so, we rely on the context to disambiguate which rule we are referring to. In particular, the label S is unambiguous whenever we use it as the name for a rule application in a derivation.

Switch rules*

$$\textbf{Sm}\ \frac{\Gamma, \mathbf{j} \leq A \vdash \mathbf{j} \leq \mathbf{m}, \Delta}{\Gamma \vdash A \leq \mathbf{m}, \Delta} \qquad \frac{\Gamma, A \leq \mathbf{m} \vdash \mathbf{j} \leq \mathbf{m}, \Delta}{\Gamma \vdash \mathbf{j} \leq A, \Delta}\ \textbf{Sj}$$

$$\textbf{Smm}\ \frac{\Gamma, \mathbf{j} \leq A \vdash \mathbf{j} \leq B, \Delta}{\Gamma, B \leq \mathbf{m} \vdash A \leq \mathbf{m}, \Delta} \qquad \frac{\Gamma, A \leq \mathbf{m} \vdash B \leq \mathbf{m}, \Delta}{\Gamma, \mathbf{j} \leq B \vdash \mathbf{j} \leq A, \Delta}\ \textbf{Sjj}$$

$$\textbf{SmT}\ \frac{\Gamma, \mathbf{j} \leq \mathbf{T} \vdash \mathbf{j} \leq A, \Delta}{\Gamma, A \leq \mathbf{m} \vdash \mathbf{T} \leq \mathbf{m}, \Delta} \qquad \frac{\Gamma, \mathbf{T} \leq \mathbf{m} \vdash A \leq \mathbf{m}, \Delta}{\Gamma, \mathbf{j} \leq A \vdash \mathbf{j} \leq \mathbf{T}, \Delta}\ \textbf{SjT}$$

$$\textbf{STm}\ \frac{\Gamma, \mathbf{j} \leq A \vdash \mathbf{j} \leq \mathbf{T}, \Delta}{\Gamma, \mathbf{T} \leq \mathbf{m} \vdash A \leq \mathbf{m}, \Delta} \qquad \frac{\Gamma, A \leq \mathbf{m} \vdash \mathbf{T} \leq \mathbf{m}, \Delta}{\Gamma, \mathbf{j} \leq \mathbf{T} \vdash \mathbf{j} \leq A, \Delta}\ \textbf{STj}$$

$$\textbf{STT'm}\ \frac{\Gamma, \mathbf{j} \leq \mathbf{T'} \vdash \mathbf{j} \leq \mathbf{T}, \Delta}{\Gamma, \mathbf{T} \leq \mathbf{m} \vdash \mathbf{T'} \leq \mathbf{m}, \Delta} \qquad \frac{\Gamma, \mathbf{T'} \leq \mathbf{m} \vdash \mathbf{T} \leq \mathbf{m}, \Delta}{\Gamma, \mathbf{j} \leq \mathbf{T} \vdash \mathbf{j} \leq \mathbf{T'}, \Delta}\ \textbf{SjTT'}$$

*Side condition: For all the switch rules except **Sm** and **Sj**, **j** and **m** do not appear in Γ or Δ. **j** (resp. **m**) in **Sm** (resp. **Sj**) must not appear in the conclusion of the rule.

Remark 1 (Analysis of switch rules). For each instantiation r of $R \in \{\mathbf{STT'm}, \mathbf{SjTT'}\}$, the instantiations of $\mathbf{j} \leq \mathbf{T''}$ and $\mathbf{T''} \leq \mathbf{m}$ (where $\mathbf{T''} \in \{\mathbf{T}, \mathbf{T'}\}$) are *nonparametric* and *locally congruent* in r (see Definition 1). For each instantiation r of any other switch rule R, the instantiations of $\mathbf{j} \leq C$ and $C \leq \mathbf{m}$ (where $C \in \{A, B\}$) are *nonparametric* and *locally congruent* in r (see Definition 1).

Adjunction rules

$$\Diamond \dashv \blacksquare\ \frac{\Gamma \vdash \Diamond\mathbf{j} \leq \mathbf{m}, \Delta}{\Gamma \vdash \mathbf{j} \leq \blacksquare\mathbf{m}, \Delta} \qquad \frac{\Gamma \vdash \mathbf{j} \leq \blacksquare\mathbf{m}, \Delta}{\Gamma \vdash \Diamond\mathbf{j} \leq \mathbf{m}, \Delta}\ \Diamond \dashv \blacksquare^{-1}$$

$$\blacklozenge \dashv \Box\ \frac{\Gamma \vdash \mathbf{j} \leq \Box\mathbf{m}, \Delta}{\Gamma \vdash \blacklozenge\mathbf{j} \leq \mathbf{m}, \Delta} \qquad \frac{\Gamma \vdash \blacklozenge\mathbf{j} \leq \mathbf{m}, \Delta}{\Gamma \vdash \mathbf{j} \leq \Box\mathbf{m}, \Delta}\ \blacklozenge \dashv \Box^{-1}$$

Structural rules for $\top$ and $\bot$

$$\top_{\Box}\ \frac{\Gamma \vdash \top \leq \mathbf{m}, \Delta}{\Gamma, \mathbf{j} \leq \top \vdash \mathbf{j} \leq \Box\mathbf{m}, \Delta} \qquad \frac{\Gamma \vdash \mathbf{j} \leq \bot, \Delta}{\Gamma, \bot \leq \mathbf{m} \vdash \Diamond\mathbf{j} \leq \mathbf{m}, \Delta}\ \bot_{\Diamond}$$

The adjunction rules above encode the fact that unary modalities $\Diamond, \blacksquare$ and $\blacklozenge, \Box$ constitute pairs of adjoint operators. The structural rules $\top_{\Box}$ (resp. $\bot_{\Diamond}$) above encodes the fact that $\Box$ preserve $\top$ (resp. $\Diamond$ preserves $\bot$).

The logical rules below encode the minimal order-theoretic properties and the arity of propositional and modal connectives.

Logical rules for propositional connectives*

$$\wedge_P\ \frac{\Gamma, \mathbf{j} \leq A_i \vdash \Delta}{\Gamma, \mathbf{j} \leq A_1 \wedge A_2 \vdash \Delta} \qquad \frac{\Gamma \vdash \mathbf{j} \leq A, \Delta \quad\quad \Gamma \vdash \mathbf{j} \leq B, \Delta}{\Gamma \vdash \mathbf{j} \leq A \wedge B, \Delta}\ \wedge_S$$

$$\vee_P\ \frac{\Gamma \vdash A \leq \mathbf{m}, \Delta \quad\quad \Gamma \vdash B \leq \mathbf{m}, \Delta}{\Gamma \vdash A \vee B \leq \mathbf{m}, \Delta} \qquad \frac{\Gamma, A_i \leq \mathbf{m} \vdash \Delta}{\Gamma, A_1 \vee A_2 \leq \mathbf{m} \vdash \Delta}\ \vee_S$$

*Side condition: labelled formula in the conclusion of any logical rule are in display.

We consider the following logical rules for modalities, where $\Box_S$ and $\Diamond_P$ are invertible, but $\Box_P$ and $\Diamond_S$ are not. This choice facilitates a smoother analysis of the rules and therefore it is preferable whenever the goal is to provide a canonical cut elimination.

Logical rules for modalities*

$$\Box_P\ \frac{\Gamma \vdash A \leq \mathbf{m}, \Delta}{\Gamma, \mathbf{j} \leq \Box A \vdash \mathbf{j} \leq \Box \mathbf{m}, \Delta} \qquad \frac{\Gamma, A \leq \mathbf{m} \vdash \mathbf{j} \leq \Box \mathbf{m}, \Delta}{\Gamma \vdash \mathbf{j} \leq \Box A, \Delta}\ \Box_S$$

$$\Diamond_P\ \frac{\Gamma, \mathbf{j} \leq A \vdash \Diamond \mathbf{j} \leq \mathbf{m}, \Delta}{\Gamma \vdash \Diamond A \leq \mathbf{m}, \Delta} \qquad \frac{\Gamma \vdash \mathbf{j} \leq A, \Delta}{\Gamma, \Diamond A \leq \mathbf{m} \vdash \Diamond \mathbf{j} \leq \mathbf{m}, \Delta}\ \Diamond_S$$

*Side conditions: **m** (resp. **j**) must not occur in the conclusion of $\Box_S$ (resp. $\Diamond_P$).
Labelled formulas in the conclusion of any logical rule are in display.

Remark 2. The invertible version of $\wedge_P$, $\vee_S$, f $\Box_P$ and $\Diamond_S$ are as follows:

$$\wedge_P\ \frac{\Gamma, \mathbf{j} \leq A, \mathbf{j} \leq B \vdash \Delta}{\Gamma, \mathbf{j} \leq A \wedge B \vdash \Delta} \qquad \frac{\Gamma, A \leq \mathbf{m}, B \leq \mathbf{m} \vdash \Delta}{\Gamma, A \vee B \leq \mathbf{m} \vdash \Delta}\ \vee_S$$

$$\Box_P\ \frac{\Gamma, \mathbf{j} \leq \Box A \vdash A \leq \mathbf{m}, \mathbf{j} \leq \Box \mathbf{m}, \Delta}{\Gamma, \mathbf{j} \leq \Box A \vdash \mathbf{j} \leq \Box \mathbf{m}, \Delta} \qquad \frac{\Gamma, \Diamond A \leq \mathbf{m} \vdash \mathbf{j} \leq A, \Diamond \mathbf{j} \leq \mathbf{m}, \Delta}{\Gamma, \Diamond A \leq \mathbf{m} \vdash \Diamond \mathbf{j} \leq \mathbf{m}, \Delta}\ \Diamond_S$$

Invertible rules can be used whenever the goal is to facilitate backwards-looking proof searches. In this case, the initial rules have to be generalized accordingly.

The table below collects the analytic rules, both in the format of display calculi and in the present format, generated by reading off the ALBA outputs of the corresponding axioms reported on in Sect. 2.3.

modal axiom	display rule	labelled rule
(4) $\Diamond\Diamond A \vdash \Diamond A$	$4\ \dfrac{\hat{\Diamond} X \vdash Y}{\hat{\Diamond}\hat{\Diamond} X \vdash Y}$	$4\ \dfrac{\Gamma \vdash \Diamond \mathbf{j} \leq \mathbf{m}, \Delta}{\Gamma, \mathbf{h} \leq \Diamond \mathbf{j} \vdash \Diamond \mathbf{h} \leq \mathbf{m}, \Delta}$
(T) $\Box A \vdash A$	$\dfrac{X \vdash \check{\Box} Y}{X \vdash Y}\ T$	$\dfrac{\Gamma \vdash \mathbf{j} \leq \Box \mathbf{m}, \Delta}{\Gamma \vdash \mathbf{j} \leq \mathbf{m}, \Delta}\ T$
(B) $A \vdash \Box\Diamond A$	$\dfrac{\hat{\Diamond} X \vdash Y}{X \vdash \check{\Box} Y}\ B$	$\dfrac{\Gamma \vdash \Diamond \mathbf{j} \leq \mathbf{m}, \Delta}{\Gamma \vdash \mathbf{j} \leq \Box \mathbf{m}, \Delta}\ B$
(D) $\Box A \vdash \Diamond A$	$\mathrm{D}\ \dfrac{\hat{\Diamond}\hat{\blacklozenge} X \vdash Y}{X \vdash Y}$	$\mathrm{D}\ \dfrac{\Gamma, \mathbf{k} \leq \blacklozenge \mathbf{j} \vdash \Diamond \mathbf{k} \leq \mathbf{m}, \Delta}{\Gamma \vdash \mathbf{j} \leq \mathbf{m}, \Delta}$
(C) $\Diamond\Box A \vdash \Box\Diamond A$	$C\ \dfrac{\hat{\Diamond}\hat{\blacklozenge} X \vdash Y}{\hat{\blacklozenge}\hat{\Diamond} X \vdash Y}$	$C\ \dfrac{\Gamma, \mathbf{k} \leq \blacklozenge \mathbf{j} \vdash \Diamond \mathbf{k} \leq \mathbf{m}, \Delta}{\Gamma, \mathbf{h} \leq \Diamond \mathbf{j} \vdash \blacklozenge \mathbf{h} \leq \mathbf{m}, \Delta}$

Where $\mathbf{k}$ in rules C and D must not appear in Γ, Δ. For any $\Sigma \subseteq \{(T), (4), (B), (D), (C)\}$, we let $\mathbf{A.L}\Sigma$ be the calculus defined by the rules in the sections above plus the additional rules in the table above corresponding to the axioms in Σ. We let $\mathbf{A.L} := \mathbf{A.L}\varnothing$.

4 Properties of the Calculus A.LΣ

4.1 Soundness

In the present section, we show that, for any $\Sigma \subseteq \{(T), (4), (B), (D), (C)\}$, the rules of $\mathbf{A.L}\Sigma$ are sound on the class $\mathsf{K}^\delta(\Sigma) := \{\mathbb{A}^\delta \mid \mathbb{A} \models \Sigma\}$. Firstly, let us recall that, as usual, any $\mathbf{A.L}$-sequent $\Gamma \vdash \Delta$ is to be interpreted as "any assignment of the variables in $\mathsf{Prop} \cup \mathsf{NOM} \cup \mathsf{CNOM}$ under which all inequalities in Γ are satisfied also satisfies some inequality in Δ"; in symbols: $\forall \overline{p} \forall \overline{\mathbf{j}} \forall \overline{\mathbf{m}} (\bigwith \Gamma \Longrightarrow \bigparr \Delta)$.

As to the basic calculus $\mathbf{A.L}$, the soundness of the initial rules, cut rules, adjunction rules, and logical rules for propositional connectives is straightforward. The soundness of the switch rules hinges on the fact that nominal and co-nominal variables range over completely join-generating and completely meet-generating subsets of $\mathbb{A}^\delta$ for any $\mathcal{L}$-algebra $\mathbb{A}$. For example, the soundness of $\mathbf{Sjm}$ follows from the following chain of equivalences:

$$
\begin{array}{lll}
& \forall \overline{p} \forall \mathbf{j} \forall \overline{\mathbf{k}} \forall \overline{\mathbf{n}} (\bigwith \Gamma \,\&\, \mathbf{j} \leq A \Rightarrow \mathbf{j} \leq B \parr \bigparr \Delta) & \text{validity of premise} \\
\text{iff} & \forall \overline{p} \forall \overline{\mathbf{k}} \forall \overline{\mathbf{n}} (\bigwith \Gamma \Rightarrow \forall \mathbf{j} (\mathbf{j} \leq A \Rightarrow \mathbf{j} \leq B) \parr \bigparr \Delta) & \text{uncurrying + side condition} \\
\text{iff} & \forall \overline{p} \forall \overline{\mathbf{k}} \forall \overline{\mathbf{n}} (\bigwith \Gamma \Rightarrow A \leq B \parr \bigparr \Delta) & \text{c. join generation} \\
\text{iff} & \forall \overline{p} \forall \overline{\mathbf{k}} \forall \overline{\mathbf{n}} (\bigwith \Gamma \Rightarrow \forall \mathbf{m} (B \leq \mathbf{m} \Rightarrow A \leq \mathbf{m}) \parr \bigparr \Delta) & \text{c. meet generation} \\
\text{iff} & \forall \overline{p} \forall \mathbf{m} \forall \overline{\mathbf{k}} \forall \overline{\mathbf{n}} (\bigwith \Gamma \,\&\, B \leq \mathbf{m} \Rightarrow A \leq \mathbf{m} \parr \bigparr \Delta) & \text{currying}
\end{array}
$$

Since $\mathbf{m}$ does not occur in Γ and Δ, the rule is also invertible. The verification of the soundness of the remaining switch rules is similar.

The soundness and invertibility of the introduction rules for the modal connectives hinge on the fact that the operation $\Diamond^\sigma$ (resp. $\Box^\pi$) is completely join-preserving (resp. completely meet preserving) and on the complete- join-generation and meet-generation properties of the subsets of $\mathbb{A}^\delta$ on which nominal and co-nominal variables are interpreted. For example, the soundness and invertibility of $\Box_S$ is verified via the following chain of equivalences:

$$
\begin{array}{lll}
& \forall \overline{p} \forall \mathbf{j} \forall \mathbf{m} \forall \overline{\mathbf{k}} \forall \overline{\mathbf{n}} (\bigwith \Gamma \,\&\, A \leq \mathbf{m} \Rightarrow \mathbf{j} \leq \Box \mathbf{m} \parr \bigparr \Delta) & \text{validity of premise} \\
\text{iff} & \forall \overline{p} \forall \mathbf{j} \forall \overline{\mathbf{k}} \forall \overline{\mathbf{n}} (\bigwith \Gamma \Rightarrow \forall \mathbf{m} (A \leq \mathbf{m} \Rightarrow \mathbf{j} \leq \Box \mathbf{m}) \parr \bigparr \Delta) & \text{uncurrying + side condition} \\
\text{iff} & \forall \overline{p} \forall \mathbf{j} \forall \overline{\mathbf{k}} \forall \overline{\mathbf{n}} (\bigwith \Gamma \Rightarrow \forall \mathbf{m} (A \leq \mathbf{m} \Rightarrow \blacklozenge \mathbf{j} \leq \mathbf{m}) \parr \bigparr \Delta) & \text{adjunction} \\
\text{iff} & \forall \overline{p} \forall \mathbf{j} \forall \overline{\mathbf{k}} \forall \overline{\mathbf{n}} (\bigwith \Gamma \Rightarrow \blacklozenge \mathbf{j} \leq A \parr \bigparr \Delta) & \text{c. meet-generation} \\
\text{iff} & \forall \overline{p} \forall \mathbf{j} \forall \overline{\mathbf{k}} \forall \overline{\mathbf{n}} (\bigwith \Gamma \Rightarrow \mathbf{j} \leq \Box A \parr \bigparr \Delta) & \text{adjunction}
\end{array}
$$

The soundness of $\Box_P$ immediately follows from the monotonicity of $\Box^\pi$. Indeed, fix an assignment of variables in $\mathsf{Prop} \cup \mathsf{NOM} \cup \mathsf{CNOM}$ under which all inequalities in Γ and $\mathbf{j} \leq \Box A$ are satisfied. If such assignment also satisfies $A \leq \mathbf{n}$, then, by monotonicity, $\mathbf{j} \leq \Box A \leq \Box \mathbf{n}$, as required. The proof for the rules $\Diamond_P$ and $\Diamond_S$ is similar.

As to the extended calculus $\mathbf{A.L}\Sigma$, the soundness of rule (4) is verified by the following chain of computations holding on the canonical extension of any $\mathcal{L}$-algebra $\mathbb{A}$ such that $\mathbb{A} \models \Diamond\Diamond p \leq \Diamond p$:

$$\begin{array}{lll}
 & \forall\overline{p}\forall\overline{\mathbf{k}}\forall\mathbf{j}\forall\mathbf{m}\forall\overline{\mathbf{n}}(\with\Gamma \Rightarrow \Diamond\mathbf{j} \leq \mathbf{m} \parr \bigparr\Delta) & \text{validity of premise}\\
\text{then} & \forall\overline{p}\forall\overline{\mathbf{k}}\forall\mathbf{j}\forall\mathbf{m}\forall\overline{\mathbf{n}}(\with\Gamma \Rightarrow \Diamond\Diamond\mathbf{j} \leq \mathbf{m} \parr \bigparr\Delta) & \text{axiom (4)}\\
\text{iff} & \forall\overline{p}\forall\overline{\mathbf{k}}\forall\mathbf{j}\forall\mathbf{m}\forall\overline{\mathbf{n}}(\with\Gamma \Rightarrow \forall\mathbf{h}(\mathbf{h} \leq \Diamond\mathbf{j} \Rightarrow \Diamond\mathbf{h} \leq \mathbf{m}) \parr \bigparr\Delta) & \text{c. join-generation}\\
\text{iff} & \forall\overline{p}\forall\overline{\mathbf{k}}\forall\mathbf{j}\forall\mathbf{m}\forall\mathbf{h}\forall\overline{\mathbf{n}}(\with\Gamma \,\&\, \mathbf{h} \leq \Diamond\mathbf{j} \Rightarrow \Diamond\mathbf{h} \leq \mathbf{m} \parr \bigparr\Delta) & \text{currying} + \mathbf{h} \text{ fresh}
\end{array}$$

The key step in the computation above is the one which makes use of the assumption of axiom (4) being valid on $\mathbb{A}$. Indeed, by the general theory of correspondence for LE-logics (cf. [7]), axiom (4) is canonical, hence the assumption implies that (4) is valid also on $\mathbb{A}^\delta$. Then, as is shown in the computation concerning axiom (4) in Sect. 2.3, the validity of (4) in $\mathbb{A}^\delta$ is equivalent to the condition $\forall\mathbf{j}\forall\mathbf{m}\,(\Diamond\mathbf{j} \leq \mathbf{m} \Rightarrow \Diamond\Diamond\mathbf{j} \leq \mathbf{m})$ holding in $\mathbb{A}^\delta$, which justifies this key step. The verification of the remaining additional rules hinges on similar arguments and facts (in particular, all axioms we consider are canonical), so in what follows we only report on the corresponding computations.

$$\begin{array}{lll}
 & \forall\overline{p}\forall\overline{\mathbf{k}}\forall\mathbf{j}\forall\mathbf{m}\forall\overline{\mathbf{n}}(\with\Gamma \Rightarrow \mathbf{j} \leq \Box\mathbf{m} \parr \bigparr\Delta) & \text{validity of premise}\\
\text{then} & \forall\overline{p}\forall\overline{\mathbf{k}}\forall\mathbf{j}\forall\mathbf{m}\forall\overline{\mathbf{n}}(\with\Gamma \Rightarrow \mathbf{j} \leq \mathbf{m} \parr \bigparr\Delta) & \text{axiom (T)}
\end{array}$$

$$\begin{array}{lll}
 & \forall\overline{p}\forall\overline{\mathbf{k}}\forall\mathbf{j}\forall\mathbf{m}\forall\overline{\mathbf{n}}(\with\Gamma \Rightarrow \Diamond\mathbf{j} \leq \mathbf{m} \parr \bigparr\Delta) & \text{validity of premise}\\
\text{then} & \forall\overline{p}\forall\overline{\mathbf{k}}\forall\mathbf{j}\forall\mathbf{m}\forall\overline{\mathbf{n}}(\with\Gamma \Rightarrow \mathbf{j} \leq \Box\mathbf{m} \parr \bigparr\Delta) & \text{axiom (B)}
\end{array}$$

$$\begin{array}{lll}
 & \forall\overline{p}\forall\overline{\mathbf{k}}\forall\mathbf{j}\forall\mathbf{m}\forall\overline{\mathbf{n}}(\with\Gamma \Rightarrow \mathbf{j} \leq \Box\mathbf{m} \parr \bigparr\Delta) & \text{validity of premise}\\
\text{then} & \forall\overline{p}\forall\overline{\mathbf{k}}\forall\mathbf{j}\forall\mathbf{m}\forall\overline{\mathbf{n}}(\with\Gamma \Rightarrow \Diamond\mathbf{j} \leq \mathbf{m} \parr \bigparr\Delta) & \text{axiom (B}^{-1}\text{)}
\end{array}$$

The soundness of the rule (C) is verified by the following chain of computations holding on the canonical extension of any $\mathcal{L}$-algebra $\mathbb{A}$ such that $\mathbb{A} \models \Diamond\Box p \leq \Box\Diamond p$:

$$\begin{array}{lll}
 & \forall\overline{p}\forall\overline{\mathbf{h}'}\forall\mathbf{j}\forall\mathbf{m}\forall\mathbf{k}\forall\overline{\mathbf{n}}(\with\Gamma \,\&\, \mathbf{k} \leq \blacklozenge\mathbf{j} \Rightarrow \Diamond\mathbf{k} \leq \mathbf{m} \parr \bigparr\Delta) & \text{validity of premise}\\
\text{iff} & \forall\overline{p}\forall\overline{\mathbf{h}'}\forall\mathbf{j}\forall\mathbf{m}\forall\overline{\mathbf{n}}(\with\Gamma \Rightarrow \forall\mathbf{k}(\mathbf{k} \leq \blacklozenge\mathbf{j} \Rightarrow \Diamond\mathbf{k} \leq \mathbf{m}) \parr \bigparr\Delta) & \text{uncurrying + side cond.}\\
\text{iff} & \forall\overline{p}\forall\overline{\mathbf{h}'}\forall\mathbf{j}\forall\mathbf{m}\forall\overline{\mathbf{n}}(\with\Gamma \Rightarrow \forall\mathbf{k}(\mathbf{k} \leq \blacklozenge\mathbf{j} \Rightarrow \mathbf{k} \leq \blacksquare\mathbf{m}) \parr \bigparr\Delta) & \text{adjunction}\\
\text{iff} & \forall\overline{p}\forall\overline{\mathbf{h}'}\forall\mathbf{j}\forall\mathbf{m}\forall\overline{\mathbf{n}}(\with\Gamma \Rightarrow \blacklozenge\mathbf{j} \leq \blacksquare\mathbf{m} \parr \bigparr\Delta) & \text{c. join-generation}\\
\text{iff} & \forall\overline{p}\forall\overline{\mathbf{h}'}\forall\mathbf{j}\forall\mathbf{m}\forall\overline{\mathbf{n}}(\with\Gamma \Rightarrow \Diamond\blacklozenge\mathbf{j} \leq \mathbf{m} \parr \bigparr\Delta) & \text{adjunction}\\
\text{then} & \forall\overline{p}\forall\overline{\mathbf{h}'}\forall\mathbf{j}\forall\mathbf{m}\forall\overline{\mathbf{n}}(\with\Gamma \Rightarrow \blacklozenge\Diamond\mathbf{j} \leq \mathbf{m} \parr \bigparr\Delta) & \text{axiom (C)}\\
\text{iff} & \forall\overline{p}\forall\overline{\mathbf{h}'}\forall\mathbf{j}\forall\mathbf{m}\forall\overline{\mathbf{n}}(\with\Gamma \Rightarrow \Diamond\mathbf{j} \leq \Box\mathbf{m} \parr \bigparr\Delta) & \text{adjunction}\\
\text{iff} & \forall\overline{p}\forall\overline{\mathbf{h}'}\forall\mathbf{j}\forall\mathbf{m}\forall\overline{\mathbf{n}}(\with\Gamma \Rightarrow \forall\mathbf{i}(\mathbf{i} \leq \Diamond\mathbf{j} \Rightarrow \mathbf{i} \leq \Box\mathbf{m}) \parr \bigparr\Delta) & \text{c. join-generation}\\
\text{iff} & \forall\overline{p}\forall\mathbf{i}\forall\overline{\mathbf{h}'}\forall\mathbf{j}\forall\mathbf{m}\forall\overline{\mathbf{n}}(\with\Gamma \,\&\, \mathbf{i} \leq \Diamond\mathbf{j} \Rightarrow \mathbf{i} \leq \Box\mathbf{m} \parr \bigparr\Delta) & \text{currying}\\
\text{iff} & \forall\overline{p}\forall\mathbf{i}\forall\overline{\mathbf{h}'}\forall\mathbf{j}\forall\mathbf{m}\forall\overline{\mathbf{n}}(\with\Gamma \,\&\, \mathbf{i} \leq \Diamond\mathbf{j} \Rightarrow \blacklozenge\mathbf{i} \leq \mathbf{m} \parr \bigparr\Delta) & \text{adjunction}
\end{array}$$

Instantiating **i** as **h** completes the proof. The key step in the computation above is the one which makes use of the assumption of axiom (C) being valid on $\mathbb{A}$. Indeed, by the general theory of correspondence for LE-logics (cf. [7]), axiom (C) is canonical, hence the assumption implies that (C) is valid also on $\mathbb{A}^\delta$. Then, as is shown in the computation concerning axiom (C) in Sect. 2.3, the validity of (C) in $\mathbb{A}^\delta$ is equivalent to the condition $\forall\mathbf{j}\forall\mathbf{m}\,(\Diamond\blacklozenge\mathbf{j} \leq \mathbf{m} \Rightarrow \blacklozenge\Diamond\mathbf{j} \leq \mathbf{m})$ holding in $\mathbb{A}^\delta$, which justifies this key step.

The soundness of the rules (D), (T), (B) is verified in a similar way.

4.2 Syntactic Completeness

In the present section, we show that all the axioms and rules of the basic logic are derivable in **A.L**, and that for any $\Sigma \subseteq \{(\mathrm{T}), (4), (\mathrm{B}), (\mathrm{D}), (\mathrm{C})\}$, the axioms and rules of $\mathbf{L}\Sigma$ are derivable in $\mathbf{A.L}\Sigma$.[6]

The sequents $p \vdash p$, $\bot \vdash p$, $p \vdash \top$, $p \vdash p \vee q$ $(q \vdash p \vee q)$, and $p \wedge q \vdash p$ $(q \wedge p \vdash q)$ are trivially derivable with one single application of the rule $Id_{\mathbf{j}\leq q}$, $\bot_w$, $\top_w$, $\vee_S$, and $\wedge_P$, respectively. The derivability of the rules $\frac{\varphi \vdash \chi \;\; \chi \vdash \psi}{\varphi \vdash \psi}$, $\frac{\varphi \vdash \chi \;\; \psi \vdash \chi}{\varphi \vee \psi \vdash \chi}$, and $\frac{\chi \vdash \varphi \;\; \chi \vdash \psi}{\chi \vdash \varphi \wedge \psi}$ can be shown by derivations in which the cut rules, $\vee_P$, and $\wedge_S$, respectively, are applied. The two derivations below show the rules concerning the connectives $\Box$ and $\Diamond$:

$$
\Box_S\;\dfrac{\Box_P\;\dfrac{\psi \leq \mathbf{m} \vdash \varphi \leq \mathbf{m}}{\mathbf{j} \leq \Box\varphi, \psi \leq \mathbf{m} \vdash \mathbf{j} \leq \Box\mathbf{m}}}{\mathbf{j} \leq \Box\varphi \vdash \mathbf{j} \leq \Box\psi}
\qquad
\dfrac{\dfrac{\mathbf{j} \leq \varphi \vdash \mathbf{j} \leq \psi}{\Diamond\psi \leq \mathbf{m}, \mathbf{j} \leq \varphi \vdash \Diamond\mathbf{j} \leq \mathbf{m}}\;\Diamond_S}{\Diamond\psi \leq \mathbf{m} \vdash \Diamond\varphi \leq \mathbf{m}}\;\Diamond_P
$$

To show the admissibility of the substitution rule $\frac{\varphi \vdash \psi}{\varphi(\chi/p) \vdash \psi(\chi/p)}$, a straightforward induction on the derivation height of $\varphi \vdash \psi$ suffices.

As to the axioms and rules of the basic logic **L**, below, we derive the axioms encoding the distributivity of $\Diamond$ over $\vee$ and $\bot$ in **A.L**. The distributivity of $\Box$ over $\wedge$ and $\top$ is derived similarly.

$$
\Diamond_P\;\dfrac{\Diamond\dashv\blacksquare^{-1}\;\dfrac{\dfrac{\vee_P\;\dfrac{S\;\dfrac{\dfrac{\dfrac{\dfrac{Id_{\mathbf{i}\leq A}\;\dfrac{}{\mathbf{i} \leq A \vdash \mathbf{i} \leq A}}{\mathbf{i} \leq A, \Diamond A \leq \mathbf{m} \vdash \Diamond\mathbf{i} \leq \mathbf{m}}\;\Diamond_S}{\mathbf{i} \leq A, \Diamond A \vee \Diamond B \leq \mathbf{m} \vdash \Diamond\mathbf{i} \leq \mathbf{m}}\;\vee_S}{\mathbf{i} \leq A, \Diamond A \vee \Diamond B \leq \mathbf{m} \vdash \mathbf{i} \leq \blacksquare\mathbf{m}}\;\Diamond\dashv\blacksquare}{\Diamond A \vee \Diamond B \leq \mathbf{m}, \blacksquare\mathbf{m} \leq \mathbf{n} \vdash A \leq \mathbf{n}} \quad \dfrac{\dfrac{\dfrac{\dfrac{Id_{\mathbf{i}\leq B}\;\dfrac{}{\mathbf{i} \leq B \vdash \mathbf{i} \leq B}}{\mathbf{i} \leq B, \Diamond B \leq \mathbf{m} \vdash \Diamond\mathbf{i} \leq \mathbf{m}}\;\Diamond_S}{\mathbf{i} \leq B, \Diamond A \vee \Diamond B \leq \mathbf{m} \vdash \Diamond\mathbf{i} \leq \mathbf{m}}\;\vee_S}{\mathbf{i} \leq B, \Diamond A \vee \Diamond B \leq \mathbf{m} \vdash \mathbf{i} \leq \blacksquare\mathbf{m}}\;\Diamond\dashv\blacksquare}{\Diamond A \vee \Diamond B \leq \mathbf{m}, \blacksquare\mathbf{m} \leq \mathbf{n} \vdash B \leq \mathbf{n}}\;S}{\Diamond A \vee \Diamond B \leq \mathbf{m}, \blacksquare\mathbf{m} \leq \mathbf{n} \vdash A \vee B \leq \mathbf{n}}}{\mathbf{h} \leq A \vee B, \Diamond A \vee \Diamond B \leq \mathbf{m} \vdash \mathbf{h} \leq \blacksquare\mathbf{m}}\;S}{\mathbf{h} \leq A \vee B, \Diamond A \vee \Diamond B \leq \mathbf{m} \vdash \Diamond\mathbf{h} \leq \mathbf{m}}}{\Diamond A \vee \Diamond B \leq \mathbf{m} \vdash \Diamond(A \vee B) \leq \mathbf{m}}
\qquad
\Diamond_P\;\dfrac{\Diamond\dashv\blacksquare^{-1}\;\dfrac{\dfrac{Id_{\mathbf{j}\leq\bot}\;\dfrac{}{\mathbf{j} \leq \bot \vdash \mathbf{j} \leq \bot}}{\bot \leq \mathbf{m}, \mathbf{j} \leq \bot \vdash \mathbf{j} \leq \blacksquare\mathbf{m}}\;\Diamond\bot}{\bot \leq \mathbf{m}, \mathbf{j} \leq \Diamond\bot \vdash \mathbf{j} \leq \mathbf{m}}}{\bot \leq \mathbf{m} \vdash \Diamond\bot \leq \mathbf{m}}
$$

The syntactic completeness for the other axioms and rules of **L** can be shown in a similar way. In particular, the admissibility of the substitution rule can be proved by induction in a standard manner.

As to the axiomatic extensions of **L**, Let us consider the axiom (4) $\Diamond\Diamond A \vdash \Diamond A$. Using ALBA we generate the first order correspondent $\forall\mathbf{j}\forall\mathbf{m}\,(\Diamond\mathbf{j} \leq \mathbf{m} \Rightarrow \Diamond\Diamond\mathbf{j} \leq \mathbf{m})$. Further processing the axiom (4) using ALBA we obtain the equivalent first order correspondent (in the so-called 'flat form'): $\forall\mathbf{j}\forall\mathbf{h}\forall\mathbf{m}\,(\Diamond\mathbf{j} \leq \mathbf{m} \Rightarrow (\mathbf{h} \leq \Diamond\mathbf{j} \Rightarrow \Diamond\mathbf{h} \leq \mathbf{m}))$, which can be written as a structural rule in the language of ALBA labelled calculi as follows:

$$
4\;\dfrac{\Gamma \vdash \Diamond\mathbf{j} \leq \mathbf{m}, \Delta}{\Gamma, \mathbf{h} \leq \Diamond\mathbf{j} \vdash \Diamond\mathbf{h} \leq \mathbf{m}, \Delta}
$$

We now provide a derivation of the axiom (4) in the basic labelled calculus **A.L** expanded with the previous structural rule 4.

---

[6] That is, we show that, for any $\mathcal{L}$-axiom $s = A \vdash B \in \{(\mathrm{T}), (4), (\mathrm{B}), (\mathrm{D}), (\mathrm{C})\}$, a derivation exists in $\mathbf{A.L}\{s\}$ of the sequent $\mathbf{j} \leq A \vdash \mathbf{j} \leq B$, or equivalently of $B \leq \mathbf{m} \vdash A \leq \mathbf{m}$.

$$
\begin{array}{ll}
\mathbf{j} \le A \vdash \mathbf{j} \le A & \mathrm{Id}_A \\ \hline
\mathbf{j} \le A, \Diamond A \le \mathbf{m} \vdash \Diamond \mathbf{j} \le \mathbf{m} & \Diamond_S \\ \hline
\mathbf{j} \le A, \Diamond A \le \mathbf{m}, \mathbf{h} \le \Diamond \mathbf{j} \vdash \Diamond \mathbf{h} \le \mathbf{m} & 4 \\ \hline
\mathbf{j} \le A, \Diamond A \le \mathbf{m}, \mathbf{h} \le \Diamond \mathbf{j} \vdash \mathbf{h} \le \blacksquare \mathbf{m} & \Diamond \dashv \blacksquare \\ \hline
\mathbf{j} \le A, \Diamond A \le \mathbf{m}, \blacksquare \mathbf{m} \le \mathbf{m}' \vdash \Diamond \mathbf{j} \le \mathbf{m}' & \mathrm{S} \\ \hline
\Diamond A \le \mathbf{m}, \blacksquare \mathbf{m} \le \mathbf{m}' \vdash \Diamond A \le \mathbf{m}' & \Diamond_P \\ \hline
\mathbf{j}' \le \Diamond A, \Diamond A \le \mathbf{m} \vdash \mathbf{j}' \le \blacksquare \mathbf{m} & \mathrm{S} \\ \hline
\mathbf{j}' \le \Diamond A, \Diamond A \le \mathbf{m} \vdash \Diamond \mathbf{j}' \le \mathbf{m} & \Diamond \dashv \blacksquare^{-1} \\ \hline
\Diamond A \le \mathbf{m} \vdash \Diamond \Diamond A \le \mathbf{m} & \Diamond_P
\end{array}
$$

Analogously, we provide a derivation of the axioms (T), (B), (D), and (C) in the basic labelled calculus **A.L** expanded with the structural rules T, B, D, and C, respectively.

$$
\begin{array}{ll}
A \le \mathbf{m} \vdash A \le \mathbf{m} & \mathrm{Id}_A \\ \hline
\mathbf{j} \le \Box A, A \le \mathbf{m} \vdash \mathbf{j} \le \Box \mathbf{m} & \Box_P \\ \hline
\mathbf{j} \le \Box A, A \le \mathbf{m} \vdash \mathbf{j} \le \mathbf{m} & T \\ \hline
\mathbf{j} \le \Box A \vdash \mathbf{j} \le A & S
\end{array}
\qquad
\begin{array}{ll}
\mathbf{j} \le A \vdash \mathbf{j} \le A & \mathrm{Id}_A \\ \hline
\mathbf{j} \le A, \Diamond A \le \mathbf{m} \vdash \Diamond \mathbf{j} \le \mathbf{m} & \Diamond_S \\ \hline
\mathbf{j} \le A, \Diamond A \le \mathbf{m} \vdash \mathbf{j} \le \Box \mathbf{m} & B \\ \hline
\mathbf{j} \le A \vdash \mathbf{j} \le \Box \Diamond A & \Box_S
\end{array}
$$

$$
\begin{array}{ll}
A \le \mathbf{n} \vdash A \le \mathbf{n} & \mathrm{Id}_A \\ \hline
\mathbf{j} \le \Box A, A \le \mathbf{n} \vdash \mathbf{j} \le \Box \mathbf{n} & \Box_P \\ \hline
\mathbf{j} \le \Box A, A \le \mathbf{n} \vdash \blacklozenge \mathbf{j} \le \mathbf{n} & \blacklozenge \dashv \Box \\ \hline
\mathbf{k} \le \blacklozenge \mathbf{j}, \mathbf{j} \le \Box A \vdash \mathbf{k} \le A & \mathrm{S} \\ \hline
\mathbf{k} \le \blacklozenge \mathbf{j}, \mathbf{j} \le \Box A, \Diamond A \le \mathbf{m} \vdash \Diamond \mathbf{k} \le \mathbf{m} & \Diamond_S \\ \hline
\mathbf{j} \le \Box A, \Diamond A \le \mathbf{m} \vdash \mathbf{j} \le \mathbf{m} & D \\ \hline
\Diamond A \le \mathbf{m} \vdash \Box A \le \mathbf{m} & \mathrm{S}
\end{array}
\qquad
\begin{array}{ll}
A \le \mathbf{o} \vdash A \le \mathbf{o} & \mathrm{Id}_A \\ \hline
\mathbf{j} \le \Box A, A \le \mathbf{o} \vdash \mathbf{j} \le \Box \mathbf{o} & \Box_P \\ \hline
\mathbf{j} \le \Box A, A \le \mathbf{o} \vdash \blacklozenge \mathbf{j} \le \mathbf{o} & \blacklozenge \dashv \Box \\ \hline
\mathbf{k} \le \blacklozenge \mathbf{j}, \mathbf{j} \le \Box A \vdash \mathbf{k} \le A & S \\ \hline
\mathbf{k} \le \blacklozenge \mathbf{j}, \mathbf{j} \le \Box A, \Diamond A \le \mathbf{m} \vdash \Diamond \mathbf{k} \le \mathbf{m} & \Diamond_S \\ \hline
\mathbf{h} \le \Diamond \mathbf{j}, \mathbf{j} \le \Box A, \Diamond A \le \mathbf{m} \vdash \blacklozenge \mathbf{h} \le \mathbf{m} & C \\ \hline
\mathbf{h} \le \Diamond \mathbf{j}, \mathbf{j} \le \Box A, \Diamond A \le \mathbf{m} \vdash \mathbf{h} \le \Box \mathbf{m} & \blacklozenge \dashv \Box^{-1} \\ \hline
\mathbf{j} \le \Box A, \Diamond A \le \mathbf{m}, \Box \mathbf{m} \le \mathbf{n} \vdash \Diamond \mathbf{j} \le \mathbf{n} & S \\ \hline
\Diamond A \le \mathbf{m}, \Box \mathbf{m} \le \mathbf{n} \vdash \Diamond \Box A \le \mathbf{n} & \Diamond_P \\ \hline
\mathbf{i} \le \Diamond \Box A, \Diamond A \le \mathbf{m} \vdash \mathbf{i} \le \Box \mathbf{m} & S \\ \hline
\mathbf{i} \le \Diamond \Box A \vdash \mathbf{i} \le \Box \Diamond A & \Box_S
\end{array}
$$

4.3 Conservativity

In the present section, we argue that, for any $\Sigma \subseteq \{(T), (4), (B), (D), (C)\}$ and all $\mathcal{L}$-formulas A and B, if the **A.L**-sequent $\mathbf{j} \le A \vdash \mathbf{j} \le B$ is derivable in $\mathbf{A.L}\Sigma$, then the $\mathcal{L}$-sequent $A \vdash B$ is an $\mathbf{L}.\Sigma$-theorem.

Indeed, because the rules of **A.L** are sound in the class $\mathsf{K}^\delta(\Sigma) = \{\mathbb{A}^\delta \mid \mathbb{A} \models \Sigma\}$, the assumption implies that $\mathbb{A}^\delta \models \forall \overline{p} \forall \mathbf{j}(\mathbf{j} \le A \Rightarrow \mathbf{j} \le B)$ which, by complete join-generation, is equivalent to $\mathbb{A}^\delta \models \forall \overline{p}(A \le B)$. Since, as discussed in Sect. 2.2, $\mathbf{L}.\Sigma$ is complete w.r.t. $\mathsf{K}^\delta(\Sigma)$, this implies that $A \vdash B$ is a theorem of $\mathbf{L}.\Sigma$, as required.

4.4 Cut Elimination and Subformula Property

As usual in the tradition of display calculi, we first characterize a new class of calculi. The actual definition is given in Appendix A. Here we just mention that a *proper labelled calculus* is a proof systems satisfying the conditions C_2–C_8 of Definition 6. Then, in the appendix, we prove the following general result, namely a canonical cut elimination theorem *à la* Belnap for any calculus in this class:

Theorem 1. *Any proper labelled calculus enjoys cut-elimination. If also the condition* C_1 *in Definition 6 is satisfied, then the proof system enjoys the subformula property.*

Finally, we obtain that the calculi $\mathbf{A.L}\Sigma$ introduced in Sect. 3 enjoys cut elimination and subformula property thanks to the following result, proved in Sect. 5.

Corollary 1. $\mathbf{A.L}\Sigma$ *is a proper labelled calculus.*

5 A.LΣ Is a Proper Labelled Calculus

In this section we first show that $\mathbf{A.L}\Sigma$ has the display property and then we provide a proof of Corollary 1 as stated in Sect. 4.4.

Lemma 1. *Let* s *be any derivable sequent in* $\mathbf{A.L}\Sigma$*. Then, up to renaming of the variables, every nominal or conominal occurring in* s *occurs in it exactly twice and with opposite polarity.*

Proof. The proof is by induction on the height of the derivation. The statement is trivially true for the axioms. All the rules in the calculi with the exception of the cut-rules either (a) introduce two occurrences of a new label, (b) eliminate two occurrences of the same existing label, or (c) keep the labels in the sequent intact. Before applying the same reasoning to $\mathrm{Cut}_{\mathbf{j}\leq A}$ and $\mathrm{Cut}_{A\leq\mathbf{m}}$, we must take care of renaming nominals and conominals in $\Gamma, \Gamma', \Delta, \Delta'$ to avoid conflicts. Thus, up to substitution, any variable occurring in a derivable sequent occurs exactly twice.

To prove that $\mathbf{A.L}\Sigma$ has the display property (see Definition 5) we need the following definition:

Definition 4. *Let* $s = \Gamma \vdash \Delta$ *be any sequent. For any structure* $\sigma = \mathbf{j} \leq \mathbf{T}$*, (resp.* $\sigma = \mathbf{T} \leq \mathbf{m}$*) in* Γ*, we say that it has a* **j**-twin *(resp.* **m**-twin*) iff there exists exactly one occurrence of* **j** *(resp.* **m***) in* Δ*.*

Proposition 1. *For any derivable sequent* $s = \Gamma \vdash \Delta$ *and any structure* $\sigma = \mathbf{j} \leq \mathbf{T}$ *(resp.* $\sigma = \mathbf{T} \leq \mathbf{m}$*) in* Γ *there exists a sequent* $s' = \Gamma' \vdash \Delta'$ *which is interderivable with* s *such that* $\sigma \in \Gamma'$ *and* σ *has a* **j**-*twin (resp.* **m**-*twin) in* s'*.*

Proof. A sequent s is said to have the **j**-*twin property* (resp. **m**-twin property) iff every structure of the form $\mathbf{j} \leq \mathbf{T}$ (resp. $\mathbf{T} \leq \mathbf{m}$) has a **j**-twin (resp. **m**-twin) in Δ. Notice that the conclusions of initial rules satisfy all twin properties trivially. We say that an inference rule preserves the twin property if the premises having a certain twin property implies that the conclusion has the same twin property.

All the rules in the basic calculus which do not involve switching of structural terms (i.e. all the rules except for the rules **STm, SmT, STj, SjT, STT′m, SjTT′**) and the rules T, B, and C preserve the twin property, since, for each nominal **j** or conominal **m** in the rule, one of the following holds:

1. each occurrence of $\mathbf{j}$ or $\mathbf{m}$ remains on the same side in the conclusion as it was in the premise (adjunction rules, rules for propositional connectives, T, B);
2. exactly two occurrences of $\mathbf{j}$ (resp. $\mathbf{m}$) are eliminated (cut, $\mathbf{Sm}$, $\mathbf{Sj}$, $\Box_S$, $\Diamond_P$, D, C);[7]
3. $\mathbf{j}$ (resp. $\mathbf{m}$) does not occur in the premise, and exactly two occurrences of $\mathbf{j}$ (resp. $\mathbf{m}$) are added in the conclusion, one in the antecedent and one in the consequent of the conclusion ($\top_\Box$, $\bot_\Diamond$, $\Box_P$, $\Diamond_S$, C).

Notice that rule C occurs both in items 2 and 3 and, in particular, we can assume that the nominal $\mathbf{h}$ in the conclusion of C is fresh (see the proof of Lemma 2).

In the case one of the $\mathbf{STm}$, $\mathbf{SmT}$, $\mathbf{STj}$, $\mathbf{SjT}$, $\mathbf{STT'm}$, $\mathbf{SjTT'}$ rules is applied, the twin property can be broken, as nominals (resp. conominal) contained in the structural terms $\mathbf{T}, \mathbf{T'}$ switch side without their twin nominals (resp. conominals) doing the same. Let $R = \frac{\Gamma_1 \vdash \Delta_1}{\Gamma_2 \vdash \Delta_2}$ be an application of any of these rules. Let $\sigma \in \Gamma_1 \cap \Gamma_2$ be a structure which has a twin in Δ_1 but not in Δ_2. This is only possible if σ is of the form $\mathbf{i} \leq \mathbf{T}_1$ (resp. $\mathbf{T}_1 \leq \mathbf{n}$) and there exists a σ' of the form $\mathbf{T} \leq \mathbf{m}$ (resp. $\mathbf{j} \leq \mathbf{T}$) for some $\mathbf{T}$ containing $\mathbf{i}$ (resp. $\mathbf{n}$) which will appear in Γ_2 with $\mathbf{j}$ (resp. $\mathbf{m}$) introduced fresh. We can switch the term $\mathbf{T}$ back to the right by using the switch rule applicable to the conclusion of R, which gives a sequent of the same form as $\Gamma_1 \vdash \Delta_1$, but with a fresh conominal (resp. nominal) in the place of $\mathbf{m}$ (resp. $\mathbf{j}$). Here, we show the proof for the rules $\mathbf{STT'm}$ and $\mathbf{SmT}$, the proof for other rules being similar.

$$\mathbf{STT'm}\,\dfrac{\mathbf{SjTT'}\,\dfrac{\Gamma, \mathbf{j} \leq \mathbf{T'} \vdash \mathbf{j} \leq \mathbf{T}, \Delta}{\Gamma, \mathbf{T} \leq \mathbf{m} \vdash \mathbf{T'} \leq \mathbf{m}, \Delta}}{\Gamma, \mathbf{j'} \leq \mathbf{T'} \vdash \mathbf{j'} \leq \mathbf{T}, \Delta} \qquad \dfrac{\dfrac{\Gamma, A \leq \mathbf{m} \vdash \mathbf{T} \leq \mathbf{m}, \Delta}{\Gamma, \mathbf{j} \leq \mathbf{T} \vdash \mathbf{j} \leq A, \Delta}\,\mathbf{STj}}{\Gamma, A \leq \mathbf{m'} \vdash \mathbf{T} \leq \mathbf{m'}, \Delta}\,\mathbf{SmT}$$

Let $s' = \Gamma_3 \vdash \Delta_3$ be any sequent derived from $\Gamma_2 \vdash \Delta_2$ such that $\sigma \in \Gamma_3$. If $\sigma' \notin \Gamma_3$, it must have been switched to the right by a switch rule, as all other rules leave pure structures on the left intact. However, whenever a switch rule is applied, the term $\mathbf{T}$ occurs on the right of the sequent and has a twin. Thus, if the last rule of the derivation is such a switch rule, then σ has a twin in s'. If $\sigma' \in \Gamma_3$ and σ' has a twin structure, then we can switch $\mathbf{T}$ to right by the application of an appropriate switch rule leading to a interderivable sequent in which σ has a twin. Thus, reasoning by induction, it is enough to show that σ' has a twin when it occurs for the first time (by the application of the switch rule). This is immediate from the fact that the label occurring in $\sigma' \in \Gamma_3$ in any switch rule is of the form $\mathbf{j} \leq \mathbf{T}$ or $\mathbf{T} \leq \mathbf{m}$ where $\mathbf{j}$ or $\mathbf{m}$ is fresh and also occurs in Δ_3. Thus switch rules preserve the twin property.

Let $R = \frac{\Gamma_1 \vdash \Delta_1}{\Gamma_2 \vdash \Delta_2}$ be an application of rule (4). Let $\sigma \in \Gamma_1 \cap \Gamma_2$ be a structure which has a twin in Δ_1 but not in Δ_2. This is only possible if σ is of the form $\mathbf{j} \leq \mathbf{T} \in \Gamma$ for some structure $\mathbf{T}$. In this case, after the application of 4 we can reintroduce the relevant twin structure $\Diamond\mathbf{j} \leq \mathbf{m}$ in the following way.

$$\mathbf{STT'n}\,\dfrac{\dfrac{4\,\dfrac{\mathbf{j} \leq \mathbf{T} \in \Gamma \vdash \Diamond\mathbf{j} \leq \mathbf{m}}{\mathbf{j} \leq \mathbf{T} \in \Gamma, \mathbf{h} \leq \Diamond\mathbf{j} \vdash \Diamond\mathbf{h} \leq \mathbf{m}}}{\mathbf{j} \leq \mathbf{T} \in \Gamma, \mathbf{h} \leq \Diamond\mathbf{j} \vdash \mathbf{h} \leq \blacksquare\mathbf{m}}\,\Diamond \dashv \blacksquare}{\mathbf{j} \leq \mathbf{T} \in \Gamma, \blacksquare\mathbf{m} \leq \mathbf{n} \vdash \Diamond\mathbf{j} \leq \mathbf{n}}$$

[7] Given the result in Lemma 1, notice that this is the same as saying that *all* occurrences of $\mathbf{j}$ (resp. $\mathbf{m}$) are eliminated, except in the case of cut rules, which have two premises. But, if the cut formula has a $\mathbf{j}$-twin (resp. $\mathbf{m}$-twin) in Γ in the one premise and in Δ' in the other, the conclusion $\Gamma, \Gamma' \vdash \Delta, \Delta'$ will still have the $\mathbf{j}$-twin (resp. $\mathbf{m}$-twin) property.

The rest of the proof now follows by the same inductive algorithm.

We can now prove the display property of the calculus.

Proposition 2. *If $s = \Gamma \vdash \Delta$ is a derivable* **A.L**Σ*-sequent, then every structure σ occurring in s is* displayable *(see Definition 5).*

Proof. We prove the display property only for the structures of the form $\mathbf{j} \leq \mathbf{T}$ and $\mathbf{j} \leq A$. The proof for structures of the form $\mathbf{T} \leq \mathbf{m}$ and $A \leq \mathbf{m}$ are dual. Let σ be a structure of the form $\mathbf{j} \leq \mathbf{T}$ or $\mathbf{j} \leq A$. By Lemma 1, $\mathbf{j}$ occurs exactly twice in any derivable sequent. Let σ' be the structure containing the other occurrence of $\mathbf{j}$. If σ' is a labelled formula we are done. The set of well-formed pure structures containing $\mathbf{j}$ is the following: $\{\mathbf{j} \leq \mathbf{m}, \Diamond\mathbf{j} \leq \mathbf{m}, \blacklozenge\mathbf{j} \leq \mathbf{m}, \mathbf{j} \leq \Box\mathbf{m}, \mathbf{j} \leq \blacksquare\mathbf{m}, \mathbf{i} \leq \Diamond\mathbf{j}, \mathbf{i} \leq \blacklozenge\mathbf{j}\}$. For any derivable sequent $s = \Gamma \vdash \Delta$, it is easy to verify the following (by inspection of all the rules in **A.L**):

(a) if $\Diamond\mathbf{j} \leq \mathbf{n}$ (resp. $\mathbf{i} \leq \Box\mathbf{m}$) occurs in s, then it occurs in Δ, given that $\Diamond_S$ and $\bot_\Diamond$ (resp. $\Box_P$ and $\top_\Box$) are the only rules introducing such a structure;
(b) if $\blacklozenge\mathbf{j} \leq \mathbf{n}$ (resp. $\mathbf{i} \leq \blacksquare\mathbf{m}$) occurs in s, then it occurs in Δ, given that $\blacklozenge$ and $\blacksquare$ can be introduced only via adjunction rules to structures $\Diamond\mathbf{j} \leq \mathbf{n}$ and $\mathbf{i} \leq \Box\mathbf{m}$ which only occur in Δ by (a);
(c) if any pure structure $t \in \{\mathbf{i} \leq \Diamond\mathbf{j}, \mathbf{i} \leq \blacklozenge\mathbf{j}, \Box\mathbf{m} \leq \mathbf{n}, \blacksquare\mathbf{m} \leq \mathbf{n}\}$ occurs in s, then it occurs in Γ, given that they can only be introduced via a switch rule (namely, **STm**, **STj**, **STT′m**, **Sj**, **TT′**).

Case (a) and (b). If $\mathbf{j}$ (resp. $\mathbf{m}$) occurs in some pure structure $\Diamond\mathbf{j} \leq \mathbf{n}$ or $\blacklozenge\mathbf{j} \leq \mathbf{n}$ (resp. $\mathbf{i} \leq \Box\mathbf{m}$ or $\mathbf{i} \leq \blacksquare\mathbf{m}$), then it occurs in Δ and it is displayable through a single application of an adjunction rule.

Case (c). If $\mathbf{j}$ (resp. $\mathbf{m}$) occurs in some pure structure $\mathbf{i} \leq \Diamond\mathbf{j}$, $\mathbf{i} \leq \blacklozenge\mathbf{j}$ (resp. $\Box\mathbf{m} \leq \mathbf{n}$, $\blacksquare\mathbf{m} \leq \mathbf{n}$), then it occurs in Γ. If we can apply a switch rule moving the relevant complex term to Δ, we reduce ourselves to the previous cases (a) or (b) and we are done. We will consider the case where a pure structure $\mathbf{i} \leq \Diamond\mathbf{j}$ occurs on the left. All the other cases are treated analogously. By Proposition 1, there exists a sequence $s' = \Gamma' \vdash \Delta'$ such that s' is interderivable with s, $\mathbf{i} \leq \Diamond\mathbf{j}$ occurs in Γ' and has an $\mathbf{i}$-twin (resp. $\mathbf{n}$-twin) in s'. Thus, $\mathbf{i}$ occurs in Δ' by definition. By (a) and (b), $\mathbf{i}$ can occur in Δ' in one of the structures $\mathbf{i} \leq A$, $\mathbf{i} \leq \mathbf{m}$ $\Diamond\mathbf{i} \leq \mathbf{m}$, or $\blacklozenge\mathbf{i} \leq \mathbf{m}$. If $\mathbf{i}$ occurs in $\mathbf{i} \leq A$ or $\mathbf{i} \leq \mathbf{m}$, we can apply a switch rule with $\mathbf{i} \leq \Diamond\mathbf{j}$ to reduce ourselves to previous cases. If $\mathbf{i}$ occurs in the structure of the form $\Diamond\mathbf{i} \leq \mathbf{m}$ or $\blacklozenge\mathbf{i} \leq \mathbf{m}$, we first apply adjunction and then proceed in the same way. This concludes the proof of display property for the basic calculus.

In case of the axiomatic extensions, we need to argue that the added rules preserve the property in Proposition 1. Notice that all the additional rules preserve the Lemma 1.

In case of rules *(T)*, *(B)*, and *(C)* no variables nominals or conominals switch side thus the twin structures are preserved by these rules. *(4)*: The rule 4 moves $\Diamond\mathbf{j}$ from the right to the left. Thus, the only structure in Γ which loses the twin property is some structure σ in Γ containing $\mathbf{j}$. However, just like in the case of the switch rules we can switch $\mathbf{j}$ back to the right by applying an adjunction followed by a switch rule on $\mathbf{h}$ as follows.

$$\dfrac{\dfrac{\Gamma, \mathbf{h} \le \Diamond \mathbf{j} \vdash \Diamond \mathbf{h} \le \mathbf{m}}{\Gamma, \mathbf{h} \le \Diamond \mathbf{j} \vdash \mathbf{h} \le \blacksquare \mathbf{m}}\ \Diamond \dashv \blacksquare}{\Gamma, \blacksquare \mathbf{m} \le \mathbf{n} \vdash \Diamond \mathbf{j} \le \mathbf{n}}\ \mathrm{STT'm}$$

Thus, the twin of σ reappears. We can now argue that the existence of an interderivable sequent with a twin is preserved in a derivation containing an application of this rule in similar way as we did with the switch rules.

(D): In case of rule D, as the premise contains the structure $\mathbf{k} \le \blacklozenge \mathbf{j}$, there must be another structure $\sigma \in \Gamma$ containing $\mathbf{j}$ of the form $\mathbf{j} \le A$ or $\mathbf{j} \le \mathbf{T}$. The premise must be interderivable with a sequent in which σ has a twin. However, in the conclusion $\mathbf{k}$ does not occur and $\mathbf{j} \le \Box \mathbf{m}$ occurs on the right. Thus, σ has a twin in the conclusion. All other nominals and conominals in the premise stay on the same side in the conclusion. Thus, rule D preserves the property of existence of interederivable sequent with the twin property.

To show that $\mathbf{A}.\mathbf{L}\Sigma$ is a proper labelled calculus, we need to verify that $\mathbf{A}.\mathbf{L}\Sigma$ satisfies each condition in Definition 6. The verification of the conditions C6 and C7 requires to preliminarily show the following:

Lemma 2 (Preservation of principal formulas' approximants). *If the sequents s and s' occur in the same branch b of an $\mathbf{A}.\mathbf{L}\Sigma$-derivation π, $\mathbf{j} \le A \in s$ and $\mathbf{i} \le A \in s'$ (resp. $A \le \mathbf{m} \in s$ and $A \le \mathbf{n} \in s'$) are* congruent *in b (see Definition 5), $\mathbf{j} \le A$ (resp. $A \le \mathbf{m}$) is principal in s and $\mathbf{i} \le A$ (resp. $A \le \mathbf{n}$) is in display in s', then the term occurring in $\mathbf{i} \le A$ (resp. $A \le \mathbf{n}$) in π can be renamed in a way such that $\mathbf{i} = \mathbf{j}$ (resp. $\mathbf{n} = \mathbf{m}$), and the new derivation π' is s.t. $\pi' \equiv \pi$ modulo a renaming of some nominal or conominal.*

Proof. Assume that $\mathbf{j} \le A \in s$ and $\mathbf{i} \le A \in s'$ (resp. $A \le \mathbf{m} \in s$ and $A \le \mathbf{n} \in s'$) are congruent in the branch b of a derivation π, $\mathbf{j} \le A$ (resp. $A \le \mathbf{m}$) is principal in s and $\mathbf{i} \le A$ (resp. $A \le \mathbf{n}$) is in display in s'. This means that there is a sub-branch $b' \subseteq b$ connecting s and s' and the height of s is strictly smaller than the height of s' in π, given that $\mathbf{j} \le A$ (resp. $A \le \mathbf{m}$) is principal in s. If $\mathbf{j} \le A$ stays parametric in b', then $\mathbf{j} = \mathbf{i}$ (resp. $\mathbf{m} = \mathbf{n}$). If not, it means that $\mathbf{j} \le A$ (resp. $A \le \mathbf{m}$) is nonparametric in an even number of applications of switch rules occurring in b', given that $\mathbf{j} \le A$ and $\mathbf{i} \le A$ (resp. $A \le \mathbf{m}$ and $A \le \mathbf{n}$) are both approximated by a nominal (resp. conominal) and a single application of a switch rules changes the nominal (resp. conominal) approximating a formula into a conominal (resp. nominal). W.l.o.g. we can confine ourselves to consider a branch b' with exactly two applications of switch rule involving $\mathbf{j} \le A$ and $\mathbf{i} \le A$ (resp. $\mathbf{m} \le A$) as nonparametric structures. If a rule R is applied in b', R is not a switch rule, and x is a nominal or conominal occurring in a nonparametric strucure in the conclusion of R but not necessarily occurring in the premise(s) of R, then we can pick x fresh in the entire sub-branch $b'' \subseteq b'$ connecting the conclusion of R with the sequent s. In the case of $\mathbf{A}.\mathbf{L}\Sigma$, such rule R is either $\Box_P$, $\Diamond_S$, C, or 4. Therefore, $\mathbf{j}$ (resp. $\mathbf{m}$) occurs neither in the premise nor in any parametric structure in the conclusion of S, where S is any application of a switch rules in the branch b' and s.t. $\mathbf{i} \le A$ (resp. $A \le \mathbf{n}$) is a nonparametric structure in the conclusion of S. So, the side conditions of switch rules are satisfied, and in any application of such S we can switch $\mathbf{i}$ for $\mathbf{j}$ (resp. $\mathbf{n}$ for $\mathbf{m}$).

We now proceed to check all the conditions C_1-C_8 defining proper labelled calculi.

Proof. The fact that **A.L**Σ satisfies condition C_5' is proved in Proposition 2. To show that **A.L**Σ satisfies condition C_8 we need to consider all possible principal cuts.

Below we consider cut formulas introduced by initial rules and we exhibit a new cut-free proof that is an axiom as well (notice that axioms are defined with empty contexts). The principal cut formula is $\mathbf{j} \leq \top$ or $\mathbf{j} \leq p$, $R \in \{\mathrm{Id}_{\mathbf{j}\leq p}, \mathrm{Id}_{\top}, \top_w\}$, and x, y, z are instantiated accordingly to R (the proof for $\bot \leq \mathbf{m}$ or $p \leq \mathbf{m}$, and $R \in \{\mathrm{Id}_{p\leq m}, \mathrm{Id}_{\bot}, \bot_w\}$ is similar and it is omitted).

$$\mathrm{Cut}_{\mathbf{j}\leq\top}\ \dfrac{\dfrac{\pi_1}{\mathbf{j} \leq x \vdash \mathbf{j} \leq y}\,R \qquad \dfrac{\pi_2}{\mathbf{j} \leq y \vdash \mathbf{j} \leq z}\,R}{\mathbf{j} \leq x \vdash \mathbf{j} \leq z} \quad \rightsquigarrow \quad \dfrac{\pi_i}{\mathbf{j} \leq x \vdash \mathbf{j} \leq z}\,R$$

Below we consider formulas introduced by logical rules. The side conditions on logical rules impose that the parametric structures are in display or can be displayed with a single application of an adjunction rule, therefore we can provide the following proof transformations where the newly generated cuts are well-defined (in particular the cut formulas are in display) and of lower complexity (the new cut formulas are immediate subformulas of the original cut formulas).

The principal cut formula is $\Box A$ and A is $\mathbf{m}$-labelled formula (see Definition 3).

$$\mathrm{Cut}_{\mathbf{j}\leq\Box A}\ \dfrac{\dfrac{\begin{matrix}\vdots\,\pi_1\\ \Gamma, A \leq \mathbf{m} \vdash \mathbf{j} \leq \Box\mathbf{m}, \Delta\end{matrix}}{\Gamma \vdash \mathbf{j} \leq \Box A, \Delta}\,\Box_S \qquad \Box_P\,\dfrac{\begin{matrix}\vdots\,\pi_2\\ \Gamma' \vdash A \leq \mathbf{m}, \Delta'\end{matrix}}{\Gamma', \mathbf{j} \leq \Box A \vdash \mathbf{j} \leq \Box\mathbf{m}, \Delta'}}{\Gamma, \Gamma' \vdash \Delta, \mathbf{j} \leq \Box\mathbf{m}, \Delta'} \ \rightsquigarrow\ \blacklozenge \dashv \Box^{-1}\ \dfrac{\dfrac{\begin{matrix}\vdots\,\pi_2\\ \Gamma' \vdash A \leq \mathbf{m}, \Delta'\end{matrix} \qquad \dfrac{\begin{matrix}\vdots\,\pi_1\\ \Gamma, A \leq \mathbf{m} \vdash \mathbf{j} \leq \Box\mathbf{m}, \Delta\end{matrix}}{\Gamma, A \leq \mathbf{m} \vdash \blacklozenge\mathbf{j} \leq \mathbf{m}, \Delta}\,\blacklozenge \dashv \Box}{\Gamma, \Gamma' \vdash \Delta, \blacklozenge\mathbf{j} \leq \mathbf{m}, \Delta'}\,\mathrm{Cut}_{A\leq\mathbf{m}}}{\Gamma, \Gamma' \vdash \Delta, \mathbf{j} \leq \Box\mathbf{m}, \Delta'}$$

The principal cut formula is $\Box A$ and A is a $\mathbf{j}$-labelled formula (see Definition 3): in this case the original cut is as in the proof above. The principal cut elimination transformation requires we perform a new cut on the immediate labelled subformulas $A \leq \mathbf{m}$, but, given A is a $\mathbf{j}$-labelled formula by assumption, only $\mathrm{Cut}_{\mathbf{j}\leq A}$ can be applied, so we need first to apply the appropriate switch rules changing the approximants of the immediate subformulas A on both branches. In branch π_1 of the original proof, we can apply adjunction and derive $\Gamma, A \leq \mathbf{m} \vdash \blacklozenge\mathbf{j} \leq \mathbf{m}$ where $\blacklozenge\mathbf{j} \leq \mathbf{m}$ is the twin structure (see Definition 4) of $A \leq \mathbf{m}$, so we can apply the rule **STj**. In branch π_2, Proposition 1 ensures that $\Gamma' \vdash A \leq \mathbf{m}, \Delta'$ is interderivable with $\Gamma'', x \leq \mathbf{m} \vdash A \leq \mathbf{m}, \Delta''$ for some x formula or x pure structure, where $x \leq \mathbf{m}$ is the twin structure of $A \leq \mathbf{m}$. Therefore, we can apply the appropriate switch rule changing the approximant of A. The proof transformation is detailed below.

$$\rightsquigarrow\quad \blacklozenge \dashv \Box^{-1}\ \dfrac{\dfrac{\dfrac{\dfrac{\dfrac{\dfrac{\begin{matrix}\vdots\,\pi_1\\ \Gamma, A \leq \mathbf{m} \vdash \mathbf{j} \leq \Box\mathbf{m}, \Delta\end{matrix}}{\Gamma, A \leq \mathbf{m} \vdash \blacklozenge\mathbf{j} \leq \mathbf{m}, \Delta}\,\blacklozenge \dashv \Box}{\Gamma, \mathbf{i} \leq \blacklozenge\mathbf{j} \vdash \mathbf{i} \leq A, \Delta}\,\mathbf{STj} \qquad \dfrac{\dfrac{\begin{matrix}\vdots\,\pi_2\\ \Gamma' \vdash A \leq \mathbf{m}, \Delta'\end{matrix}}{\Gamma'', x \leq \mathbf{m} \vdash A \leq \mathbf{m}, \Delta''}\,\text{Prop. 1}}{\Gamma'', \mathbf{i} \leq A \vdash \mathbf{i} \leq x, \Delta''}\,\mathrm{S}}{\Gamma, \Gamma'', \mathbf{i} \leq \blacklozenge\mathbf{j} \vdash \mathbf{i} \leq x, \Delta, \Delta''}\,\mathrm{Cut}_{\mathbf{i}\leq A}}{\Gamma, \Gamma'', x \leq \mathbf{m} \vdash \blacklozenge\mathbf{j} \leq \mathbf{m}, \Delta, \Delta''}\,\mathrm{S}}{\Gamma, \Gamma' \vdash \blacklozenge\mathbf{j} \leq \mathbf{m}, \Delta, \Delta'}\,\text{Prop. 1}}{\Gamma, \Gamma' \vdash \mathbf{j} \leq \Box\mathbf{m}, \Delta, \Delta'}$$

The principal cut formula is $A \wedge B$ and both immediate subformulas are **j**-labelled formulas (the case in which at least one immediate subformula is an **m**-labelled formula is analogous to the proof transformation step for $\Diamond A$ and it is omitted. The case for $A \vee B$ are similar and they are omitted).

$$\mathrm{Cut}_{\mathbf{j} \leq A_1 \wedge A_2}\dfrac{\dfrac{\dfrac{\vdots \pi_1}{\Gamma \vdash \mathbf{j} \leq A_1, \Delta} \quad \dfrac{\vdots \pi_2}{\Gamma \vdash \mathbf{j} \leq A_2, \Delta}}{\Gamma \vdash \mathbf{j} \leq A_1 \wedge A_2, \Delta}\wedge_S \quad \wedge_P\dfrac{\dfrac{\vdots \pi_3}{\Gamma', \mathbf{j} \leq A_i \vdash \Delta'}}{\Gamma', \mathbf{j} \leq A_1 \wedge A_2 \vdash \Delta'}}{\Gamma, \Gamma' \vdash \Delta, \Delta'} \rightsquigarrow \mathrm{Cut}_{\mathbf{j} \leq A_i}\dfrac{\dfrac{\vdots \pi_i}{\Gamma \vdash \mathbf{j} \leq A_i, \Delta} \quad \dfrac{\vdots \pi_3}{\Gamma', \mathbf{j} \leq A_i \vdash \Delta'}}{\Gamma, \Gamma' \vdash \Delta, \Delta'}$$

To show that **A**.**L**Σ satisfies all the other conditions is immediate and it requires inspecting all the rules in **A**.**L**Σ.

6 Conclusions

In the present paper, we have showcased a methodology for introducing labelled calculi for nonclassical logics (LE-logics) in a uniform and principled way, taking the basic normal lattice-based modal logic **L** and some of its axiomatic extensions as a case study. This methodology hinges on the use of semantic information to generate calculi which are guaranteed by their design to enjoy a set of basic desirable properties (soundness, syntactic completeness, conservativity, cut elimination and subformula property). Interestingly, the methodology showcased in the present paper naturally imports the one developed and applied in [15] in the context of proper display calculi to the proof-theoretic format of labelled calculi. Specifically, just like the algorithm ALBA, the main tool in unified correspondence theory, was used in [1, 15] to generate proper display calculi for basic (D)LE-logics in arbitrary signatures and their axiomatic extensions defined by analytic inductive axioms, in the present paper, ALBA is used to generate labelled calculi for **L** and 31 of its axiomatic extensions. Similarly to extant labelled calculi in the literature (viz. those introduced in [16]), the language of the calculi introduced in the present paper manipulate a language which properly extends the original language of the logic, and includes *labels*. However, the language of these labels is the same language manipulated by ALBA, the intended interpretation of which is provided by a suitable *algebraic* environment, rather than by a relational one; specifically, by the canonical extensions of the algebras in the class canonically associated with the given logic. Just like the use of canonical extensions as a semantic environment for unified correspondence theory has allowed for the mechanization and uniform generalization of correspondence arguments from classical normal modal logic to the much wider setting of normal LE-logics without relying on the availability of any particular relational semantics, this same semantic setting allows for the uniform generation of labelled calculi for LE-logics in a way that does not rely on a particular relational semantics. However, via general duality theoretic facts, these calculi will be sound also w.r.t. any relational semantic environment for the given logic, and can also provide a "blueprint" for the introduction of labelled calculi designed to capture the logics of specific classes of relational structures (cf. [20]). In future work, we will generalize the current results to arbitrary LE-signatures, and establish systematic connections, via formal translations, between proper display calculi and labelled calculi for LE-logics.

A Proper Labelled Calculi

In what follows we provide a formal definition of the display property and proper labelled calculi.

Definition 5 (Display). *A nominal* $\mathbf{j}$ *(resp. conominal* $\mathbf{m}$*) is always* in display *in a labelled formula* $\mathbf{j} \leq A$ *(resp.* $A \leq \mathbf{m}$*), and is* in display *in a pure structure* t *iff* $t = \mathbf{j} \leq \mathbf{T}$ *(resp.* $t = \mathbf{T} \leq \mathbf{m}$*) for some term* $\mathbf{T}$ *such that* $\mathbf{j}$ *does not occur in* $\mathbf{T}$.

A pure structure $\mathbf{j} \leq \mathbf{T}$ *(resp.* $\mathbf{T} \leq \mathbf{m}$*) is* in display *in a sequent* s *if* $\mathbf{j}$ *(resp.* $\mathbf{m}$*) is in display in each structure of* s *in which it occurs.*

A labelled formula $\mathbf{j} \leq A$ *(resp.* $A \leq \mathbf{m}$*) is* in display *in a sequent* s *if* $\mathbf{j}$ *(resp.* $\mathbf{m}$*) is in display in each structure of* s *in which it occurs.*

A proof system enjoys the display property *iff for every derivable sequent* $s = \Gamma \vdash \Delta$ *and every structure* $\sigma \in s$*, the sequent* s *can be equivalently transformed, using the rules of the system, into a sequent* s' *s.t.* σ *occurs in display in* s' *(in this case we might say that* σ *is* displayable*).*[8]

Definition 6 (Proper labelled calculi). *A proof system is a* proper labelled calculus *if it satisfies the following list of conditions:*

C_1: Preservation of Formulas. *Each formula occurring in a premise of an inference* r *is a subformula of some formula in the conclusion of* r.

C_2: Shape-Alikeness of Parameters and Formulas/Terms in Congruent Structures. *(i) Congruent parameters are occurrences of the same (meta-)structure (i.e. instantiations of structure metavariables in the application of a rule* R *except for switch rules); (ii) instantiations of labelled formulas in the application of switch rules (in the case of* $\mathbf{A.L}\Sigma$*, occurrences of the form* $\mathbf{j} \leq C$ *and* $C \leq \mathbf{m}$*) are congruent and the formulas in these occurrences instantiate the same formula metavariable (namely* C*); (iii) instantiations of pure structures in the application of switch rules (in the case of* $\mathbf{A.L}\Sigma$*, occurrences of the form* $\mathbf{j} \leq \mathbf{T}''$ *and* $\mathbf{T}'' \leq \mathbf{m}$*) are congruent and exactly one term in these occurrences instantiate the same term metavariable (namely* $\mathbf{T}''$*).*

C_3: Non-proliferation of Parameters and Congruent Structures. *(i) Each parameter in an inference* r *is congruent to at most one constituent in the conclusion of* r*. (ii) Each nonparametric structure in the instantiation* r *of a switch rule* R *is congruent to at most one nonparametric structure in the conclusion of* r.

C_4: Position-Alikeness of Parameters and Congruent Structures. *Congruent parameters and congruent structures occur in the same position (i.e. either in precedent position or in succedent position) in their respective sequents.*

C_5: Display of Principal Constituents. *If a labelled formula* a *is principal in the conclusion of an inference* r*, then* a *is in display.*

C'_5: Display-Invariance of Axioms and Structural Rules. *If a structure* σ *occurs in the conclusion* s *of a structural rule* $\frac{s_1, \ldots, s_n}{s} R$ *(where* R *is an axiom whenever the set of*

[8] Notice that we are not requiring that every meta-structure is displayable.

premises is empty), then either σ occurs in display in s, or a structure σ' and a sequent s' exist s.t. σ' is in display in s', and s' is derivable from s via application of switch and adjunction rules only, and σ and σ' are congruent in this derivation. Moreover, if the rule R is an axiom, then $\dfrac{}{s'}\,R'$ *is an axiom of the calculus as well.*

C_6: Closure Under Substitution for Succedent Parts. *Each rule is closed under simultaneous substitution of (sets of) arbitrary structures for congruent labelled formulas occurring in succedent position.*

C_7: Closure Under Substitution for Precedent Parts. *Each rule is closed under simultaneous substitution of (sets of) arbitrary structures for congruent labelled formulas occurring in precedent position.*
where $\overline{\sigma}$ is a multi-set of structures and $\overline{\sigma}/a$ means that $\overline{\sigma}$ are substituted for a.
This condition caters for the step in the cut-elimination procedure in which the cut needs to be "pushed up" over rules in which the cut-formula in succedent position is parametric.

C_8: Eliminability of Matching Principal Constituents. *This condition requests a standard Gentzen-style checking, which is now limited to the case in which both cut formulas are* principal, *i.e. each of them has been introduced with the last rule application of each corresponding subdeduction. In this case, analogously to the proof Gentzen-style, condition C_8 requires being able to transform the given deduction into a deduction with the same conclusion in which either the cut is eliminated altogether, or is transformed in one or more applications of the cut rule(s), involving proper subformulas of the original cut-formula.*

We now provide the proof of Theorem 1 stated in Sect. 4.4.

Proof. The proof is close to the proofs in [12] and [22, Section 3.3, Appendix A]. As to the principal move (i.e. both labelled cut formulas are principal), condition C_8 guarantees that this cut application can be eliminated. As to the parametric moves (i.e. at least one labelled cut formula is parametric), we are in the following situation:

$$
\dfrac{\begin{array}{c}\vdots\,\pi_1\\ (\Pi \vdash \Sigma)[a]^{suc}\end{array}
\qquad
\begin{array}{c}
\begin{array}{ccc}\vdots\,\pi_{2.1} & & \vdots\,\pi_{2.n}\\ (\Gamma_1 \vdash \Delta_1)[a_{u_1}]^{pre} & \cdots & (\Gamma_n \vdash \Delta_n)[a_{u_n}]^{pre}\end{array}\\
\ddots\;\vdots\;\cdot^{\cdot^{\cdot}}\;\pi_2\\
(\Gamma \vdash \Delta)[a]^{pre}\end{array}}
{\Pi, \Gamma \vdash \Sigma, \Delta}\;\text{Cut}
$$

where we assume that the cut labelled formula a is parametric in the conclusion of π_2 (the other case is symmetric), and $(\Gamma \vdash \Delta)[a]^{pre}$ (resp. $(\Pi \vdash \Sigma)[a]^{suc}$) means that a occur in precedent (resp. succedent) position in $\Gamma \vdash \Delta$ (resp. $\Pi \vdash \Sigma$).

Conditions C_2-C_4 make it possible to follow the history of that occurrence of a, since these conditions enforce that the history takes the shape of a tree, of which we consider each leaf. Let a_{u_i} (abbreviated to a_u from now on) be one such uppermost-occurrence in the history-tree of the parametric cut term a occurring in π_2, and let $\pi_{2.i}$ be the subderivation ending in the sequent $\Gamma_i \vdash \Delta_i$, in which a_u is introduced.

Wansing's parametric case (1) splits into two subcases: (1a) a_u is introduced in display; (1b) a_u is not introduced in display. Condition C_5' guarantees that (1b) can only be the case when a_u has been introduced via an axiom.

If (1a), then we can perform the following transformation:

$$\dfrac{\dfrac{\vdots\, \pi_1}{(\Pi \vdash \Sigma)[a]^{suc}} \quad \dfrac{\begin{matrix}\vdots\, \pi_{2.i}\\ (\Gamma_i \vdash \Delta_i)[a_u]^{pre}\\ \vdots\, \pi_2\end{matrix}}{(\Gamma \vdash \Delta)[a]^{pre}}}{\Pi, \Gamma \vdash \Sigma, \Delta}\,\text{Cut} \quad \rightsquigarrow \quad \begin{matrix}\dfrac{\begin{matrix}\vdots\, \pi_1\\ (\Pi \vdash \Sigma)[a]^{suc}\end{matrix} \quad \begin{matrix}\vdots\, \pi_{2.i}\\ (\Gamma_i \vdash \Delta_i)[a_u]^{pre}\end{matrix}}{\Pi, \Gamma_i \vdash \Sigma, \Delta_i}\,\text{Cut'}\\ \vdots\, \pi_2\,[\{\Pi, \Sigma\}/a]\\ \Pi, \Gamma \vdash \Sigma, \Delta\end{matrix}$$

where $\pi_2\,[\{\Pi, \Sigma\}/a]$ is the derivation obtained by π_2 by substituting Π, Σ for a in π_2.[9] Notice that the assumption that a is parametric in the conclusion of π_2 and that a_u is principal implies that π_2 has more than one node, and hence the transformation above results in a cut application of strictly lower height. Moreover, condition C_7 implies that Cut' is well defined and the substitution of $\{\Pi, \Sigma\}$ for a in π_2 gives rise to an admissible derivation $\pi_2\,[\{\Pi, \Sigma\}/a]^{pre}$ in the calculus (use C_6 for the symmetric case). If (1b), i.e. if a_u is the principal formula of an axiom, the situation is illustrated below in the derivation on the left-hand side:

$$\dfrac{\dfrac{\vdots\, \pi_1}{(\Pi \vdash \Sigma)[a]^{suc}} \quad \dfrac{\begin{matrix}(\Gamma_i \vdash \Delta_i)[a_u]^{pre}\\ \vdots\, \pi_2\end{matrix}}{(\Gamma \vdash \Delta)[a]^{pre}}}{(\Pi, \Gamma \vdash \Sigma, \Delta}\,\text{Cut} \quad \rightsquigarrow \quad \begin{matrix}\dfrac{\begin{matrix}\vdots\, \pi_1\\ (\Pi \vdash \Sigma)[a]^{suc}\end{matrix} \quad (\Gamma_i' \vdash \Delta_i')[a_u]^{pre}}{\Pi, \Gamma' \vdash \Sigma, \Delta'}\,\text{Cut'}\\ \vdots\, \pi'\\ (\Gamma_i \vdash \Delta_i)[\{\Pi, \Sigma\}/a]\\ \vdots\, \pi_2\,[\{\Pi, \Sigma\}/a]\\ \Pi, \Gamma \vdash \Sigma, \Delta\end{matrix}$$

where $(\Gamma_i \vdash \Delta_i)[a_u]^{pre}[a]^{suc}$ is an axiom. Then, condition C_5' implies that some sequent $(\Gamma_i' \vdash \Delta_i')[a_u]^{pre}[a]^{suc}$ exists, which is display-equivalent to the first axiom, and in which a_u occurs in display. This new sequent can be either identical to $(\Gamma_i \vdash \Delta_i)[a_u]^{pre}[a]^{suc}$, in which case we proceed as in case (1a), or it can be different, in which case, condition C_5' guarantees that it is an axiom as well. Further, if π is the derivation consisting of applications of adjunction and switch rules which transform the latter axiom into the former, then let $\pi' = \pi\,[\{\Pi, \Sigma\}/a_u]$. As discussed when treating (1a), the assumptions imply that π_2 has more than one node, so the transformation described above results in a cut application of strictly lower height. Moreover, condition C_7 implies that Cut' is well defined and substituting $\{\Pi, \Sigma\}$ for a_u in π_2 and in π gives rise to admissible derivations $\pi_2\,[\{\Pi, \Sigma\}/a_u]$ and π' in the calculus (use C_6 for the symmetric case).

As to Wansing's case (2), assume that a_u has been introduced as a parameter in the conclusion of $\pi_{2.i}$ by an application r of the rule R.

[9] Notice that the writing $\pi_2\,[\{\Pi, \Sigma\}/a]$ does not mean that Π and Σ remain untouched in π_2, namely it does not mean that every sequent in π_2 is of the form $\Pi, \Gamma'' \vdash \Sigma, \Delta''$ for some Γ'', Δ''. Indeed, structures in Π or in Σ might play the role of active structures in some applications of switch rules occurring in π_2, if any.

Therefore, the transformation below yields a derivation where π_1 is not used at all and the cut is not applied.

$$\dfrac{\dfrac{\vdots\ \pi_1}{(\Pi \vdash \Sigma)[a]^{suc}} \qquad \dfrac{\dfrac{\vdots\ \pi_{2.i}}{(\Gamma_i \vdash \Delta_i)[a_u]^{pre}}}{\dfrac{\vdots\ \pi_2}{(\Gamma \vdash \Delta)[a]^{pre}}}}{\Pi, \Gamma \vdash \Sigma, \Delta}\ \text{Cut} \quad \rightsquigarrow \quad \dfrac{\dfrac{\vdots\ \pi'_{2.i}}{(\Gamma_i \vdash \Delta_i)[\{\Pi, \Sigma\}/a_u]}}{\dfrac{\vdots\ \pi_2\ [\{\Pi, \Sigma\}/a]}{\Pi, \Gamma \vdash \Sigma, \Delta}}$$

From this point on, the proof proceeds like in [22].

References

1. Chen, J., Greco, G., Palmigiano, A., Tzimoulis, A.: Syntactic completeness of proper display calculi. ACM Trans. Comput. Log. **23**, 1–46 (2022)
2. Conradie, W., Frittella, S., Manoorkar, K., Nazari, S., Palmigiano, A., Tzimoulus, A., Wijnberg, N.M.: Rough concepts. Inf. Sci. **561**, 371–413 (2021)
3. Conradie, W., Frittella, S., Palmigiano, A., Piazzai, M., Tzimoulis, A., Wijnberg, N.M.: Categories: how i learned to stop worrying and love two sorts. In: Väänänen, J., Hirvonen, Å., de Queiroz, R. (eds.) WoLLIC 2016. LNCS, vol. 9803, pp. 145–164. Springer, Heidelberg (2016). https://doi.org/10.1007/978-3-662-52921-8_10
4. Conradie, W., Frittella, S., Palmigiano, A., Piazzai, M., Tzimoulis, A., Wijnberg, N.M.: Toward an epistemic-logical theory of categorization. EPTCS, vol. 251 (2017)
5. Conradie, W., Ghilardi, S., Palmigiano, A.: Unified correspondence. In: Baltag, A., Smets, S. (eds.) Johan van Benthem on Logic and Information Dynamics. OCL, vol. 5, pp. 933–975. Springer, Cham (2014). https://doi.org/10.1007/978-3-319-06025-5_36
6. Conradie, W., Palmigiano, A.: Algorithmic correspondence and canonicity for distributive modal logic. Ann. Pure Appl. Logic **163**(3), 338–376 (2012)
7. Conradie, W., Palmigiano, A.: Algorithmic correspondence and canonicity for non-distributive logics. Ann. Pure Appl. Logic **170**(9), 923–974 (2019)
8. Conradie, W., Palmigiano, A., Robinson, C., Wijnberg, N.: Non-distributive logics: from semantics to meaning. In: Rezus, A. (ed.) Contemporary Logic and Computing. Landscapes in Logic, vol. 1, pp. 38–86. College Publications (2020)
9. Davey, B.A., Priestley, H.A.: Introduction to Lattices and Order, 2nd edn. Cambridge University Press, New York (2002)
10. De Domenico, A., Greco, G.: Algorithmic correspondence and analytic rules. In: Advances in Modal Logic, vol. 14, pp. 371–389. College Publications (2022)
11. Dunn, J.M., Gehrke, M., Palmigiano, A.: Canonical extensions and relational completeness of some structural logics. J. Symb. Log. **70**(3), 713–740 (2005)
12. Frittella, S., Greco, G., Kurz, A., Palmigiano, A., Sikimić, V.: Multi-type display calculus for dynamic epistemic logic. J. Log. Comput. **26**(6), 2017–2065 (2016)
13. Ganter, B., Wille, R.: Formal Concept Analysis: Mathematical Foundations. Springer, Heidelberg (2012)
14. Gehrke, M., Harding, J.: Bounded lattice expansions. J. Algebra **238**, 345–371 (2001)
15. Greco, G., Ma, M., Palmigiano, A., Tzimoulis, A., Zhao, Z.: Unified correspondence as a proof-theoretic tool. J. Log. Comput. **28**(7), 1367–1442 (2016)
16. Negri, S.: Proof analysis in modal logic. J. Philos. Log. **34**, 507–544 (2005)
17. Negri, S., Dyckhoff, R.: Proof analysis in intermediate logics. Arch. Math. Logic **51**(1), 71–92 (2012)

18. Pawlak, Z.: Rough sets. IJCIS **11**(5), 341–356 (1982)
19. Simpson, A.K.: The proof theory and semantics of intuitionistic modal logic. Ph.D. dissertation, University of Edinburgh (1994)
20. van der Berg, I., De Domenico, A., Greco, G., Manoorkar, K., Palmigiano, A., Panettiere, M.: Labelled calculi for the logics of rough concepts. In: Banerjee, M., Sreejith, A.V. (eds.) ICLA 2023. LNCS, vol. 13963, pp. 172–188. Springer, Heidelberg (2023). https://doi.org/10.1007/978-3-031-26689-8_13
21. Wansing, H.: Displaying Modal Logic. Kluwer (1998)
22. Wansing, H.: Sequent systems for modal logics. In: Gabbay, D.M., Guenthner, F. (eds.) Handbook of Philosophical Logic, vol. 8, pp. 61–145. Springer, Dordrecht (2002). https://doi.org/10.1007/978-94-010-0387-2_2

Two Ways to Scare a Gruffalo

Shikha Singh[1(✉)], Kamal Lodaya[2], and Deepak Khemani[1]

[1] Department of Computer Science and Engineering, IIT Madras, Chennai 600036, India
cs16d008@smail.iitm.ac.in, khemani@cse.iitm.ac.in

[2] Independent Researcher, Bengaluru 560064, India

Abstract. This paper applies and extends the results from [22] on *agent-update frames* and their logic. Several interesting examples of actions for forgery and deception, agent-upgrade and agent-downgrade are considered. Going on from the earlier paper, a second interesting children's story is modelled using these ideas. A dynamic epistemic logic is defined with all these actions and provided with a complete axiomatization. Decision procedures for satisfiability and model checking follow. A planning-oriented approach is also discussed.

Keywords: Agent-update · Deception · Forgery · Completeness and decidability · Epistemic planning

1 Introduction

In [22], the authors modelled Julia Donaldson's children's story *The Gruffalo* [14] in dynamic epistemic logic [28]. The technical enhancement required extending the update modality, specified by an action frame $U = (E, O_i, pre, post)$ [3], with a product update operation providing the updated Kripke model. We extended the semantics with agent-update frames $U = (E, O_i, O_i^+, O_i^-, pre, post)$, which allow adding and deleting agents at states of the Kripke model. The different viewpoints of agents provided in an action frame are used to model deception. An extended sum-product update operation underlies this development.

Our work showed us the flexibility and extendibility provided by action frames. Our extensions were extremely general, and dealing with how to ascribe beliefs to the updated agents was a challenge. The axioms we came up with seemed quite ad hoc.

In *The Gruffalo*, a mouse runs into a fox, owl and snake, all intent on eating it. It deceives them by claiming it is friends with a terrible gruffalo, and they all run away. In a magical twist, the gruffalo appears and wants to eat the mouse. The mouse claims everyone is scared of it and takes the gruffalo to fox, owl and snake, each of whom run away on seeing them. The gruffalo is convinced this must be because of the mouse, and it too runs away.

Remark 1 (Historical). The story by Donaldson [14] is close to one in the *Arabian Nights* [7]. Donaldson's story is simpler and easier to model in pure doxastic

M. Banerjee and A. V. Sreejith (Eds.): ICLA 2023, LNCS 13963, pp. 48–67, 2023.
https://doi.org/10.1007/978-3-031-26689-8_4

logic. It appears this Arabian nights story is derived from a simpler one in the Buddhist Jataka tales, which go at least as far back as the 3rd century [13].

In this paper we model a sequel: Julia Donaldson's *The Gruffalo's Child* [15]. Here is the story in brief. There are several agents in the story. We begin with four: the gruffalo (child) g, mouse m, fox f and owl o, and all four believe in each others' agency. For simplicity, we do not consider a snake agent from the book. The gruffalo has been brought up to believe that the mouse is big and bad, which it is skeptical about. In the context of the story, "big and bad" means that it eats gruffalos. The gruffalo first meets the fox (instead of the snake in the story) and then the owl, both of whom reiterate this belief. The gruffalo remains skeptical. It then meets the mouse who is not big at all, and it sees that it can eat the mouse. The mouse utilizes the rising Moon to project a big shadow of itself. This reverses the gruffalo's belief and it runs away.

Remark 2 (Historical). A similar idea (a solar eclipse) was used by Tintin in *Prisoners of the Sun* [16].

To model the story we introduce agent-upgrade and agent-downgrade operations: the mouse is downgraded in the gruffalo's eyes, and then the mouse upgrades itself. We see them as variants of agent-addition and agent-deletion. There is a different way to present belief upgrade and downgrade without changing the set of agents [4,24,25] using ordered Kripke models. Although we do not study this, our work suggests that the two approaches may be inter-translatable.

Our paper [22] and the present paper began with a problem in AI planning. This is seen as model checking from a logic perspective [10,11], and has been studied in DEL [17,18]. We sketch how formulating it in logic suggests ways to tackle it.

Remark 3 (On stories). Amarel [1] suggested using the folk problem of missionaries and cannibals to study planning problems in artificial intelligence. Smullyan's books, starting from [23], are masterpieces of logic puzzles of various kinds. The book of [27] is an inspiring account of modelling epistemic puzzles as stories. Woods's books on fiction [31,32] explore the paradox that Sherlock Holmes lived in 221B Baker Street in the 19th century, and that he didn't since he didn't exist then.

Here is an outline of the paper.

Section 2 gives some basic definitions of models, as well as the agent-update semantics of [22]. In Sect. 3 we define a few kinds of agent-updates and a logic with which we can use them. In particular we show how forgery can be modelled in addition to deception, as well as new operations of agent upgrade and downgrade. These updates are used to model the story from *The Gruffalo's Child* [15]. In Sect. 4 we prove the usual theoretical results: completeness of the proof system, algorithms for satisfiability and model checking. Section 5 has a discussion suggesting a more planning-oriented approach.

We want to thank Hans van Ditmarsch, Anantha Padmanabha, R. Ramanujam and Yanjing Wang for discussions on the earlier paper [22] which led to the writing of this paper.

2 Models and Logic

We begin with Kripke structures.

Definition 1 (Kripke model). $M = (S, \{R_i \mid i \in A\}, V)$, *where model* M *consists of a set of possible worlds* S *and accessibility relations* $R_i \subseteq S \times S$ *for every agent* $i \in A$, *and a valuation function* $V : Prop \to 2^S$ *assigns states to a proposition.* sR_it *abbreviates* $\langle s,t\rangle \in R_i$ *and it means that at a world* s, *agent* i *believes possible that the world may be* t. *When an agent relation is reflexive, symmetric and transitive the worlds are said to be indistinguishable by the agent. A pointed Kripke model is written as* (M, s) *where* $s \in S$ *is a* designated *state.*

In the figures, a directed arrow labelled with i from world s to world t depicts sR_it and an undirected line between two worlds, say s and t, labelled with i, represents arrows for sR_it and tR_is.

We will assume a fixed set of propositions *Prop* throughout this article. When used as an input to an algorithm, the size of a Kripke model is the sum of the number of states $|S|$, the number of agents $|A|$, the sizes of the accessibility relation $|R_i|$ of every agent i and the size of the valuation, presented in some convenient manner such as a bitvector of states for every proposition. The asymptotically dominant component will be the sizes of the relations, which can be quadratic in the number of states. The size of the valuation is only linear in the number of states. Thus the input is of size $O(|A||S|^2)$.

2.1 Updating Kripke Models with Actions

We present our agent-updates in the style of Baltag, Moss and Solecki's *action frames* [3], further developed in [26].

We formally define *Agent Update frames* on a countable set of potential agents $\mathcal{A}$ and a finite $A \subseteq \mathcal{A}$ of agents in a model [22]. The logic EL will be defined in Definition 4.

Definition 2 (Agent-update frame on $A \subseteq \mathcal{A}$). *An* agent-update frame *is a finite structure* $U = (E, \{O_i \mid i \in A\}, \{O_i^+ \mid i \in \mathcal{A}\}, \{O_i^- \mid i \in A\}, pre)$ *with a finite set of events* E, observability *relations for each agent:* $O_i, O_i^+, O_i^- \subseteq E \times E$, *the former two being transitive, together with function* $pre : E \to EL$ *which assigns a* precondition *for each event.* uO_iv *means that agent* i *perceives event* u *as event* v. uO_i^+v *means that event* u *adds agent* i, *we collect such added agents* i *in the set* $Add(u)$. uO_i^-v *means that event* u *deletes agent* i, *and* $Del(u)$ *is the collection of such deleted agents. A* pointed agent-update frame *is written as* (U, u) *where* $u \in E$ *is a designated event.*

A pointed frame (U, u) with $u \in E$ specifies the semantics of an action which updates a Kripke model, applied at event u where the precondition $pre(u)$ holds. See Definition 3 below.

In pictures, in addition to the traditional (solid) arrows (here denoted as O_i) in an action frame on A, we have two other types of arrows: *sum* arrows, dashed, for O_i^+, which can range over new agents outside A, and *del* arrows, dotted, for O_i^- on A in the agent-update frames. Where required, the precondition of an event is shown alongside. Otherwise the precondition at an event u can be taken as $pre(u) = \top$.

We use letters a, b, g, h, i, j, k to denote agents, s, t to denote worlds in Kripke frames, and u, v, w, x to denote events in the agent update frames throughout the paper. We will use R_X for a subset of agents X to abbreviate the transitive closure of $\bigcup\{R_j \mid j \in X\}$.

A *skip* event, represented as an event with $\top$ precondition and self-loops for all agents A, denotes no change. It will be frequently seen in agent frames.

The updated model after an action is formalized as a product of a Kripke model with an action frame [3].

We defined sum-product update [22] to describe belief update for the existing agents and to ascribe beliefs to the newly added agents, and drop beliefs of the deleted agents. During model transformation, for an existing agent a, the possible worlds for an agent in the updated model are inherited from the possible worlds it considered earlier. In world (s, u) (after execution of event u in world s) of the product model, another world (t, v) is possible if and only if t is possible from s, and v is possible from u. For the agent i being added due to an agent adding event u ($i \in Add(u)$), the worlds that i considers possible at (s, u) are *observer-dependent*.

The beliefs of the existing agents are determined by product, the beliefs of the newly added/deleted agents are determined by sum/difference. We describe the transformation of a model on A when an agent-update frame on $\mathcal{A}$ is applied to it, and we call it *sum-product update*. This is product update for agents in A, along with sum and difference for agents in $Add()$ and $Del()$ respectively. An agent's deletion takes priority over its addition.

Definition 3 (Sum-product update). *Given a pointed Kripke model* (M, s) *on agents* A *and a pointed agent-update frame* (U, e) *with* $U = (E, O, O^+, O^-, pre)$ *on agents* $\mathcal{A}$, *the updated pointed Kripke model* $(M * U, (s, e))$, *is defined as:* $(S', \{R'_a \mid a \in A'\}, V')$ *on the updated set of agents* A' *(those* a *such that* R'_a *is nonempty), where:*

1. $S' = \{(s, u) \mid M, s \models pre(u)\} \cap (S \times E)$
2. $V'(p) = \{(s, u) \in S' \mid s \in V(p)\}$
3. R'_a *is the transitive closure of* $(Q_a^{unf} \cup Q_a^{asc} \cup Q_a^{inh})$, *where:*
   unforgotten: $(s, u)Q_a^{unf}(t, v) \iff sR_a t$ *and* $uO_a v$ *and not* $uO_a^- v$
   ascribed: $(s, u)Q_a^{asc}(s, v) \iff uO_a^+ v$, *for* $a \in (Add(u) \setminus Del(u))$
   inherited: $(s, u)Q_a^{inh}(t, u) \iff sR_{Obs(u)} t$, *for* $a \in (Add(u) \setminus Del(u))$

2.2 Logic

We define our agent-update logic using the BNF below. Let $p \in Prop$ be a proposition, Y, X, H be disjoint subsets of $\mathcal{A}$ and i an element. We add specific agent-changing operators U given in the BNF below to obtain the language

AUL. The sublanguage without these operators is called EL. The book of Van Ditmarsch, Van der Hoek and Kooi [28] presents various dynamic epistemic logics.

Definition 4 (Formulas of updates and language AUL).

$$U ::= skip \mid p\ for\ X \mid p\ dcv\ X \mid +Y\ for\ X \mid -Y\ for\ X \mid \Uparrow Y\ for\ X \mid \Downarrow Y\ for\ X \mid H : +Y\ dcv\ X \mid H : -Y\ dcv\ X \mid H : \Uparrow Y\ dcv\ X \mid H : \Downarrow Y\ dcv\ X$$

$$\phi ::= p \mid \neg\phi \mid (\phi \wedge \phi) \mid P_i\phi \mid \langle U\rangle\phi$$

The modality $P_i\phi$ is read as "agent i possibly believes ϕ". The dual modality $B_i\phi = \neg P_i\neg\phi$ is read as "agent i believes ϕ". The other modalities are action modalities, $\langle U\rangle\phi$ is read as "after possible update U, ϕ holds". The dual modality is $[U]\phi = \neg\langle U\rangle\neg\phi$. The updates will be explained through examples in Sect. 3.

Each action operator U is provided a specific action frame $F(U)$. More specifically, given these fixed frames (defined in Sect. 3), the semantics of AUL can be defined as follows, using Definition 3 for sum-product update. We use u for the designated event of the update.

Definition 5 (Truth at a world in a model). *Given a formula ϕ, at a pointed Kripke model (M, s), the assertion "formula ϕ is true at world s in model M" is abbreviated as $M, s \models \phi$ and recursively defined as:*

- $M, s \models \top$ *(always),*
- $M, s \models p \Leftrightarrow s \in V(p)$,
- $M, s \models \neg\phi \Leftrightarrow$ *not* $(M, s) \models \phi$,
- $M, s \models (\phi \wedge \psi) \Leftrightarrow (M, s) \models \phi$ *and* $(M, s) \models \psi$, *and*
- $M, s \models P_i\phi \Leftrightarrow$ *for some* t, sR_it *and* $(M, t) \models \phi$
- $M, s \models \langle U\rangle\phi$ *iff* $(M * F(U), (s, u)) \models \phi$

A formula is valid *if it is true in all models at all worlds. It is* satisfiable *if it is true in some model at some world.*

We work only with *transitive* relations R_i, hence $B_i\phi \implies B_iB_i\phi$ is a valid formula. It says that positive belief is introspective. In our models, $\neg B_i\phi \implies B_i\neg B_i\phi$ is not a valid formula. It says that negative belief is introspective. Chellas has a textbook treatment of modal logic [9] which describes such correspondences of valid formulas with properties of Kripke frames.

Independently of Wang *et al.* [30] which has the same idea, we model existence of agents at a world using presence of that agent's accessibility at the world. We sometimes use the "agency" of an agent i, by which we mean: An agent i exists at a world s in model M iff $(M, s) \models P_i\top$. An agent i exists for another agent j at a world s if i's agency holds at all the worlds t reachable by j from s. Formally, $B_jP_i\top$.

2.3 Proof System

The proof system gives axioms and inference rules to prove valid formulas. There are 8 axioms below and 2 standard inference rules. Several axioms for the update operators will be presented in Sect. 3.

1. all instances of propositional tautologies
2. $B_a(\phi \implies \psi) \implies (B_a\phi \implies B_a\psi)$
3. $B_a\phi \implies B_aB_a\phi$
4. $[U](\phi \implies \psi) \implies ([U]\phi \implies [U]\psi)$
5. $[U]p \Leftrightarrow (pre(u) \implies p)$
6. $[U]\neg\phi \Leftrightarrow (pre(u) \implies \neg[U]\phi)$
7. $[U](\phi \wedge \psi) \Leftrightarrow ([U]\phi \wedge [U]\psi)$
8. $\langle skip\rangle\phi \iff \phi$
9. From ϕ and $\phi \implies \psi$, infer ψ
10. From ϕ, infer $B_a\phi$

3 Agent-Update Actions and Their Logic

In this section, we will examine different kinds of agent-update actions.

We first identify a set of agents whose beliefs remain unchanged at an event in an agent-update frame.

Definition 6 (Observer). *The set of* observers $Obs(u)$ *at an event* u *in an agent-update frame is those* j *with agency at* u *such that* $uO_jv \iff v = u$.

A subset of these are deceivers. In brief, the deceived come to believe the situation depicted at event v observable from u. But at u, the deceivers' beliefs are unchanged.

Definition 7 (Deceiver). *In an agent-update frame if event* v *is observable at* u *by* X (uO_Xv), *the set of* deceivers $Dcvr(u, v)$ *is observers at* u, $Dcvr(u, v) \subseteq Obs(u)$, *whereas the* deceived $Dcvd(u, v)$ *are those* $D \subseteq X$ *disjoint from* $Dcvr(u, v)$ *such that* $D \cup Dcvr(u, v)$ *are observers at* v.

Agents from A which are added, deleted, observed at u or deceived at v, or to whom information is communicated participate in an action. We call other agents *remote*.

The following axioms are validities. The first axiom says that no beliefs change for remote agents. The next axiom is the epistemic action axiom (we call it belief-action) which is common in the literature [2,3,28]. It says that for agents which are observers at the designated event u, beliefs after the update can be reduced to beliefs before the update.

11. $\langle U\rangle P_k\phi \iff P_k\langle skip\rangle\phi$, for $k \in A \setminus (Add() \cup Del() \cup Obs(u) \cup Dcvd(u, v))$
12. $\langle U\rangle P_j\phi \iff P_j\langle U\rangle\phi$, for $j \in Obs(u)$

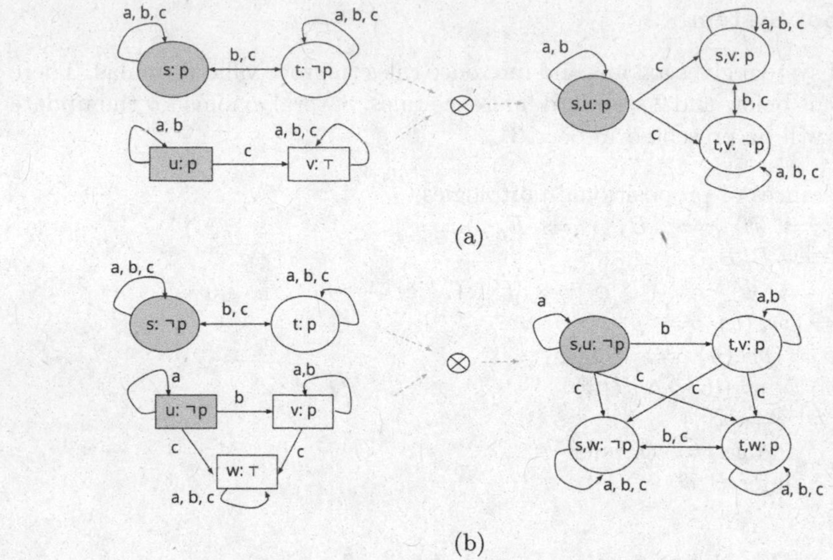

Fig. 1. (a) Alice informing Bob, $F(p\ for\ \{a,b\})$; (b) Alice lying to Bob, $F(a : p\ dcv\ b)$

3.1 Private Update and Lying

Example 1. Suppose Alice knows the truth value of proposition p and Bob does not know. The actions of Alice telling Bob the truth value of p ($p\ for\ \{a,b\}$) and Alice lying to Bob about the truth value of p ($a : p\ dcv\ b$) are depicted in Fig. 1 [29].

In Fig. 1a, top left is a Kripke model, bottom left is an action frame with $Obs(u) = \{a,b\}$ and on the right is the product Kripke model. Alice telling Bob that p is true is modelled with a single event with precondition p, such that both Alice and Bob believe p after the update. Another agent c is unaffected.

Whereas in Fig. 1b above left is a Kripke model, below left is an action frame with $Obs(u) = \{a\}, Dcvr(u,v) = \{a\}$ and $Dcvd(u,v) = \{b\}$, on the right is the product Kripke model. Alice lying to Bob that p is true is modelled using v with precondition p, representing perception of Bob, while event u with precondition $\neg p$ and with an outgoing Bob-arrow to event v is the perception of Alice. Agent c remains unaffected.

The next axiom expresses a validity about information communicated during an update. The remote agent axiom covers the other agents.

13. $\langle p\ for\ X\rangle P_j\phi \iff (p \wedge P_j\langle p\ for\ X\rangle\phi)$, for $j \in X = Obs(u)$

Next we have the axioms for lying. The first reduces to the previous truthful update axioms. This is a pattern which we will repeatedly see in deception. The belief-action axiom covers agents in $H = Dcvr(u) = Obs(u)$ and the remote agent axiom covers the rest.

14. $\langle H : p\ dcv\ X\rangle P_j\phi \iff (\neg p \wedge P_j\langle p\ for\ X \cup H\rangle\phi)$, for $j \in X = Dcvd(u,v)$

3.2 Forgery with Deception and Without

The first agent-update operator we consider is agent-addition $H : +Y\ dcv\ X$ which is deceptive [20,21]. For our example we consider a generalization where the deceivers H do not reveal themselves but use a forged message pretending to be from I.

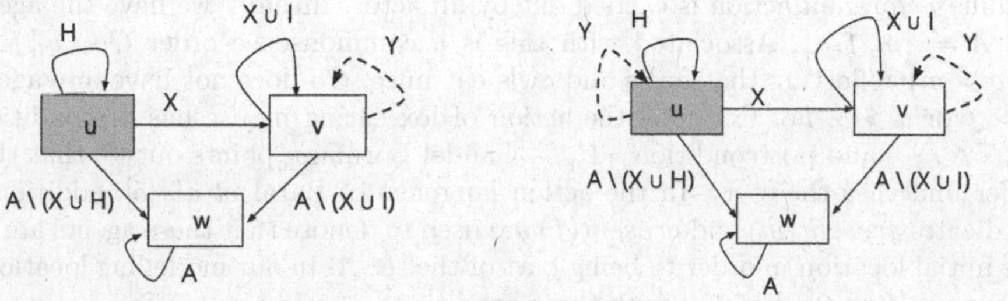

(a) $F(H\ forge\ I : +Y\ dcv\ X)$ (b) $H\ forge\ I : +Y\ for\ X$

Fig. 2. Forgery

Example 2. Figure 2a is from *The final problem* of Sherlock Holmes [12] with $H = \{Moriarty\}$, $X = \{Dr.Watson\}$, *and* $I = \{Innkeeper\}$.

Moriarty wants to deceive Dr Watson away from Holmes by saying that there is a lady who is ill at their inn and needs his help. Such an attempt would not succeed because Watson would not believe Moriarty. So Moriarty forges a letter from the innkeeper, and Watson gets deceived. Thus Watson believes the innkeeper knows of the existence of the lady, whereas Moriarty knows that the innkeeper knows nothing.

We consider lying $H : +Y\ dcv\ X$ plausible when Y are new agents, thus the deceived are credulous. The set of agents is now $A' = A \cup Y$. When all the Y are new fictitious agents (we restrict to $Y \cap A = \emptyset$), the next axiom is valid. For observers and remote agents, we do not repeat the axioms.

15. $[H\ forge\ I : +Y\ dcv\ X]B_l\bot$, for $l \in Y = Add(v)$ (so $\langle H\ forge\ I : +Y\ dcv\ X\rangle P_l\phi \iff \bot$)
16. $\langle H\ forge\ I : +Y\ dcv\ X\rangle P_j\phi \Leftrightarrow P_j\langle +Y\ for\ X \cup I\rangle\phi$, for $j \in X = Dcvd(u, v)$

The second axiom above is a belief-action which reduces the deceptive agent-addition operator to a private agent-addition operator which is described next.

Example 3. Figure 2b illustrates a variant of the Sherlock Holmes story if such a lady did exist. For example, Moriarty could send a lady agent to the inn who could then have pretended to be ill. The forged message from the innkeeper would say a lady is arriving at the inn and has requested a physician's help. The innkeeper is not aware of the existence of the lady.

17. $\langle H\ forge\ I : +Y\ for\ X\rangle P_l\phi \Leftrightarrow \phi \vee \bigvee_{h\in H} P_h\langle H\ forge\ I : +Y\ for\ X\rangle\phi$, for $l \in Y = Add(u) = Add(v)$ (so $\langle H\ forge\ I : +Y\ for\ X\rangle P_l\top$ is valid)
18. $\langle H\ forge\ I : +Y\ for\ X\rangle P_j\phi \Leftrightarrow P_j\langle +Y\ for\ X \cup I\rangle\phi$, for $j \in X = Obs(v)$

3.3 Agent-Deletion with Deception and Without

In modelling the gruffalo story we use commonsense conditions from AI, which include *actors*: an action is carried out by an actor. Initially we have the agent set $A = \{m, f, o\}$. Associated with this is a commonsense order $Co = \{f > m, o > m\}$ reflecting that foxes and owls eat mice. Co does not have any agent $a > f$ or $a > o$. For example, the action of fox eating mouse has precondition $P_m\top \wedge P_f\top$ and postcondition $\neg P_m\top$. Candel Bormann points out [8] that this order underlies the story. In the action language of Baral *et al.* [5], additional predicates $present(m)$ and $present(f)$ are used to denote that these agents are at the initial location in order to being part of the set A. In our modelling locations play no role so we dispense with these conditions.

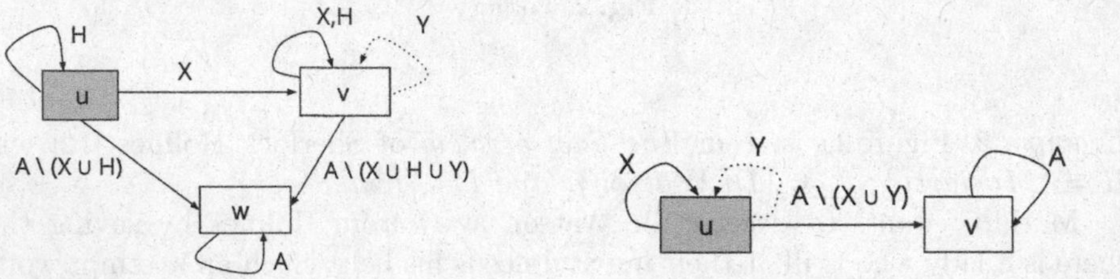

(a) $(F(H : -Y\ dcv\ X), u)$ for *deceptive agent-deletion* update

(b) $(F(-Y\ for\ X), u)$ for *private agent-deletion* update

Fig. 3. Agent-deletion

In Fig. 3a, deceivers H, whose beliefs about agency of Y (which are neither deceiver nor deceived) are unchanged at u, deceive X into believing that agents $Y \subseteq A \setminus (X \cup H)$ have been privately deleted at v for themselves and for H.

In a private deletion, agents Y are selectively deleted for observers in $X \subseteq A$ ($Y \subseteq A \setminus X$), at event u in an agent update frame. The remaining agents are oblivious at v.

Example 4. In Fig. 3b, the dotted self-loop for Y at u could stand for the mouse $i \in Y$ being eaten by the fox f, observed by others $X (f \in X)$, the rest of the animals being oblivious of the meal. The beliefs of the rest of the animals in $A \setminus X$ are unchanged, v is a skip. In particular, the animals in $A \setminus X$ believe in the agency of i at v.

Based on the commonsense order, we have the actions $Act = \{f : -m\ for\ X, o : -m\ for\ X\}$, for subsets $X \subseteq A$. We will make up the action syntax as we go along, it is copied from the AUL update modalities and only meant to informally refer to the actions. The actors are f and o respectively (which do the eating). The actions $-f\ for\ X$ and $-o\ for\ X$ of fox or owl being eaten are not in *Act* since they do not respect the commonsense order: there is no agent eating them.

Because no i-arrows remain after i-deletion, the next axiom is valid. The next two axioms follow from belief-action axiom.

The next three axioms for deceptive agent-deletion follow from belief-action axiom.

19. $\langle H : -Y\ dcv\ X\rangle P_i\phi \Leftrightarrow P_i\langle skip\rangle\phi$, for $i \in Y = Del(v) \cup A \setminus (Del(v) \cup Obs(u) \cup Dcvd(u,v))$
20. $\langle H : -Y\ dcv\ X\rangle P_h\phi \Leftrightarrow P_h\langle H : -Y\ dcv\ X\rangle\phi$, for $h \in H = Dcvr(u,v) = Obs(u)$
21. $\langle H : -Y\ dcv\ X\rangle P_j\phi \Leftrightarrow P_j\langle -Y\ for\ (X \cup H)\rangle\phi$, for $j \in X = Dcvd(u,v)$

Here is the key axiom for private agent-deletion.

22. $[-Y\ for\ X]B_i\bot$, for $i \in Y - Del(u)$. So $\langle -Y\ for\ X\rangle P_i\phi \iff \bot$.

Fox Tries to Convince Gruffalo. The initial situation in the *Gruffalo's child* story is modelled with M_0 with a designated world s as is shown in Fig. 4. $(M_0, s) \models P_g\top \wedge P_f\top \wedge P_o\top \wedge P_m\top \wedge \neg P_m p$. By the proposition is meant $p \iff \neg P_g\top$, that is, a "big bad mouse" is one which eats gruffalos.

Example 5. In *The Gruffalo's child*, the first move is fox telling the gruffalo g of a mouse which likes to eat fox. This move is modelled as a combination of the addition and deletion actions. We write it as a $f : (-g)\ for\ g$ action. At $v1$, the fox believes the mouse believes in eating gruffalos, which we represent as a deletion of gruffalo at $x1$. At $u1$, the gruffalo does not buy the belief. Another agent, the owl, is oblivious of this interaction at $w1$.

Owl Tries to Convince Gruffalo

Example 6. The mouse runs into the owl after deceiving fox and makes a deceptive move again, $o : (-g)\ for\ g$, as before a combination of an addition and a deletion in Fig. 5. At $v2$ the owl believes that the mouse believes in eating gruffalos. At $u2$ the gruffalo does not accept believing this, with fox being oblivious of the interaction.

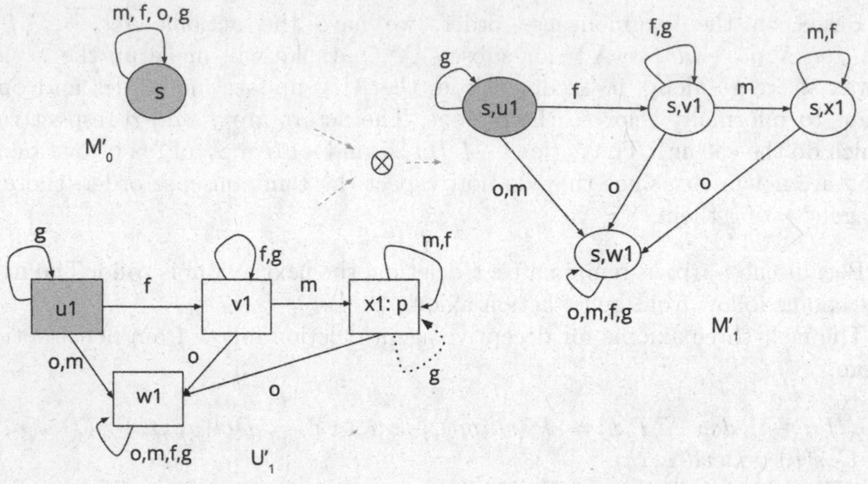

Fig. 4. Fox tries to convince gruffalo of a gruffalo-eating mouse

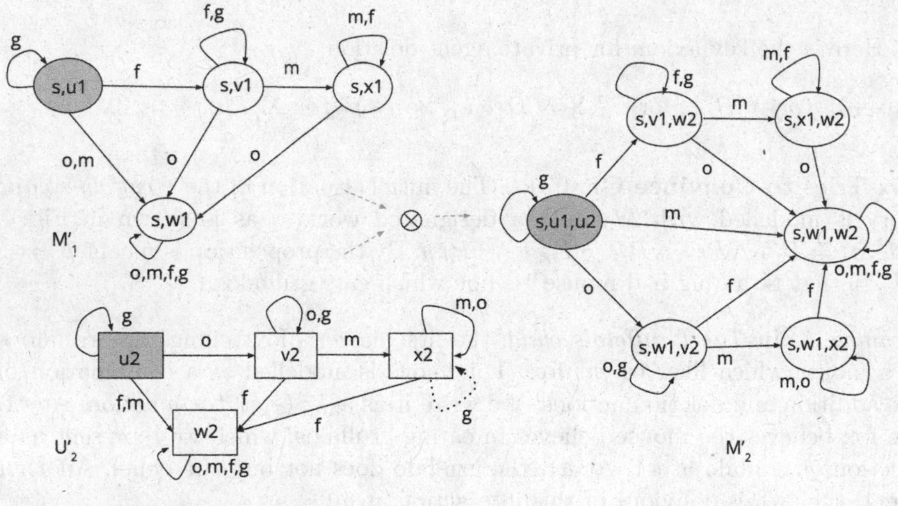

Fig. 5. Owl tries to convince gruffalo that there is a gruffalo-eating mouse

3.4 Private Agent-Addition: +Y for X

In Fig. 6, agents Y are selectively added at event u for observers $X \subseteq A$ $(Y \cap X = \emptyset)$ in an agent frame. Agents in Y can be outside A. Event v is a skip event that does not change anything for anyone. At u, agents in $A \setminus X$ believe that event v occurs; they consider that all agents in A are observers at v.

Example 7. In Fig. 6, let a new agent owl $i \in Y$ appear in the action $+Y$ *for* X as indicated by the dashed arrow. The actor is Y, so we could write it as Y : $+Y$ *for* X. The mouse m is present on the scene at u, it constitutes $X(m \in X)$.

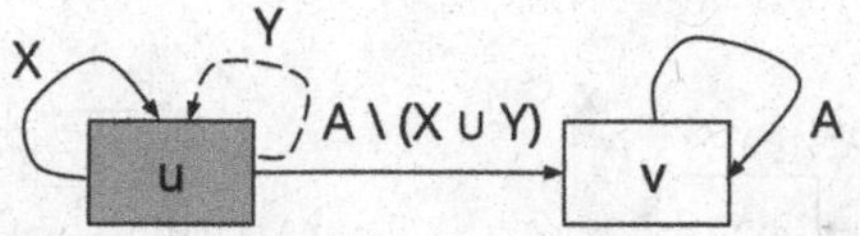

Fig. 6. $(F(+Y\ for\ X), u)$ for *private agent-addition* update

Other animals such as the fox f in $A \setminus X$ are unaware of the agency of i at this moment. They believe that nothing happens at v (a skip).

The agent set $A = \{m, f, g\}$ is expanded to $A' = \{m, f, g, i\}$. The commonsense order Co is unchanged, it has $g > m$ from the earlier introduction of the gruffalo by fox. An action $a : +i\ for\ X$ with $i > a$ does not respect the commonsense order, how would a commandeer such a performance?

When all the Y are new agents (so we restrict to $Y \cap A = \emptyset$), the next axiom is a valid equivalence. The next two axioms follow from the belief-action axiom.

23. $\langle +Y\ for\ X\rangle P_i\phi \Leftrightarrow \phi \vee \bigvee_{j\in X} P_j\langle +Y\ for\ X\rangle\phi$, for $i \in Y = Add(u)$ (so $\langle +Y\ for\ X\rangle\top$ is valid)

3.5 Downgrade and Deceptive Downgrade

Figure 7a shows an agent-downgrade action. In the literature with ordered Kripke models [4,24], such updates typically refer to a proposition. For example an action $\Downarrow p$ would place worlds satisfying p below worlds that do not satisfy p. Our interest in [22] was in existence of agents, where we introduced agent-addition and agent-deletion operations. In this paper, we attempt integrating these ideas into commonsense situations which appear in AI modelling, which are represented by the order Co. Hence agent-downgrade (and agent-upgrade) actions will affect propositional values related to the commonsense order.

Example 8. The agent-downgrade action is motivated by our story *The Gruffalo's child.* Here $g \in X$ at event u downgrades the mouse $m \in Y$ at event v, which it had heard of from fox and owl as eating gruffalos, to one which does not eat gruffalos. That is, the commonsense order Co is updated to remove $m > g$. This has the postcondition $P_m\top$, f continues to be present but without its desire to eat m the mouse remains safe. However, notice that at event v, a self-loop for agent g is added. That is, if the high-grade mouse considered gruffalos as vermin which it had eaten up, the low-grade mouse allows gruffalos to peacefully co-exist.

The explanation above serves to reduce agent-downgrade to agent-addition, which provides a simple axiom.

24. $\langle \Downarrow Y\ for\ X\rangle P_m\phi \iff P_m\langle +X\ for\ Y\rangle\phi$, for $m \in Y = Obs(v), X = Add(v)$

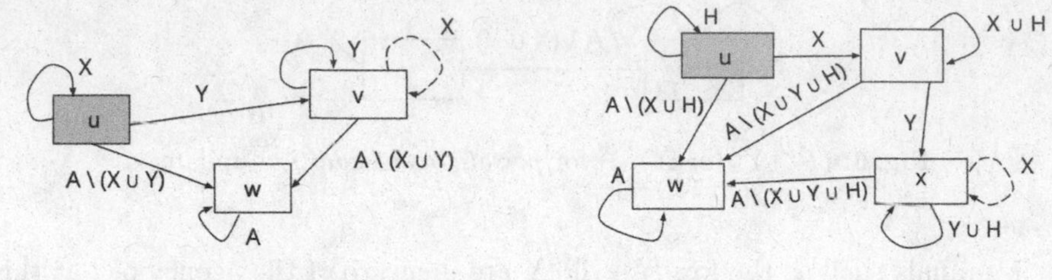

(a) F($\Downarrow Y\ for\ X$) (b) F($H : \Downarrow Y\ dcv\ X$)

Fig. 7. Agent-downgrade

The deceptive downgrade $\langle H :\Downarrow Y\ dcv\ X\rangle\phi$ removes j-deletion arrows ($j \in X$), however H does not believe that j is not capable of eating Y, as shown in Fig. 7b.

25. $\langle H :\Downarrow Y\ dcv\ X\rangle P_g\phi \iff P_g\langle\Downarrow Y\ for\ (X \cup H)\rangle\phi$, for $g \in X = Add(x)$

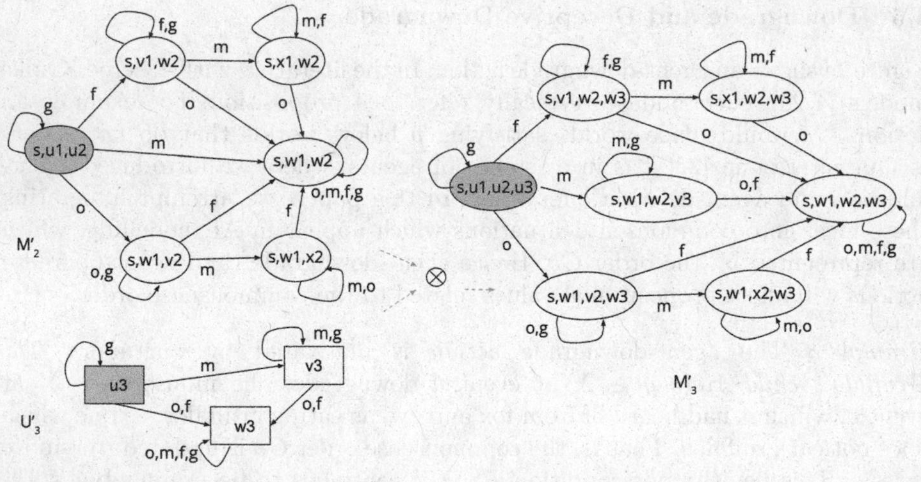

Fig. 8. Mouse appears for gruffalo

The Mouse Appears

Example 9. Further in *The Gruffalo's child*, the gruffalo runs into a mouse which is not big and bad. We model this as a downgrade $\Downarrow m\ for\ g$ about m appearing for gruffalo. m doesn't have any g-deletion arrow. See update $U3$ illustrated in Fig. 8.

3.6 Upgrade and Deceptive Upgrade

An i-upgrade for j is one which adds the possibility of j-deletion as shown in Fig. 9a.

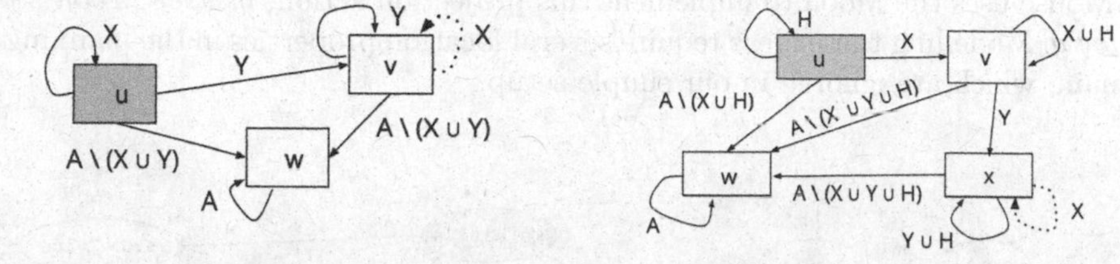

(a) F($\Uparrow Y$ *for* X) (b) F($H : \Uparrow Y$ *dcv* X)

Fig. 9. Agent-upgrade

The deceptive upgrade also adds j-deletion arrows for X. Beliefs of $A \setminus X$ as well as H about Y's capabilities will be unaffected as shown in Fig. 9b.

26. $\langle \Uparrow Y \; for \; X \rangle P_m \phi \iff P_m \langle -X \; for \; Y \rangle \phi$, for $m \in Y = Obs(v), X = Del(v)$

As usual, the axioms for deceptive upgrade use those for upgrade.

27. $\langle H : \Uparrow Y \; dcv \; X \rangle P_g \phi \iff P_g \langle \Uparrow Y \; for \; (X \cup H) \rangle \phi$, for $g \in X = Del(x)$

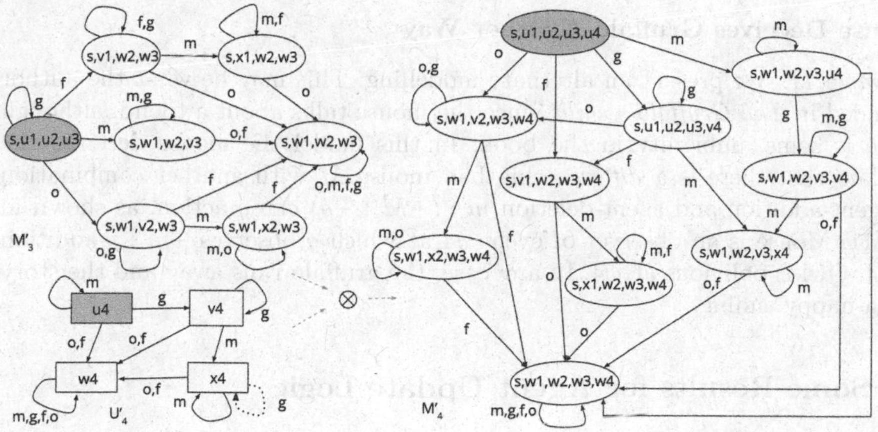

Fig. 10. Mouse deceives gruffalo that it is the big bad mouse, $F(m : \Uparrow m \; dcv \; g)$

Mouse Deceives Gruffalo

Example 10. Further in the story, the mouse deceives the gruffalo $m :\Uparrow m\ dcv\ g$ by showing m capable of eating g. This makes the upgraded m have a $-g$ arrow, although m itself does not believe in its upgraded capability. Owl is oblivious at $w4$. This is illustrated in Fig. 10.

Mouse uses the Moon to implement this projection action, $m\ uses\ Moon :\Uparrow m\ dcv\ g$. Modelling these ideas require several location properties in the planning domain, which are ignored in our simple setup.

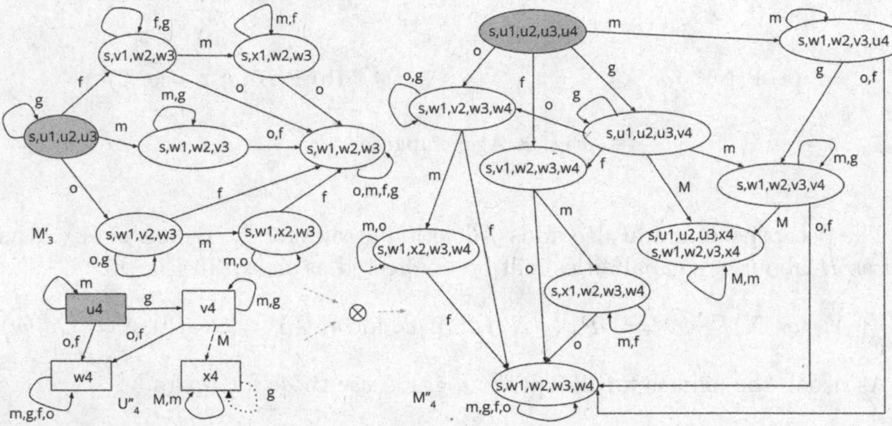

Fig. 11. Mouse deceives gruffalo that there is a big bad mouse

Mouse Deceives Gruffalo Another Way

Example 11. We present an alternate modelling. This may be what the author intended in *The Gruffalo's child*, since the mouse talks about a friend, although there is some ambiguity in the book. In this model the mouse deceives the gruffalo that there is a *different* big bad mouse M with another combination of agent-addition and agent-deletion $m : (+M : -g)\ dcv\ g$ action, as shown in Fig. 11. Mouse is an observer of event $u4$ at which g observes $+m$: p-addition at $v4$. Owl is oblivious at $w4$. In any case, the gruffalo runs away and the story has a happy ending.

4 Some Results for Agent Update Logic

Theorem 1 (Completeness). *The proof system of Sects. 2.3 and 3 is sound and complete over transitive Kripke models.*

Proof. For the proof we define the lexicographic size of a formula, following the DEL book [28, Definition 7.38]. For all formulas, this is as expected, for example $\ell(P_a\phi) = 1 + \ell(\phi)$; except only the update formula: $\ell(\langle U\rangle\phi)$ is defined as $(4 + \ell(U))\ell(\phi)$. This means that the lexicographic size of the left hand side of every bi-implication in the proof system is greater than the lexicographic size of its right hand side. For example, for axiom 23: $\langle +Y\ for\ X\rangle P_j\phi = \phi \vee \bigvee_{j\in X} P_j\langle +Y\ for\ X\rangle\phi$, it can be shown that $(4 + n + k)(1 + m) > (1 + n) + (4 + n + k)m$ where n is $|X|$, k is $|Y|$ and m is $\ell(\phi)$. Thus a reduction algorithm can apply these equivalences to go from an AUL formula to an equivalent EL formula. EL is complete over transitive Kripke models [9]. □

The reader may ask what is achieved by having a proof system with 2 general axioms and 15 specific axioms for 10 update operations (excluding *skip*), compared to the couple of axioms for a single general update operation in [22]. We will discuss this in the context of AI planning in Sect. 5.

Next we provide decision procedures mentioned in [22]. The proof of the first theorem follows from the completeness argument.

Theorem 2 (Satisfiability). *There is a polynomial space algorithm to check satisfiability of an AUL formula.*

Theorem 3 (Model checking). *Given a transitive Kripke model, checking whether an AUL formula holds at a designated state can be done in polynomial time.*

Proof. A labelling algorithm can be implemented by saving the action updates as one proceeds inwards in the formula (without performing the updates). On evaluating a belief modality which requires an agent relation in the updated model, the relation after the updates is calculated by using the saved updates. An extra multiplication by the length of the formula is needed for the number of modalities this has to be done for. This gives a polynomial time algorithm for model checking on transitive models. □

5 Planning

The planning community has traditionally worked with a fixed set of actions *Act*, and a planning problem is defined as a triple $\langle S, Act, G\rangle$ where S is the start state that is completely specified, and G is the set of goal predicates that are desired to be true. The goal predicates G may be true in many states, and any one of them may be acceptable to the planner. For example, in the *Gruffalo's child* story both the mouse running away and the gruffalo running away would satisfy the goal of the mouse being alive.

In the real world the set of actions available to an agent may virtually be unlimited, limited only by the agent's imagination. For a planner, considering a much larger set of actions may be intractable. In the real world agents normally pick a familiar sequence of actions that have been known to work in the past.

For example a traveller may choose between going by bus to the airport or hiring a taxi based on time and money constraints. But what if there were to be a taxi strike and time is running out? In that scenario the traveller may think of the option of calling up a friend to drop him to the airport, an action that one would not normally consider.

We propose that when the set of operators may be potentially unlimited, one can prescribe a graded set of partitions that are accessible to the planner, in a lazy evaluation manner. Thus the set *Act* may be partitioned into an ordered set of subsets $Act_1 \subset Act_2 \subset \ldots$. The planner can now operate with an iterative broadening algorithm in which it begins searching for a plan with the minimal set Act_1 and under certain conditions broadens it to Act_2, and so on. This broadening could be when a plan cannot be found within a reasonable time with only the set Act_1, but there could be other conditions too involving sub-goal interaction.

The above scenario is exemplified in the stories that we are considering where desperate agents seek desperate solutions, often in life threatening situations. For example, the default plan that a mouse may have is to flee in the presence of a hungry predator, but spatial proximity may prohibit that, prompting it to think of other options. In the original story by Donaldson [14], the mouse invents the Gruffalo, with fingers crossed that the predator will swallow the story. And when the unexpected happens and the relieved mouse next encounters the Gruffalo in flesh and blood, it is compelled to spin yet another yarn.

The approach that we are advocating is to not limit a planner to a fixed set of actions, but have access to graded sets of actions when a plan with fewer actions cannot be found. The actions in the extended sets may be computationally more demanding, or may have a lesser chance of success.

The goal for the planner is $P_m\top$ after one step. Informally speaking we have a set of actions Act_1 available to model check the transitive closure $\langle Act_1^+\rangle P_m\top$ (this is outside the logic AUL) at the initial state [11,17]. Since $[Act_1]\neg P_m\top$, the goal is unreachable after 1 step, the base of the transitive closure. Thus from this Act_1, the planner moves to a larger set of actions Act_2.

To reach the goal we restrict ourselves to actions that add agents from Sect. 3, that only alter matters related to agent existence. Since the possible agents form an infinite set, for a practical solution we will have to use some rules about how to go about adding agents.

First Rule. We use the following commonsense inference rule. Suppose $s \models P_f\top \wedge P_m\top$. If $s \models \bigwedge_{a>m, a\neq f\in A} \neg P_a\top$, for $Co \supseteq \{f > m\}$, then add fresh $g \notin A$ to get $A' = A \cup \{g\}$ with $Co' = Co \cup \{g > f\}$. The word "fresh" indicates that the agents outside A form one equivalence class; an arbitrary g is chosen from them, thus dividing the equivalence class into $\{g\}$ which gets added to A' and another equivalence class of the agents outside A'.

Let *Act* be the current set of actions. Consider $Act \cup \{g : -f\}$. The new action is not applicable since $P_g\top$ is false in the initial state.

So one generates a new action, $+g\ for\ f$ which has the postcondition $P_g\top$ for the sub-goal. This action does not identify an actor. Try $m : +g\ for\ f$ using agent m as actor. ($g : +g\ for\ f$ is useless because precondition $P_g\top$ is not met.) But the commonsense order $m < f < g$ says mouse cannot commandeer a gruffalo to appear for fox.

So one generates a new action $m : +g\ dcv\ f$. This action has an actor present, respects commonsense (assuming a credulous fox) and achieves the desired sub-goal. The fact that gruffalo g is fictitious helps in plausibility. So the action set expands to $Act' = Act \cup \{g : -f, m : +g\ dcv\ f\}$.

This process can be repeated for the owl. The goal $P_m\top$ is reached.

Second Rule. Here is another inference rule. If $P_m\top \wedge P_f\top \wedge \bigwedge_{f>a>m\in A} \neg P_a\top$, then add fresh $h \notin A$ with the new commonsense order $Co' = Co \cup \{f > h, h > m\}$ on the expanded agent set $A' = A \cup \{h\}$.

By the reasoning process we saw above, this will eventually lead to an action $m : +h\ dcv\ f$ being added to *Act*. For example, mouse leads the fox to believe there is a hen which is more delicious than itself. The mouse has to still find a way to escape, but for the moment the action is plausible as it preserves $P_m\top$. It requires a planning domain where the story will move to a location where mouse can escape. This is basically the action in Book 4 of the *Panchatantra* stories [6,19] where the monkey who foolishly asked a crocodile to ferry it across the river on its back, only to find itself being considered a meal, tells the crocodile it has left its most delicious heart on the shore, and exhorts the crocodile to swim to the river bank so that the heart can be recovered. This requires a planning domain where river and its bank are modellable.

Remark 4 (Historical). The *Panchatantra* stories are dated to the 3rd century. One of the stories appears in sculpture at a Nalanda temple (7th century).

Third Rule. Here is another inference rule. In a commonsense order with $m > g \in Co$, remove $m > g$ to get $Co' = Co \setminus \{m > g\}$. This is the essential idea behind the action of agent-downgrade. In the *Gruffalo's child*, the downgrade leads to $Co' = (Co \setminus \{m > g\}) \cup \{g > m\}$.

These are only suggestions towards a planning-oriented view of the agent-update logic.

6 Conclusion

Van Ditmarsch, Van der Hoek and Kooi's book on DEL [28, Section 6.1] has a discussion on action frames as syntax and semantics for a logic. In this paper, we suggested using an explicit syntax for our agent-update modalities. The bulk of the paper is a discussion on what kind of syntax works to model stories in an AI planning setting. The usual theoretical results of completeness and algorithms for satisfiability and model checking were obtained. Our syntactic view suggests an approach to synthesis which can be used in planning. A collaboration between people working in logic and AI can lead to fruitful results in this area.

References

1. Amarel, S.: On representation of problems of reasoning about action. In: Michie, D. (ed.) Machine Intelligence 3, pp. 131–171. Edinburgh University Press (1971)
2. Baltag, A., Moss, L.S.: Logics for epistemic programs. Synthese **139**(2), 165–224 (2004)
3. Baltag, A., Moss, L.S., Solecki, S.: The logic of common knowledge, public announcements, and private suspicions. In: Gilboa, I. (ed.) Proceedings of 7th TARK, Evanston, pp. 43–56. Morgan Kaufmann (1998)
4. Baltag, A., Smets, S.: A qualitative theory of dynamic interactive belief revision. In: Bonnano, G., van der Hoek, W., Wooldridge, M. (eds.) Logic and the Foundations of Game and Decision Theory (LOFT 7), pp. 9–58. Amsterdam University Press, Amsterdam (2008)
5. Baral, C., Gelfond, G., Pontelli, E., Son, T.C.: An action language for multi-agent domains. Artif. Intell. **302**, 103601 (2022)
6. Bühler, G.: Panchatantra (5 volumes). Bombay (1891)
7. Burton, R.F.: The Book of the Thousand Nights and One Night. Kama Shashtra Society (1888)
8. Bormann, D.C.: Moving possible world theory from logic to value. Poetics Today **34**(1–2), 177–231 (2013)
9. Chellas, B.F.: Modal Logic: An Introduction. Cambridge University Press, Cambridge (1980)
10. Cimatti, A., Pistore, M., Roveri, M., Traverso, P.: Weak, strong, and strong cyclic planning via symbolic model checking. Artif. Intell. **147**(1–2), 35–84 (2003)
11. Cimatti, A., Pistore, M., Traverso, P.: Automated planning. In: Handbook of Knowledge Representation, pp. 841–867. Elsevier (2008)
12. Doyle, A.C.: The memoirs of Sherlock Holmes. G. Newnes Ltd. (1894)
13. Cowell, E.B., Neil, R.A.: The Jatakas or Stories of the Buddha's Former Births. Cambridge University Press, Cambridge (1907)
14. Donaldson, J.: The Gruffalo. Pan Macmillan (1999)
15. Donaldson, J.: The Gruffalo's child. Pan Macmillan (2004)
16. Hergé. Prisoners of the sun. Casterman (1949)
17. Li, Y., Quan, Yu., Wang, Y.: More for free: a dynamic epistemic framework for conformant planning over transition systems. J. Log. Comput. **27**(8), 2383–2410 (2017)
18. Löwe, B., Pacuit, E., Witzel, A.: DEL planning and some tractable cases. In: van Ditmarsch, H., Lang, J., Ju, S. (eds.) LORI 2011. LNCS (LNAI), vol. 6953, pp. 179–192. Springer, Heidelberg (2011). https://doi.org/10.1007/978-3-642-24130-7_13
19. Ryder, A.W.: The Panchatantra. University of Chicago Press, Chicago (1925)
20. Sakama, C.: A formal account of deception. In: Deceptive and Counter-Deceptive Machines, AAAI Fall Symposia, Arlington, pp. 34–41 (2015)
21. Sarkadi, S., Panisso, A., Bordini, R., McBurney, P., Parsons, S., Chapman, M.: Modelling deception using theory of mind in multi-agent systems. AI Commun. **32**(4), 287–302 (2019)
22. Singh, S., Lodaya, K., Khemani, D.: Agent-update models. arXiv preprint arXiv:2211.02452 (2022)
23. Smullyan, R.M.: What is the Name of this Book? Prentice-Hall, Hoboken (1978)
24. Van Benthem, J.: Dynamic logic for belief revision. J. Appl. Nonclass. Logic **17**(2), 129–155 (2007)

25. Van Benthem, J.: Modal logic for open minds. CSLI (2010)
26. Van Benthem, J., van Eijck, J., Kooi, B.: Logics of communication and change. Inform. Comput. **204**(11), 1620–1662 (2006)
27. Van Ditmarsch, H., Kooi, B.: One Hundred Prisoners and a Light Bulb. Springer, Cham (2015). https://doi.org/10.1007/978-3-319-16694-0
28. Van Ditmarsch, H., van Der Hoek, W., Kooi, B.: Dynamic Epistemic Logic. Synthese Library, vol. 337. Springer, Dordrecht (2008). https://doi.org/10.1007/978-1-4020-5839-4
29. van Ditmarsch, H., van Eijck, J., Sietsma, F., Wang, Y.: On the logic of lying. In: van Eijck, J., Verbrugge, R. (eds.) Games, Actions and Social Software. LNCS, vol. 7010, pp. 41–72. Springer, Heidelberg (2012). https://doi.org/10.1007/978-3-642-29326-9_4
30. Wang, Y., Wei, Y., Seligman, J.: Quantifier-free epistemic term-modal logic with assignment operator. Ann. Pure Appl. Log. **173**, 103071 (2022)
31. Woods, J.: The Logic of Fiction: A Philosophical Sounding of Deviant Logic. Mouton (1974)
32. Woods, J.: Truth in Fiction: Rethinking its Logic. Springer, Cham (2018). https://doi.org/10.1007/978-3-319-72658-8

Determinacy Axioms and Large Cardinals

Sandra Müller(✉)

Institut für Diskrete Mathematik und Geometrie,
TU Wien, Wiedner Hauptstraße 8-10/104, 1040 Wien, Austria
sandra.mueller@tuwien.ac.at
https://dmg.tuwien.ac.at/sandramueller/

Abstract. The study of inner models was initiated by Gödel's analysis of the constructible universe. Later, the study of canonical inner models with large cardinals, e.g., measurable cardinals, strong cardinals or Woodin cardinals, was pioneered by Jensen, Mitchell, Steel, and others. Around the same time, the study of infinite two-player games was driven forward by Martin's proof of analytic determinacy from a measurable cardinal, Borel determinacy from ZFC, and Martin and Steel's proof of levels of projective determinacy from Woodin cardinals with a measurable cardinal on top. First Woodin and later Neeman improved the result in the projective hierarchy by showing that in fact the existence of a countable iterable model, a mouse, with Woodin cardinals and a top measure suffices to prove determinacy in the projective hierarchy. This opened up the possibility for an optimal result stating the equivalence between local determinacy hypotheses and the existence of mice in the projective hierarchy. This article outlines the main concepts and results connecting determinacy hypotheses with the existence of mice with large cardinals as well as recent progress in the area.

Keywords: Determinacy · Infinite game · Large cardinal

1 Introduction

The standard axioms of set theory, Zermelo-Fraenkel set theory with Choice (ZFC), do not suffice to answer all questions in mathematics. While this follows abstractly from Kurt Gödel's famous incompleteness theorems, we nowadays know numerous concrete examples for such questions. A large number of problems in set theory, for example, regularity properties such as Lebesgue measurability and the Baire property are not decided – for even rather simple (for example, projective) sets of reals – by ZFC. Even many problems outside of set

Supported by L'ORÉAL Austria, in collaboration with the Austrian UNESCO Commission and in cooperation with the Austrian Academy of Sciences - Fellowship *Determinacy and Large Cardinals* and the Austrian Science Fund (FWF) under Elise Richter grant number V844, international project number I6087, and START grant number Y1498.

M. Banerjee and A. V. Sreejith (Eds.): ICLA 2023, LNCS 13963, pp. 68–78, 2023.
https://doi.org/10.1007/978-3-031-26689-8_5

theory have been showed to be unsolvable, meaning neither their truth nor their failure can be proven from ZFC. This includes the Whitehead Problem (group theory, [49]), the Borel Conjecture (measure theory, [22]), Kaplansky's Conjecture on Banach algebras (analysis, [8]), and the Brown-Douglas-Fillmore Problem (operator algebras, [11]). A major part of set theory is devoted to attacking this problem by studying various extensions of ZFC and their properties. One of the main goals of current research in set theory is to identify *the "right" axioms for mathematics* that settle these problems. This, in part philosophical, problem is attacked with technical mathematical methods by analyzing various extensions of ZFC and their properties. **Determinacy assumptions** are canonical extensions of ZFC that postulate the existence of winning strategies in natural two-player games. Such assumptions are known to imply regularity properties, and enhance sets of real numbers with a great deal of canonical structure. Other natural and well-studied extensions of ZFC are given by the hierarchy of **large cardinal axioms**. Determinacy assumptions, large cardinal axioms, and their consequences are widely used and have many fruitful implications in set theory and even in other areas of mathematics such as algebraic topology [7], topology [6,13,38], algebra [10], and operator algebras [11]. Many applications, in particular, proofs of consistency strength lower bounds, exploit the interplay of large cardinals and determinacy axioms. Thus, understanding the connections between determinacy assumptions and the hierarchy of large cardinals is vital to **answer questions left open by ZFC itself**. The results outlined in this article are closely related to this overall goal.

To explore the connections between large cardinals and determinacy at higher levels, the study of other hierarchies, for example, with more complex inner models called **hybrid mice**, has been very fruitful. **Translation procedures** are needed to translate these hybrid models, whose strength comes from descriptive set theoretic features, back to standard inner models while making use of their hybrid nature to obtain stronger large cardinals in the translated model. They are therefore a key method **connecting descriptive set theory with inner model theory**. One of the results surveyed in this article is a new translation procedure extending work of Sargsyan [41], Steel [53], and Zhu [61]. This new translation procedure yields a countably iterable inner model with a cardinal λ that is both a limit of Woodin cardinals and a limit of strong cardinals [30]. So it improves Sargsyan's construction in [41] in two ways: It can be used to obtain infinitely many instead of finitely many strong cardinals and the models it yields are countably iterable – a crucial property of mice. This translation procedure can be applied to prove a **conjecture of Sargsyan** on the consistency strength of the Axiom of Determinacy when all sets are universally Baire [30], a central and widely used property of sets of reals introduced implicitly in [47] and explicitly in [12]. In fact, the new translation procedure can be applied in a much broader context. Moreover, it provides the basis for translation procedures resulting in more complex patterns of strong cardinals, for example, a strong cardinal that is a limit of strong cardinals.

Recent seminal results of Sargsyan and Trang [44–46], see also the review [29], as well as Larson and Sargsyan [20, 42] suggest that we are at a *turning point* in the search for natural constructions of canonical models with a Woodin limit of Woodin cardinals and thereby for proving better lower bounds for natural set theoretic hypotheses.

2 Determinacy for Games of Length ω and Large Cardinals

In 1953, Gale and Stewart [14] developed a basic theory of infinite games. For notational simplicity, we identify reals in $\mathbb{R}$ with ω-sequences of natural numbers in ${}^{\omega}\omega$. Gale and Stewart considered, for every set of reals A, a two-player game $G(A)$ of length ω, where player I and player II alternate playing natural numbers $n_0, n_1, \ldots$, as follows:

$$\begin{array}{c|cccc} \text{I} & n_0 & & n_2 & \ldots \\ \hline \text{II} & & n_1 & & n_3 \ldots \end{array}$$

They defined that player I wins the game $G(A)$ if and only if the sequence $x = (n_0, n_1, \ldots)$ of natural numbers produced during a run of the game $G(A)$ is an element of A; otherwise, player II wins. We call A the payoff set of $G(A)$. The game $G(A)$ (or the set A itself) is called determined if and only if one of the two players has a winning strategy, meaning that there is a method by which they can win in the game described above, no matter what their opponent does. The **Axiom of Determinacy** (AD) is the statement that all sets of reals are determined.

Already in [14], the authors were able to prove that every open and every closed set of reals is determined under ZFC. But they also proved that determinacy for all sets of reals contradicts the Axiom of Choice. This leads to the natural question as to how the picture looks for definable sets of reals which are more complicated than open and closed sets. After some partial results by Wolfe [59] and Davis [9], Martin was able to prove in 1975 [24] that every Borel set of reals is determined (again using ZFC).

In the meantime, the development of so-called **large cardinal axioms** was proceeding in set theory, and Solovay was able to prove regularity properties, a known consequence of determinacy, for a specific pointclass, assuming the existence of a measurable cardinal, instead of a determinacy axiom. Finally, Martin was able to prove a direct connection between large cardinals and determinacy axioms: He showed, in 1970, that the existence of a measurable cardinal implies determinacy for every analytic set of reals [23]. Eight years later, Harrington established that this result is, in some sense, optimal, by proving that determinacy for all analytic sets of reals implies that $0^{\#}$, a countable active iterable canonical inner model which can be obtained from a measurable cardinal, exists [15]. Here, an **iterable canonical inner model**, or **mouse**, is a fine structural model that is, in some sense, iterable. This notion goes back to Jensen [16]. Together with Martin's argument mentioned above, this yields an equivalence between the two statements. The construction of such canonical inner models and their connection with determinacy was later extended in work of Dodd, Jensen, Martin, Mitchell, Neeman, Schimmerling, Schindler, Solovay, Steel, Woodin, Zeman, and others (see, e.g., [25,26,48,55,60]; see the preface of [32] or Larson's history of determinacy [19] for a more detailed overview). In the projective hierarchy, this led to the following fundamental theorem. Here, $M_0^\#(x)$ denotes $x^\#$, a version of $0^\#$ relativized to a real x, and $M_n^\#(x)$ denotes a minimal countable active mouse with n Woodin cardinals constructed above x.

Theorem 1 (Harrington, Martin, Neeman, Woodin [15,23,32,33,36]**).** Let n be a natural number. Then the following are equivalent:

1. All $\boldsymbol{\Pi}^1_{n+1}$ sets are determined, and
2. for all $x \in {}^\omega\omega$, $M_n^\#(x)$ exists and is ω_1-iterable.

The proof that the determinacy of sets in the projective hierarchy implies the existence of mice with finitely many Woodin cardinals in this exact level-by-level correspondence first appeared in [27,32] and is originally due to Woodin. As shown in [27], the underlying methods can be used to obtain similar results for certain hybrid mice in the $L(\mathbb{R})$-hierarchy. These tight connections are, at first, very surprising, as they show that two ostensibly completely different notions, from distinct areas of set theory – determinacy from descriptive set theory, and inner models with large cardinals from inner model theory – are, in fact, the same.

3 Determinacy for Games Longer Than ω

It turns out that the correspondence between determinacy and inner models with large cardinals does not stop at games of length ω. For every ordinal α and set $A \subseteq {}^\alpha\omega$, we can define a game $G(A)$ of length α with payoff set A, as follows:

$$\begin{array}{c|cccccc} \mathrm{I} & n_0 & & n_2 & \ldots & n_\omega & & \ldots \\ \hline \mathrm{II} & & n_1 & & n_3 \ldots & & n_{\omega+1} & \ldots \end{array}$$

The players alternate playing natural numbers n_i for $i < \alpha$, and we again say that player I wins the game if and only if the sequence $x = (n_0, n_1, \dots)$ of length α they produce is an element of A; otherwise, player II wins. In landmark results, Neeman [37] developed powerful techniques to prove the determinacy of projective games longer than ω from large cardinals. A first step in this direction is, for example, the following result:

Theorem 2 (Neeman, [37]**).** Let $n \in \omega$ and suppose that $M_{\omega+n}^\#(x)$ exists for all reals $x \in {}^\omega\omega$. Then all games of length ω^2 with $\boldsymbol{\Pi}^1_{n+1}$ payoff are determined.

This result in fact holds for games of fixed length α, for all countable ordinals α, instead of games of length ω^2. The following theorem complements Neeman's results for projective games of length ω^2:

Theorem 3 (Aguilera, Müller, [2,28]) Let n be a natural number and suppose that all games of length ω^2 with $\boldsymbol{\Pi}^1_{n+1}$ payoff are determined. Then, for every $x \in {}^\omega\omega$, there is a model $\mathcal{M}$ of ZFC, with $\omega + n$ Woodin cardinals, such that $x \in \mathcal{M}$.

At this level, the interplay of determinacy and large cardinals is already understood quite well (see also [1,3]). For games of length ω^α with analytic payoff, for countable ordinals α, similar results have previously been established by Trang [57], building on unpublished results of Woodin, using canonical models of determinacy with a generalized Solovay measure. The Solovay measure is also called a **supercompact measure for** ω_1 as it witnesses a degree of supercompactness for ω_1.

When considering much stronger notions of determinacy, the picture is less clear. For example, it was already shown by Mycielski in 1964 that determinacy for all games of length ω_1 is inconsistent with Zermelo-Fraenkel set theory (ZF). Nevertheless, there are subclasses of games of length ω_1 that are still known to be determined under large cardinal assumptions.

An intermediate step are games that do not have a fixed countable length but still end after countably many rounds. In 2004, Neeman showed in groundbreaking work, from large cardinals, that certain games that are not of fixed countable length are still determined. These so-called **games of continuously coded length**, which go back to Steel [50], are defined as follows: For any set $A \subset ({}^\omega\omega)^{<\omega_1}$ and partial function $\nu\colon {}^\omega\omega \rightharpoonup \omega$, the game $G_{\text{cont}}(\nu, A)$ is given by the following rules:

$$\begin{array}{c|cccccccc} \text{I} & y_0(0) & & y_0(2) & & \ldots\ y_\alpha(0) & & y_\alpha(2) & & \ldots \\ \hline \text{II} & & y_0(1) & & y_0(3)\ \ldots & & y_\alpha(1) & & y_\alpha(3)\ \ldots \end{array}$$

We canonically identify segments of the game of length ω as mega-rounds, and let y_α denote the real that the two players together produce in mega-round α. If $\nu(y_\alpha)$ is undefined, the game ends, and player I wins if and only if $\langle y_\xi \mid \xi \leq \alpha\rangle \in A$. Otherwise, let $n_\alpha = \nu(y_\alpha)$. Then the game ends if $n_\alpha \in \{n_\xi \mid \xi < \alpha\}$, and again, player I wins if and only if $\langle y_\xi \mid \xi \leq \alpha\rangle \in A$. If neither of these alternatives hold, the game continues.

Theorem 4 (Neeman, [37]**).** Suppose there is an iterable proper class model M, with a Woodin cardinal δ and a cardinal $\kappa < \delta$ that is $(\delta+1)$-strong in M, such that $V^M_{\delta+1}$ is countable in V. Then the game $G_{\text{cont}}(\nu, A)$ is determined for every ν in the class $\boldsymbol{\Sigma}^0_2$ and every A that is $<\omega^2 - \boldsymbol{\Pi}^1_1$ in the codes.

Here, being Γ in the codes for a pointclass Γ and a set $A \subset ({}^\omega\omega)^{<\omega_1}$ is defined via a natural coding of elements of A as reals; A is Γ *in the codes* if the set of codes of elements of A belongs to Γ. It is not known whether Theorem 4 is optimal, but results of Neeman and Steel [34] show that it cannot be very far away from optimal. I conjecture that it is indeed optimal in the following sense:

Conjecture 1. Suppose the game $G_{\text{cont}}(\nu, A)$ is determined for every ν in the class $\boldsymbol{\Sigma}^0_2$ and every A that is $<\omega^2 - \boldsymbol{\Pi}^1_1$ in the codes. Then there is a model of ZFC with a Woodin cardinal δ and a cardinal $\kappa < \delta$ that is $(\delta+1)$-strong.

A similar conjecture at a higher level is moving toward a Holy Grail of current inner model theory. More precisely, it concers the aim to prove the existence of an inner model with a Woodin cardinal that is a limit of Woodin cardinals from the determinacy of certain long games. The natural games to consider at this level have length ω_1 and their payoff set is ordinal definable using reals as parameters. The converse was shown by Woodin, using results of Neeman [37] and ideas going back to Kechris and Solovay [18].

Theorem 5 (Neeman, Woodin, [37]**).** Suppose there is an iterable proper class model with a Woodin cardinal that is a limit of Woodin cardinals and countable in V. Then there is a model of ZFC in which all ordinal definable games of length ω_1 on natural numbers with real parameters are determined.

In fact, Woodin showed that determinacy of these games of length ω_1 is equiconsistent with a seemingly weaker statement: determinacy of certain games that are **constructibly uncountable in the play**. These games are defined as follows: For a payoff set $A \subset ({}^{\omega}\omega)^{<\omega_1}$, players I and II alternate playing natural numbers to produce reals y_α.

$$\begin{array}{c|cccccccc} \text{I} & y_0(0) & & y_0(2) & \ldots\ y_\alpha(0) & & y_\alpha(2) & & \ldots \\ \hline \text{II} & & y_0(1) & & y_0(3)\ \ldots & y_\alpha(1) & & y_\alpha(3)\ \ldots & \end{array}$$

The game ends when its length reaches the first ordinal γ which is uncountable in $L[y_\alpha \mid \alpha < \gamma]$, and player I wins if and only if $\langle y_\alpha \mid \alpha < \gamma \rangle \in A$. Here $L[y_\alpha \mid \alpha < \gamma]$ denotes Gödel's Constructible Universe L relative to $(y_\alpha \mid \alpha < \gamma)$. In this case $\gamma = \omega_1^{L[y_\alpha \mid \alpha<\gamma]}$, so it makes sense to say that the game **ends at** ω_1 **in** L **of the play**. We technically define that II wins if the game lasts ω_1 (in V) rounds, but mild large cardinal assumptions yield an ordinal γ, as above, that is countable in V. Neeman proved that, for sufficiently definable payoff sets A, these games are determined, via a sophisticated extension of the methods used in the proof of Theorem 4.

Theorem 6 (Neeman, [37]**).** Suppose there is an iterable proper class model with a Woodin cardinal that is a limit of Woodin cardinals and countable in V. Then all games ending at ω_1 in L of the play with payoff sets that are $\Game(<\omega^2 - \Pi^1_1)$ in the codes are determined.

Here, $\Game$ denotes the game quantifier for games of length ω. In [35], Neeman showed the consistency of the hypotheses of Theorems 5 and 6 from large cardinals. In light of Theorem 6, Theorem 5 is a consequence of the following result of Woodin's:

Theorem 7 (Woodin, [37]**).** The following theories are equiconsistent:

1. ZFC + all ordinal definable games of length ω_1 on natural numbers with real parameters are determined.
2. ZFC + all games ending at ω_1 in L of the play with payoff sets that are $\Game(<\omega^2 - \Pi^1_1)$ in the codes are determined.

I conjecture that Theorem 5 is optimal, in the following sense:

Conjecture 2. Suppose all ordinal definable games of length ω_1 on natural numbers with real parameters are determined. Then there is a model of ZFC with a Woodin cardinal that is a limit of Woodin cardinals.

This would be the first correspondence between a natural determinacy notion and large cardinals at the level of a Woodin cardinal that is a limit of Woodin cardinals. It cannot be achieved using current methods such as the core model induction technique due to Woodin (see, for example, the review [29]), which Sargsyan and Trang [44–46] have recently shown runs into serious issues before reaching this level. In addition, by recent results of Larson and Sargsyan [20, 42], also the well-known liberal K^c construction in [4, 17] can fail if there is a Woodin cardinal that is a limit of Woodin cardinals.

Therefore, understanding the large cardinal strength of the determinacy of such uncountable games might shed light on how to canonically obtain inner models with a Woodin cardinal that is a limit of Woodin cardinals.

4 Strong Models of Determinacy for Games of Length ω

Another approach to strengthen determinacy is to keep playing games of length ω and impose additional structural properties on the model. Examples of such structural properties are "$\theta_0 < \Theta$," "Θ is regular," or the Largest Suslin Axiom, see, for example, [40, 44, 53]. Here Θ is given by

$$\Theta = \sup\{\alpha \mid \text{there is a surjection } f\colon \mathbb{R} \to \alpha\}$$

and we write θ_0 for the least ordinal α such that there is no surjection of $\mathbb{R}$ onto α which is ordinal definable from a real. While in models of the Axiom of Choice Θ is simply equal to $(2^{\aleph_0})^+$, it has very interesting behaviour in models of the Axiom of Determinacy.

Other examples of properties that can be used to obtain strong models of determinacy are "all sets of reals are Suslin" or "all sets of reals are universally Baire." Being Suslin is a generalization of being analytic. More precisely, a set of reals is *Suslin* if it is the projection of a tree on $\omega \times \kappa$ for some ordinal κ. Woodin and Steel determined the exact large cardinal strength of the theory "AD + all sets of reals are Suslin" [52, 54]:

Theorem 8 (Steel, Woodin, [52, 54]). The following theories are equiconsistent (over ZF):

1. AD + all sets of reals are Suslin,
2. ZFC + there is a cardinal λ that is a limit of Woodin cardinals and a limit of $<\lambda$-strong cardinals.

By results of Martin and Woodin, see [54, Theorems 9.1 and 9.2], assuming AD, the statement "all sets of reals are Suslin" is equivalent to the Axiom of Determinacy for games on reals ($\mathsf{AD}_{\mathbb{R}}$). Being universally Baire is a strengthening of being Suslin that was introduced implicitly in [47] and explicitly by Feng, Magidor and Woodin [12].

Definition 1 (Feng, Magidor, Woodin [12]**).** *A subset A of a topological space Y is* universally Baire *if $f^{-1}(A)$ has the property of Baire in any topological space X, where $f: X \to Y$ is continuous.*

The exact consistency strength of the statement that all sets of reals are universally Baire under determinacy was conjectured by Sargsyan, in 2014, after he was able to obtain an upper bound with Larson and Wilson [21] via an extension of Woodin's famous derived model theorem. One fact that makes their argument particularly interesting is that no model of the form $L(\mathcal{P}(\mathbb{R}))$ is a model of "AD+ all sets of reals are universally Baire." Universal Baireness is not only widely used across set theory but a crucial property in inner model theory: Universally Baire iteration strategies (canonically coded as a set of reals) can be extended from countable to uncountable iterations (see, for example, [39]). The following theorem proves **Sargsyan's conjecture** by showing that the upper bound Larson, Sargsyan, and Wilson obtained is optimal:

Theorem 9 (Larson, Sargsyan, Wilson, [21]**, Müller,** [30]**).** The following theories are equiconsistent (over ZF):

1. AD + all sets of reals are universally Baire,
2. ZFC + there is a cardinal that is a limit of Woodin cardinals and a limit of strong cardinals.

To construct and analyze the relevant models to prove the direction $Con(1.) \Rightarrow Con(2.)$ in this theorem, instead of just considering two hierarchies – determinacy axioms and inner models with large cardinals – a third hierarchy is used to reach higher levels in the other two. These three hierarchies together form what Steel calls the **triple helix** of inner model theory. The new hierarchy goes back to Woodin and Sargsyan and consists of canonical models called **hybrid mice**, or **hod mice**. These models are not only enhanced by large cardinals witnessed by extenders on their sequence, but also equipped with partial iteration strategies for themselves, see [40]. The strength of these models intuitively comes from the descriptive set theoretic complexity of these partial iteration strategies.

The name **hod mouse** comes from the fact that these mice naturally occur as the result of analyses of HOD, the hereditarily ordinal definable sets, in various models of determinacy. This analysis was pioneered by Steel and Woodin [51,56] in the model $L(\mathbb{R})$, as well as in $L[x][g]$ for a cone of reals x, where g is generic for Lévy collapsing the least inaccessible cardinal in $L[x]$ (both under a determinacy hypothesis). It was extended to larger models of determinacy by Sargsyan, Trang, and others [5,40,43,58]. In [31] we showed how to analyze HOD in $M_n(x)[g]$, for a cone of reals x, where g is generic for Lévy collapsing the least inaccessible cardinal in $M_n(x)$ (under a determinacy hypothesis).

The technical innovation behind the direction $Con(1.) \Rightarrow Con(2.)$ in Theorem 9 is a new translation procedure to translate hybrid mice into mice with a limit of Woodin and strong cardinals [30]. This required an iterability proof for models obtained via a novel backgrounded construction. In [30] it is shown that the resulting models are countably iterably, meaning that countable substructures are iterable, and, in fact, a bit more. But the following natural question is left open:

Question 1. Is there a translation procedure that yields fully iterable mice with a limit of Woodin and strong cardinals (when applied to suitable hybrid mice)?

References

1. Aguilera, J.P., Müller, S.: Projective games on the reals. Notre Dame J. Formal Logic **61**, 573–589 (2020). https://doi.org/10.1215/00294527-2020-0027
2. Aguilera, J.P., Müller, S.: The consistency strength of long projective determinacy. J. Symb. Log. **85**(1), 338–366 (2020). https://doi.org/10.1017/jsl.2019.78
3. Aguilera, J.P., Müller, S., Schlicht, P.: Long games and σ-projective sets. Ann. Pure Appl. Log. **172**, 102939 (2021). https://doi.org/10.1016/j.apal.2020.102939
4. Andretta, A., Neeman, I., Steel, J.R.: The domestic levels of K^c are iterable. Israel J. Math. **125**, 157–201 (2001). https://doi.org/10.1007/BF02773379
5. Atmai, R., Sargsyan, G.: Hod up to $AD_{\mathbb{R}} + \Theta$ is measurable. Ann. Pure Appl. Log. **170**(1), 95–108 (2019). https://doi.org/10.1016/j.apal.2018.08.013
6. Carroy, R., Medini, A., Müller, S.: Every zero-dimensional homogeneous space is strongly homogeneous under determinacy. J. Math. Log. **20**, 2050015 (2020). https://doi.org/10.1142/S0219061320500154
7. Casacuberta, C., Scevenels, D., Smith, J.H.: Implications of large-cardinal principles in homotopical localization. Adv. Math. **197**(1), 120–139 (2005). https://doi.org/10.1016/j.aim.2004.10.001
8. Dales, H.G., Woodin, W.H.: An Introduction to Independence for Analysts. London Mathematical Society Lecture Note series. Cambridge University Press, Cambridge (1987). https://doi.org/10.1017/CBO9780511662256
9. Davis, M.: Infinite games of perfect information. In: Dresher, M., Shapley, L.S., Tucker, A.W. (eds.) Advances in Game Theory, vol. 52, pp. 85–101 (1964)
10. Eklof, P.C., Mekler, A.H.: Almost Free Modules, North-Holland Mathematical Library, vol. 65. North-Holland Publishing Co. (2002)
11. Farah, I.: All automorphisms of the Calkin algebra are inner. Ann. Math. **173**(2), 619–661 (2011). https://doi.org/10.4007/annals.2011.173.2.1
12. Feng, Q., Magidor, M., Woodin, W.H.: Universally baire sets of reals. In: Judah, H., Just, W., Woodin, W.H. (eds.) Set Theory of the Continuum. Mathematical Sciences Research Institute Publications, vol. 26, pp. 203–242. Springer, Cham (1992). https://doi.org/10.1007/978-1-4613-9754-0_15
13. Fleissner, W.G.: If all normal Moore spaces are metrizable, then there is an inner model with a measurable cardinal. Trans. Amer. Math. Soc. **273**, 365–373 (1982). https://doi.org/10.1090/S0002-9947-1982-0664048-8
14. Gale, D., Stewart, F.M.: Infinite games with perfect information. In: Kuhn, H.W., Tucker, A.W. (eds.) Contributions to the Theory of Games, vol. 2, pp. 245–266 (1953)
15. Harrington, L.: Analytic determinacy and $0^\#$. J. Symb. Log. **43**, 685–693 (1978). https://doi.org/10.2307/2273508
16. Jensen, R.B.: The fine structure of the constructible hierarchy. Ann. Math. Logic **4**, 229–308 (1972). https://doi.org/10.1016/0003-4843(72)90001-0

17. Jensen, R.B., Schimmerling, E., Schindler, R., Steel, J.R.: Stacking mice. J. Symb. Log. **74**(1), 315–335 (2009). https://doi.org/10.2178/jsl/1231082314
18. Kechris, A., Solovay, R.: On the relative consistency strength of determinacy hypotheses. Trans. Amer. Math. Soc. **290**(1), 179–211 (1985). https://doi.org/10.1090/S0002-9947-1985-0787961-2
19. Larson, P.B.: A brief history of determinacy. In: Kechris, A.S., Löwe, B., Steel, J.R. (eds.) Large Cardinals, Determinacy and Other Topics, pp. 3–60 (2020). https://doi.org/10.1017/9781316863534.002
20. Larson, P.B., Sargsyan, G.: Failure of square in $\mathbb{P}_{\max}$ extensions of Chang models (2021)
21. Larson, P.B., Sargsyan, G., Wilson, T.M.: A model of the Axiom of Determinacy in which every set of reals is universally Baire (2018)
22. Laver, R.: On the consistency of Borel's conjecture. Acta Math. **137**, 151–169 (1976). https://doi.org/10.1007/BF02392416
23. Martin, D.A.: Measurable cardinals and analytic games. Fundam. Math. **66**, 287–291 (1970). https://doi.org/10.4064/fm-66-3-287-291
24. Martin, D.A.: Borel determinacy. Ann. Math. **102**(2), 363–371 (1975). https://doi.org/10.2307/1971035
25. Martin, D.A., Steel, J.R.: A proof of projective determinacy. J. Amer. Math. Soc. **2**(1), 71–125 (1989). https://doi.org/10.2307/1990913
26. Mitchell, W.J., Steel, J.R.: Fine Structure and Iteration Trees. Lecture Notes on Logistics, vol. 3. Springer-Verlag, New York (1994)
27. Uhlenbrock (now Müller), S.: Pure and Hybrid Mice with Finitely Many Woodin Cardinals from Levels of Determinacy. Ph.D. thesis, University of Münster (2016)
28. Müller, S.: The axiom of determinacy implies dependent choice in mice. Math. Logic Quarter. **65**(3), 370–375 (2019). https://doi.org/10.1002/malq.201800077
29. Müller, S.: Four papers on the large cardinal strength of PFA via core model induction. Bull. Symb. Log. **26**(1), 89–92 (2020). https://doi.org/10.1017/bsl.2020.6
30. Müller, S.: The consistency strength of determinacy when all sets are universally Baire (2021). Submitted
31. Müller, S., Sargsyan, G.: HOD in inner models with Woodin cardinals. J. Symb. Log. (2021). https://doi.org/10.1017/jsl.2021.61
32. Müller, S., Schindler, R., Woodin, W.: Mice with finitely many woodin cardinals from optimal determinacy hypotheses. J. Math. Log. **20**, 1950013 (2020). https://doi.org/10.1142/S0219061319500132
33. Neeman, I.: Optimal proofs of determinacy. Bull. Symb. Log. **1**(3), 327–339 (1995). https://doi.org/10.2307/421159
34. Neeman, I.: Games of Countable Length. In: Cooper, S.B., Truss, J.K. (eds.) Sets and Proofs, London Mathematical Society Lecture Note series, pp. 159–196 (1999). https://doi.org/10.1017/CBO9781107325944.009
35. Neeman, I.: Inner models in the region of a Woodin limit of Woodin cardinals. Ann. Pure Appl. Log. **116**(1), 67–155 (2002). https://doi.org/10.1016/S0168-0072(01)00103-8
36. Neeman, I.: Optimal Proofs of Determinacy II. J. Math. Log. **2**(2), 227–258 (2002). https://doi.org/10.1142/S0219061302000175
37. Neeman, I.: The Determinacy of Long Games, De Gruyter series in logic and its applications, vol. 7. De Gruyter (2004). https://doi.org/10.1515/9783110200065
38. Nyikos, P.J.: A provisional solution to the normal Moore space problem. Proc. Amer. Math. Soc. **78**, 429–435 (1980). https://doi.org/10.1090/S0002-9939-1980-0553389-4

39. Sargsyan, G.: Descriptive inner model theory. Bull. Symbolic Logic **19**(1), 1–55 (2013). https://doi.org/10.2178/bsl.1901010
40. Sargsyan, G.: Hod Mice and the Mouse Set Conjecture. Memoirs of the American Mathematical Society, vol. 236 (2015). https://doi.org/10.1090/memo/1111
41. Sargsyan, G.: Translation procedures in descriptive inner model theory. In: Foundations of Mathematics, American Mathematical Society, vol. 690, pp. 205–223 (2017). https://doi.org/10.1090/conm/690/13869
42. Sargsyan, G.: Announcement of recent results in descriptive inner model theory (2021)
43. Sargsyan, G., Trang, N.D.: The Largest Suslin Axiom (2016)
44. Sargsyan, G., Trang, N.D.: The exact consistency strength of the generic absoluteness for the universally Baire sets (2019)
45. Sargsyan, G., Trang, N.D.: Sealing from iterability. Trans. Amer. Math. Soc. Ser. B **8**, 229–248 (2021). https://doi.org/10.1090/btran/65
46. Sargsyan, G., Trang, N.D.: Sealing of the universally Baire sets. Bull. Symb. Log. (2021). https://doi.org/10.1017/bsl.2021.29
47. Schilling, K., Vaught, R.: Borel games and the Baire property. Trans. Amer. Math. Soc. **279**(1), 411–428 (1983). https://doi.org/10.2307/1999393
48. Schindler, R., Steel, J.R., Zeman, M.: Deconstructing inner model theory. J. Symb. Log. **67**, 721–736 (2002). https://doi.org/10.2178/jsl/1190150106
49. Shelah, S.: Infinite abelian groups, Whitehead problem and some constructions. Israel J. Math. **18**, 243–256 (1974). https://doi.org/10.1007/BF02757281
50. Steel, J.R.: Long games. In: Cabal Seminar 81–85, Lecture Notes in Mathematical 1333, pp. 56–97. Springer Verlag, Cham (1988). https://doi.org/10.1007/BFb0084970
51. Steel, J.R.: $\mathrm{HOD}^{L(\mathbb{R})}$ is a core model below Θ. Bull. Symb. Log. **1**(1), 75–84 (1995). https://doi.org/10.2307/420947
52. Steel, J.R.: An optimal consistency strength lower bound for $\mathsf{AD}_{\mathbb{R}}$ (2008)
53. Steel, J.R.: Derived models associated to mice. In: Computational Prospects of Infinity - Part I: Tutorials, vol. 14, pp. 105–193. World Scientific (2008). https://doi.org/10.1142/9789812794055_0003
54. Steel, J.R.: The derived model theorem. In: Cooper, S.B., Geuvers, H., Pillay, A., Vänänen, J. (eds.) Logic Colloquium 2006, pp. 280–327 (2009). https://doi.org/10.1017/CBO9780511605321.014
55. Steel, J.R.: An Outline of Inner Model Theory, pp. 1595–1684. Springer, Cham (2010). https://doi.org/10.1007/978-1-4020-5764-9_20
56. Steel, J.R., Woodin, W.H.: HOD as a core model. In: Kechris, A.S., Löwe, B., Steel, J.R. (eds.) Ordinal Definability and Recursion Theory, pp. 257–346 (2016). https://doi.org/10.1017/CBO9781139519694.010
57. Trang, N.D.: Generalized Solovay Measures, the HOD Analysis, and the Core Model Induction. Ph.D. thesis, University of California at Berkeley (2013)
58. Trang, N.D.: HOD in natural models of AD^+. Ann. Pure Appl. Log. **165**(10), 1533–1556 (2014). https://doi.org/10.1016/j.apal.2014.04.006
59. Wolfe, P.: The strict determinateness of certain infinite games. Pac. J. Math. **5**, 841–847 (1955). https://doi.org/10.2140/pjm.1955.5.841
60. Zeman, M.: Inner Models and Large Cardinals, De Gruyter series in logic and its applications, vol. 5. De Gruyter (2002). https://doi.org/10.1515/9783110857818
61. Zhu, Y.: Realizing an AD^+ model as a derived model of a premouse. Ann. Pure Appl. Log. **166**, 1275–1364 (2015). https://doi.org/10.1016/j.apal.2015.05.002

Big Ideas from Logic for Mathematics and Computing Education

R. Ramanujam[1,2](✉)

[1] Institute of Mathematical Sciences, Chennai, India
jam@imsc.res.in
[2] Azim Premji University, Bengaluru, India
http://www.imsc.res.in/r_ramanujam

Abstract. Despite logic and reasoning being considered central to mathematics and computing education, it plays a largely peripheral role in high school or undergraduate curricula. The language of propositional logic and formal deductions are taught in high school mathematics as well as undergraduate discrete mathematics courses. Opportunities for learning mathematical logic are few. In this paper, we address the question whether mathematical logic can meaningfully contribute to mathematics and computing education at the school level.

Along the lines of *Big Ideas of mathematics* [4] we propose a set of "Big Ideas" of mathematical logic relevant to education: for instance, the notion of truth relative to a structure, construction of models for a set of sentences, consistency of procedures used for computation, comparison of algorithms, and so on. We discuss how they can significantly enrich mathematics and computing science curriculum and pedagogy, clearing up students' common misconceptions, referring to some experiences with high school mathematics teachers and students.

Keywords: High school mathematics curriculum · Mathematics education · Algorithms · Logic

1 Logic in School Mathematics

The structure of school mathematics curricula dates back to the days of the industrial revolution. The content areas are: arithmetic, algebra, geometry, trigonometry, the differential and integral calculus, probability and statistics. Combinatorics and propositional logic are accommodated within this structure. In terms of relative space, arithmetic in the early years, algebra and geometry in the middle years, and calculus in the later years occupy the centre [15]. While there are differences across national curricula, the principal structure remains similar.

In what follows, we focus on the curriculum of the Central Board of Secondary Education (CBSE) in India, principally because of our familiarity with it and relative ignorance of syllabi elsewhere. However, even a superficial study

M. Banerjee and A. V. Sreejith (Eds.): ICLA 2023, LNCS 13963, pp. 79–91, 2023.
https://doi.org/10.1007/978-3-031-26689-8_6

of curricula across the world show similarities to CBSE. As it happens, India has 32 Boards of education (every state has its own, and there are other national ones as well); however, while there are differences between the syllabi, the structure is similar, especially with regard to the role logic plays in school mathematics. (In India secondary school education covers grades 9 to 12, when students are typically from age 13 to 17.)

In this structure, the study of logic is principally limited to geometric reasoning and using the language of propositional logic. We discuss these below.

1.1 Geometric Reasoning

Geometry occupies a significant space in the secondary school curriculum, in grades 9 and 10. In grade 9, it constitutes nearly half of the instruction period (75 out of 160 scheduled class hours). In grade 10, it reduces to less than a quarter (30 out of 160). **Proofs** are accorded importance in the study of geometry. There are theorems and proofs in other ares such as number theory and algebra, but these are not explicitly signalled and discussed as they are in geometry. Some algebraic identities are stated without proof, whereas geometric propositions are always proved.

Geometry begins with a historical introduction to Euclid. The syllabus document [13] says:

> Euclid's method of formalising observed phenomenon into rigorous mathematics with definitions, common/obvious notions, axioms/postulates, and theorems. The five postulates of Euclid. Equivalent versions of the fifth postulate. Showing the relationship between axiom and theorem.

Subsequently a number of propositions on lines, angles, triangles, quadrilaterals and circles are stated and proved. Geometry in grade 10 is similar, with the additional notions of congruence and similarity, tangents. The proofs are largely rigorous but informal. Some of the proofs, such as ones using congruence, require some depth of reasoning using many assertions proved earlier. An appendix in each book discusses the nature of deductive proofs, stressing the role of formal derivations and the distinction between verification and proof.

In this regard, geometry instruction employs logic purposefully even while not stressing on the *language* of logic or on formal deduction in an inference system.

1.2 Propositional Reasoning

The CBSE syllabus for grade 11 includes a unit titled *Mathematical reasoning.* The document [13] says:

> Mathematically acceptable statements. Connecting words/phrases – consolidating the understanding of "if and only if (necessary and sufficient) condition", "implies", "and/or", "implied by", "and", "or", "there exists"

and their use through a variety of examples related to real life and Mathematics. Validating the statements involving the connecting words – difference between contradiction, converse and contrapositive.

The textbook chapter introduces the language of propositional logic, and explains the use of connectives very well. The principal focus is on implication: getting students to appreciate the distinction between $p \implies q$ and $q \implies p$, to understand inclusive disjunction and to realise that a single counter-example suffices to falsify a universally quantified statement. Considerable effort is spent in formalising intuitive statements in the language of propositional logic. It is also important to point out that geometry is extensively used as the terrain from which propositions are picked.

For instance, an exercise ([14], Chap. 14, page 344) asks the students to: Write the following statement in *five different ways*, conveying the same meaning.

If a triangle is equiangular, then it is an obtuse angled triangle.

For someone formalising this as $p \implies q$ (which is what the book teaches the student to do), there are indeed equivalent forms to try, such as $\neg q \implies \neg p$. If a student tries to use the language of geometry starting with a triangle ABC and proceeding further, it is much less clear if she would get anywhere with this exercise.

Our remark here is limited to pointing out the use of a formal logical language in mathematical context in the syllabus, and the derivation of differently expressed properties by virtue of logical equivalence. This is an important logical exercise in which students gain some proficiency.

1.3 What is Achieved

How effective is this foray into logic for the student learning mathematics? It is clear that geometry is central to mathematics learning and students should be able to see that the theorems proved are deduced from clearly stated axioms. They also get some practice in proving some assertions themselves. Does it matter that the axioms and inference rules are never formally spelt out, nor that an informal argument is preferred over a formal deduction? Most mathematics educators are clear that informal, rigorous argument is more important for mathematics learning than formal deductions. There is extensive literature on students' reasoning and structure in proofs: [8, 10, 17], and more. Invariably, these researchers point out that developing intuition is more important for mathematics learning than formal proofs. For our discussion here, the more relevant question is whether the language of logic is playing any significant role when the student is learning geometry, or whether the student is learning deductive systems by way of geometry. The answer to both of these questions seems to be negative. [1] argues persuasively that mathematical logic is not really necessary for teaching proofs in mathematics. [6] offers nuanced arguments on different aspects of logic relevant in the teaching of proofs.

On the other hand, learning the correct use of propositional connectives and quantifiers unambiguously constitutes logic learning, and this is also indispensable for obtaining fluency in the language of mathematics. However, in practice, even this is entirely unclear [9].

To understand why logic may be ineffective in these respects, we need to look at the *placement* of this syllabus unit in the curricular structure. The other syllabus units among which boolean logic is placed involve topics such as: Trigonometric Functions, Complex Numbers, Binomial Theorem, Conic Sections, Frequency Distributions, Limits and Derivatives. In fact, in the CBSE's official textbook, the chapter on propositional logic follows the chapter on limits and derivatives. After solving exercises computing the derivatives of functions such as $\frac{x}{sin\ x}$ the student encounters the relationship between $p \implies q$ and $q \implies p$. It is hardly surprising that the student is left wondering whether this is merely an interlude between differentiation and integration (which do consume most of the time and energy of the higher secondary mathematics student). While students are happy to take any "easy" topic that comes their way, the deeper issues pass them by, and the chapter on propositional reasoning merely becomes another set of (thankfully simple) rules to be learnt and forgotten soon.

If logic has very little place in school mathematics, and even the little that is on offer is ineffective, should we conclude that logic is irrelevant for school mathematics?

We suggest in what follows that we take a slightly different view: rather than teach logic, we should consider incorporating ideas from logic into the teaching of the content areas of mathematics such as algebra, geometry, number systems, and so on.

2 Conceptual Difficulties

In this section, we point out a number of difficulties faced by students learning mathematics where logic can be of significant help. These are symptomatic rather than constituting a detailed analysis. These difficulties are acknowledged and discussed extensively in the mathematics education literature. If we have any novelty to offer it is only in the suggested use of mathematical logic for addressing these difficulties.

1. *The use of variables*: We ask middle school children to solve equations of the form $x + 3 = 5$, and they learn that x is an unknown number. Then they go on to consider equations such as $x + y = 5$, when x can be one of many numbers, somehow dependent on the value that y takes. We also ask them to "see" that $x + y = y + x$, but now x can be *any* value whatsoever. There is worse to come when we ask them to consider the line given by the equation $x = k$ where k is some constant [3].
   All these are legitimate and reasonable uses of x in mathematical contexts, but being syntactically disciplined about quantifying x appropriately depending on context would solve much of the confusion.

2. *True or false?*: Durand-Guerrier [7] discusses the proposition: "Any number that ends in 4 is divisible by 4." Some students consider such statements true as well as false, since it holds for some numbers and does not hold for some numbers.
3. *Truth in a structure*: Is $x^2 - 2 = 0$? Well, that depends on which number system you are solving the equation in. But this is a source of endless confusion to students.
The problem is not only of truth being structure dependent, but even interpretation of operations and functions being structure dependent as well. This confusion gets considerably worse when we consider statements such as "multiplication is the same as repeated addition". This works on positive integers and this is how students learn arithmetic in primary school. But this reasoning is unhelpful when they need to learn $\sqrt{2} \times \sqrt{3}$ and so on.
4. *Equation solving*: A classic problem asks students to evaluate $\frac{x^2-1}{x-1}$ at $x = 1$. The student answering this as $x = 2$ gets shocked when made to realise that the function cannot be evaluated at $x = 1$. The same student, further on, is asked to compute the limit of $\frac{x^2-1}{x-1}$ as $x \to 1$. Now the limit is indeed 2. The student is left with a bad taste in the mouth: something is surely wrong!
5. *Heuristic advice*: One of the side effects of the trouble above is the typical advice given to the student to always substitute the answer she obtains in the given equation and verify. This does help in this instance, since the student can immediately verify that it leads to division by zero.
Equation solving comes with a range of heuristics such as grouping like terms, moving all terms involving the same variable to one side, change of sign across equality symbol, and so on. However, very rarely does the student get any assurance on the reliability of these heuristics, and whether they suffice in all contexts.
6. *Functional variation*: The student learns the definition of the *sine* function as ratio of the 'opposite side' to the hypotenuse of a right angled triangle. Later on, the student gets to graph the function as an oscillating curve. What is varying here? Are we saying something about the universe of right angled triangles? This again is a cause of confusion for many students.

There are many such sources of confusion and incomprehension in school mathematics. We mention these here merely to point out how logic can be helpful in this regard.

3 Big Ideas

[4] suggested a curriculum for mathematics education based on **Big Ideas** of mathematics. The argument is that rather than teaching algebra, geometry, number systems etc we should attempt to communicate the ideas of mathematics that make it at once abstract and at the same time uniquely powerful in being widely applicable. Charles offers a list of 21 such ideas; these have been debated by other researchers [11] but everyone would agree that notions like algorithm,

functional variation, symmetry and transformation, linearity, etc. are among the central ideas of mathematics. Whether a curriculum based on such ideas can address the social as well as disciplinary objectives of mathematics education needs to be critically assessed.

Our reference to Big Ideas here is towards a different purpose. We could presumably consider a similar set of Big Ideas of Mathematical Logic. More specifically we can ask, what are the Big Ideas of Mathematical Logic of relevance to mathematics education (if any)? This distinction is important, since there are some important ideas of logic whose relevance to school mathematics is not immediately clear. For instance the *compositionality principle*, by which the meaning of sentences is entirely derived from the semantics of connectives and given meaning of primitive elements, is of central importance in logic. However, it is not clear how important this is for pedagogic purposes of school mathematics.

We propose the following partial list of Big Ideas of Mathematical Logic for mathematics education, many of which have been discussed extensively by [2].

1. Truth relative to a structure: Students are trained to consider mathematical statements to be true or false. Often some underlying structure is assumed, against which statements are being evaluated. Unfortunately, the ground shifts, so to speak, leaving many students confused. For instance, does $x^2 - 2 = 0$ have a solution? The answer is that it depends on what x ranges over. Is multiplication repeated addition? Yes, it is, over natural numbers, and hence this is true in primary school, but not in high school. Is the angle sum property true? Yes, if you consider triangles on the plane. The fact that truth is relative to the assumed structure is not easily understood.
2. Model construction: The fact that definitions are in themselves neither true nor false may seem obvious to the mature mathematics student, but is a source of confusion to school children. Indeed, it is pedagogically non-trivial to show that alternate definitions give us different objects and that it makes sense to ask whether two definitions are equivalent. All this has to do with model constructions, a logical enterprise. A direct way this appears in school mathematics is in the use of multiple but equivalent representations of the same mathematical object, where the equivalence is not easy to establish or comprehend. For instance, $\sqrt{2}$ is the solution to the equation $x^2 - 2 = 0$ over the reals, but locating that real number on the real line is not easy. It is also the length of the diagonal of the unit square, and while students are comfortable with denoting both by $\sqrt{2}$, they are puzzled when they try to extend these meanings to $2\sqrt{2}$.
3. The Syntax Semantics distinction: A student is puzzled by the fact that though $\frac{x^2-1}{x-1} = x + 1$ seems to be correct, at $x = 1$ we cannot evaluate the left hand side, whereas we can indeed evaluate the right hand side. The fact that equivalence of expressions at the syntactic level denotes equivalence of functions at the semantic level is difficult to grasp for the student. Logic makes the syntax semantics distinction explicit. We have already referred to the use of variables in school and the confusion caused therein, due to the hidden quantifiers. The use of the same symbols for arithmetical operations

whether they are interpreted over integers or the reals again causes a great deal of confusion. The syntax semantics distinction, a central notion of logic, can greatly help mathematics pedagogy as well [7].

4. Formalisation: When asked whether $x > y$ implied $x^2 > y^2$, a student answered that this was *both true and false*, since it was true for "half the numbers". In general, students hold intuitive notions of disjunction, implication and universal/existential quantification, whereas the mathematics they learn works with these in formal terms, where the definition is different. (The author has often seen frequentist interpretations of implication by school students.)
5. Consistency of rules: Students tend to learn mathematics as a collection of rules. For instance, while solving equations, one moves all "like terms" together. Many heuristics are learnt and they are largely helpful. For instance, while solving trigonometric equations, one actively looks for $sin^2x + cos^2x$ hidden in different ways. Very rarely is the question raised whether these rules are reliable, and whether they are always applicable. Doubts often creep in when encountering notions such as limits and derivatives, or infinite sums, when suddenly some of the old rules are not applicable. Again, logic has an important notion to contribute, namely the notion of consistency.
6. Proving vs Checking: Typically, proofs in school mathematics involve demonstration that a universally quantified statement is true in some implicit structure. The quantification could be over numbers, triangles, circles, and so on. (Note that even the assertion of an infinity of primes is of the form $\forall x \exists y$.) On the other hand, students are used to checking that a property holds for some object: for instance that 137 is a prime, or that perpendicular bisectors of a given triangle have a point of concurrency, and so on. Yet the former isn't called a proof whereas the latter is termed a *theorem*. These are clearly logical issues and can easily be cleared up with some understanding of logic.
7. Reasoning about algorithms: This is of sufficient importance for us to discuss it separately in the next section.

We reiterate the fact that this is an indicative list, and we make no claims to being comprehensive. Further, we emphasise that we are not advocating the teaching of these notions to students but only that mathematics curricula take cognizance of them, and that teachers learn sufficient logic so that it informs pedagogy.

4 Reasoning About Algorithms

If there is one thing that characterises school mathematics for a student, it is the learning of algorithms in one context after another, be it long division, or factorisation of quadratic expressions, or matrix inversion, or computing compound interest, or finding the standard deviation of given data, and so on. These are largely seen to be disparate, to be learnt and applied in context. It is not an exaggeration to say that the teaching and learning of these algorithms dominates school mathematics almost to the exclusion of their declarative content.

Moreover, the conceptual vs procedural debate dominates the discourse on mathematics education.

Our mention of algorithms is to point out that *reasoning* about algorithms is crucial for bridging the conceptual – procedural divide (where it exists) and that this is an indispensable role for logic. Moreover, it is such reasoning that is central to **computational thinking**, a term of great current interest. Indeed it is when algorithms are taught and learnt as procedures without any explicit reasoning about them that mathematics becomes difficult for many. The position paper [15] on teaching of mathematics in the 2005 National Curriculum Framework asserts:

> ... we have repeatedly referred to offering a multiplicity of approaches, procedures, solutions. We see this as crucial for liberating school mathematics from the tyranny of the one right answer, found by applying the one algorithm taught. When many ways are available, one can compare them, decide which is appropriate when, and in the process gain insight.

When we have a multiplicity of methods, they need to be compared and analysed to determine which method works best when, and this is the conceptual understanding of importance to mathematics learning. Ideally when students come up with algorithms, compare theirs with other algorithms and argue about them, they acquire confidence in the use of these algorithms. Logic has a great deal to contribute in this regard.

Opportunities for such reasoning are present throughout school. Very small children working with some number of counters, say 20, need to reason thus when they are asked, "how do you know you have counted them all?" and come up with grouping strategies. When a child who has distributed 20 toffees among 10 children is asked, "how many do you have?" answers readily. When asked further, "how many do each of your friends have?", needs to reason explicitly by symmetry, provided the algorithm she used for toffee distribution treats all children symmetrically.

At primary school, students get asked how many different pairs of whole numbers add to (say) 10. When a student offers the solution, there is an implicit question to her: how do you know you have counted them all? This process of verification involves reasoning, and when it becomes a habit, shows the value of formal reasoning as it helps the student catch routine mistakes. The transition from "do you know" to "how do you know it" is important for mathematics, and reasoning provides the natural vehicle for carrying the student through the transition.

One can list a range of algorithms in number theory, algebra, geometry, trigonometry, calculus, statistics and business mathematics and ask, in each case, how students reason about the correctness of the algorithm they learn. Much better, if and when they get to devise algorithms of their own, one can ask how they argue that their method works. While formal proofs of correctness may neither be feasible nor required, argumentation is important, and logic can help in this.

All this, of course, is even more relevant to computing education. Reasoning about procedures is at the heart of computational thinking [16] and the connection to logic is generally missed in school. When a small child is asked to add 2, 104 and 78, and regroups them to first get 80 first and then adds it to 104, this re-ordering is a crucial element of computational thinking. Over time a student builds up such rules, and reflection on these rules and their applicability is a logical exercise, very much necessary for meta-cognition.

Students learning programming need to also learn to build correctness arguments (even if not proofs). Reasoning about programs requires understanding the syntax semantics distinction, and checking whether a claim is true or not at different points of program execution. Logic plays an indispensable role here [5].

We point out once again that it is quite unclear how a separate logic course taught as a curricular unit will help in the ways listed above. More than the formal language of logic, integral use of these Big Ideas within the mathematics and computing curricula would be helpful.

5 Experiences

Experiential accounts of educational interventions are of very limited value. What we need is research based on strong data from classroom practices, analysed in appropriate theoretical frameworks. Our discussion here should be seen as stressing the need for such research, one that examines whether a logician's perspective may be able to influence mathematics and computing curricula positively, and if yes, understand how, in a nuanced manner.

However, two instances of discussion on some of the issues discussed in this paper seem to be worth reporting. The first of these was during a workshop for high school teachers of mathematics in 2015 and the second was in a mathematics club activity with students in the higher secondary stage (age 16–17, in grades 11–12, the last stage before students enter university). To a great extent, these were among the experiences initiated the author's foray into discussions with mathematics teachers and educators on logic.

5.1 Algebra Tricks

The workshop for high school teachers was on problem solving in algebra and geometry. In a particular session, there was discussion on problem solving techniques that had come up in different contexts, and while we were listing them, one teacher V made the remarkable assertion that there was no unifying method combining these techniques, especially in the context of algebra. He contended that they were "tricks" and best learnt thus. Several teachers disputed this, but when called upon to present a unifying method, found it hard to even argue that any such method exists, let alone present one, much to the glee of the (by now growing) camp of "algebra tricks" enthusiasts.

At this point, one of the participant teachers M wondered whether this was a problem specific to algebra, and whether the situation was different with regard to geometric reasoning. She felt that geometry was about proofs, and that should help. The author deferred to V on this, who was emphatic that geometry was "logical", entirely based on axioms and proof, and hence all techniques were unified. The ensuing discussion led to a consensus among teachers that the axiomatic method in geometry provided a sound basis for problem solving as well.

At this point, the author reminded teachers that algebra was entirely axiomatic in its development. A teacher protested that this was true of "abstract algebra" studied in the university, not algebra in school mathematics which was only about "solving equations". This led to a general discussion but it was clear that there was a sense of unease with what had been concluded some moments ago. The author gave an impromptu lecture on equational theories and decision problems related to them. Almost everyone had heard of Gödel's theorems but had not seen their relevance to the matter under discussion. Tarski's theorem on real arithmetic was entirely unfamiliar to teachers. They were impressed that an algorithm could possibly be constructed to solve high school algebra problems, but this also made them somehow uneasy.

This instance is recounted only to highlight the point made earlier that notions such as consistency and algorithms for satisfiability in equational theories are relevant for mathematics pedagogy, and hence that teachers would benefit from familiarity with these notions.

5.2 What are Foundations?

The other instance occurred when the author was visiting a school to conduct a mathematics club session for students in the last two years of school, preparatory to entering the university. The author was introduced as someone working on 'the logical foundations of computer science'. The day's plan was to play some games, introduce some game theoretic ideas. Before we could get started, student K wanted to ask something, and was clearly hesitant to pipe up. There was some discussion between him and his neighbour. After some prodding, he asked: *What does 'logical foundation' mean? Mathematics is the one beneath everything.* The side comment was a particularly enticing invitation for discussion (and game theory was happily abandoned then and there).

The discussion was initially incoherent but settled on the question of whether logic was founded on mathematics or it was the other way. The student group was overwhelmingly in support of the former opinion, and the author's prompts about logic providing the language of mathematics were ineffective. The pivotal opinion was expressed by student D who said, *How can we talk of* $\frac{d}{dx}$ *and limits as* $\wedge$ *and* $\vee$*?*. She carried the day, with others pitching in with their favourite construct such as $\sqrt{}$, integrals, and so on. The students were familiar with the language of boolean logic, and while they had some idea of quantifiers, the word logic largely meant assertions such as *law of contrapositives* and formal deductions (as seen in geometry).

Feeling obliged to talk a little on foundations of mathematics, without the rigorous setting of first order logic, the author embarked on a presentation of Peano's axioms for arithmetic, and having got the consent of students that they were reasonable, showed that even a 'weird' arithmetic, with two copies of the number line, "one sitting on top of another", would support these axioms. The idea was to show that it makes sense to define mathematical objects by their properties, but that the properties may lead us to new 'weird' objects. The students were almost spell-bound by this demonstration and the author was gratified to get a round of spontaneous applause for the 'weird' number line. (After the talk, a bothered student asked the author which was actually true, whether there was only one number line, or many more.)

Since this incident, the author has had similar discussions with undergraduate students of mathematics as well, with varied experiences. But the idea of a nonstandard model (admittedly only demonstrated, not proven) was always an *aha moment* for students.

Again, this instance is recounted not to ask for inclusion of foundations of mathematics in the curriculum, but only to point to one of the Big Ideas discussed earlier, that of *Model construction*. This idea is not only a foundational concept in mathematics but of pedagogic purpose as well, in terms of inspiring students and giving them a glimpse of how the mathematical edifice is constructed, rather than taking mathematical constructs as given.

Indeed, all the Big Ideas presented here are supported by experiences of this kind, arising from interactions with students and teachers.

6 Room for Logic

ICLA is not the appropriate forum for discussing school curricula in mathematics and computing. However, certain remarks, as they pertain to ways by which logicians can contribute meaningfully to school education, seem worth offering. We list some below in formulaic terms, without detailed justification.

- Logic has a significant role to play in school mathematics, but it is quite unclear whether this is actualised by teaching the language of boolean logic and syllogisms, or by emphasis on formal deductions in Euclidean geometry.
- Logic can be greatly helpful in clearing up a range of misconceptions that students have, and in easing students into facility with the formal language of mathematics. This is of particular importance since mathematical language is known to play a greatly alienating role for a large proportion of the student population (perhaps the majority). Formalisation and formalisability are notions relevant to understanding discourse in mathematics classrooms [12], and logicians have significant expertise in this regard.
- Mathematics teachers can greatly benefit from learning the central ideas of logic, but again it is unclear whether the standard introductory course in mathematical logic offered at the university achieves this purpose.
- *Reasoning about procedures* is central to computational thinking and this is a fundamental need for both mathematics and computing education. While this is being acknowledged in recent times, we have some way to go before we have curricular designs for schools that incorporate such reasoning.

- A range of technological tools (such as Geogebra or Mathigon) and Computer Algebra Systems have had significant impact on mathematics classrooms with pedagogic integration. Can *logic tools* play a similar role? Do we need specialised child-friendly tools for school students' logical explorations?
- As a rule, mathematics educators seem to be largely unaware of the interaction between mathematical logic and mathematics, and what logic might mean at different stages of schooling. While there is extensive literature on how students reason in mathematics classrooms, this is not examined in the light of logic in itself.

[15] talks of *compartmentalisation* as one of the major problems of the mathematics education milieu, with little interaction between mathematics teachers at the university and those in high school, or between the latter and teachers in the elementary school. The tribe of logicians is miniscule in comparison, but interaction between logicians and mathematics teachers can be of mutual benefit to both.

Acknowledgements. The writings of several researchers such as John Baldwin, Viviane Durand-Guerrier, Susanna Epp, Gila Hanna and Gian-Carlo Rota have been influential in formulating many of the ideas discussed here. I thank the school students and mathematics teachers who have interacted with me extensively and taught me with great patience about school education, and I thank Tamil Nadu Science Forum for always making room for such interactions.

References

1. Aristidou, M.: Is mathematical logic really necessary in teaching mathematical proofs? Athens J. Educ. **7**(1), 99–122 (2020)
2. Baldwin, J.: Logic across the high school curriculum. https://homepages.math.uic.edu/~jbaldwin/pub/logicnov14.pdf. Accessed 8 Aug 2020
3. Banerjee, R., Subramaniam, K.: Evolution of a teaching approach for beginning algebra. Educ. Stud. Math. **81**(2), 351–367 (2012)
4. Charles, R.: Big ideas and understandings as the foundation for elementary and middle school mathematics. J. Math. Educ. Leadersh. **7**(3), 9–24 (2005)
5. Chellougui, F., Durand-Guerrier, V., Meyer, A.: Discrete mathematics, computer science, logic, proof, and their relationships. In: Durand-Guerrier, V., Hochmuth, R., Nardi, E., Winslow, C. (eds.) Research and Development in University Mathematics Education. Routledge (2021). https://doi.org/10.4324/9780429346859
6. Durand-Guerrier, V., Boero, P., Douek, N., Epp, S.S., Tanguay, D.: Examining the role of logic in teaching proof. In: Hanna, G., de Villiers, M. (eds.) Proof and Proving in Mathematics Education. NISS, vol. 15, pp. 369–389. Springer, Dordrecht (2012). https://doi.org/10.1007/978-94-007-2129-6_16
7. Durand-Guerrier, V.: The place of logic in school curricula and advocacy for logic at school. Presented at the Panel on Logic Education, ICLA 2021, Indian Conference on Logic and Applications. https://www.isichennai.res.in/~sujata/icla2021/panelslides/durand-guerrier.pdf
8. Epp, S.: The role of logic in teaching proof. Am. Math. Mon. **110**(10), 886–899 (2003)

9. Epp, S.: The language of quantification in mathematics instruction. In: Stiff, L.V., Curcio, F.V., (eds.) Developing Mathematical Reasoning in Grades K-12. NCTM, Reston (1999)
10. Hanna, G.: Rigorous Proof in Mathematics Education. OISE Press, Toronto (1983)
11. Hurst, C.: Big challenges and big opportunities: the power of 'big ideas' to change curriculum and the culture of teacher planning. In: Anderson, J., Cavanagh, M., Prescott, A. (eds.) Curriculum in Focus: Research Guided Practice (Proceedings of the 37th Annual Conference of the Mathematics Education Research Group of Australasia), Sydney, pp. 287–294 (2014)
12. Jayasree, S., Subramaniam, K., Ramanujam, R.: Coherent formalisability as acceptability criterion for students' mathematical discourse. J. Res. Math. Educ. (2022). https://doi.org/10.1080/14794802.2022.2041469
13. National Council for Educational Research and Training: Mathematics (Classes XI - XII). https://ncert.nic.in/pdf/syllabus/desm_s_Mathematics.pdf
14. National Council for Educational Research and Training: Mathematics Textbook for Class XI. https://ncert.nic.in/textbook.php?kemh1=0-16
15. Position paper of the National Focus Group on Teaching of Mathematics, NCERT 2006. https://ncert.nic.in/pdf/focus-group/math.pdf
16. Ramanujam, R.: Computational thinking: the new buzz. At Right Angles (13), 23–31 (2022)
17. Rota, G.-C.: The phenomenology of mathematical proof. Synthese **111**, 183–196 (1997)

Modal Logic of Generalized Separated Topological Spaces

Qian Chen[1] and Minghui Ma[2(✉)]

[1] Tsinghua-Amsterdam Joint Research Centre for Logic, Department of Philosophy, Tsinghua University, Beijing, China
chenq21@mails.tsinghua.edu.cn

[2] Institute of Logic and Cognition, Department of Philosophy, Sun Yat-sen University, Guangzhou, China
mamh6@mail.sysu.edu.cn

Abstract. For each non-zero cardinal κ, we introduce a generalized separation axiom T_0^κ for topological spaces. For every integer $n > 0$, under the d-semantics which interpret $\Diamond$ as the derived set operator in a topological space, the class of all T_0^n-spaces is d-defined by the modal formula t_0^n, and we show that $\mathsf{wK4T}_0^n = \mathsf{wK4} \oplus \mathsf{t}_0^n$ is the d-logic of all T_0^n-spaces. For $\kappa \geq \aleph_0$, the class of all T_0^κ-spaces is not d-definable.

Keywords: Topological space · Separation axiom · Modal logic

1 Introduction

McKinsey and Tarski [18] proposed interpretations of the unary modal operator $\Diamond$ as the closure operator C and the derived set operator d of a topological space. These interpretations give the topological C-semantics and d-semantics for modal logic. According to the work [8], we distinguish C-logics under the C-semantics and d-logics under the d-semantics. The C-logic of all topological spaces is the modal logic S4 (cf. [18]), and the d-logic of all topological spaces is the modal logic wK4 (cf. [15]). The fundamental connection between C-semantics and d-semantics gives an embedding of the modal logic S4 into wK4 (cf. e.g. [8,16]). These discoveries drive the development of the fruitful branch of topological semantics in the study of modal logic. Many modal logics are shown to be C-logics or d-logics of various classes of topological spaces (cf. e.g. [2,3,6–11]), and topological structures are also used for modeling concrete scenarios like reasoning about knowledge and belief (cf. e.g. [20]).

In the present work, we are concerned with modal logics of topological spaces defined by separation axioms. It is quite standard in the study of topology to consider a series of separation axioms including T_0 (Kolmogoroff), T_1 (Fréchet), T_2 (Hausdorff), T_3 (Vietoris/regular) and T_4 (Tietze/normal) etc. (cf. e.g. [1,4,14,19,24]). These axioms obtain their new forms or applications in the study of

Supported by Chinese National Funding of Social Science (Grant no. 18ZDA033).

M. Banerjee and A. V. Sreejith (Eds.): ICLA 2023, LNCS 13963, pp. 92–104, 2023.
https://doi.org/10.1007/978-3-031-26689-8_7

topological spaces (cf. e.g. [5,13,17,21,22]). Few results on the logic of separation axioms can be found in the literature (cf. [23]). Here let us concentrate on the d-semantics of modal logics, and hence the d-definability of separation axioms. It is known that K4 is the d-logics of all T_d-spaces (also called $T_{1/2}$-spaces) as well as the one of all Stone spaces (cf. [9]). Since all Stone spaces are normal, it follows that T_1, T_2, T_3 and T_4 are not d-definable in modal logic. However, it is shown in [10] that T_0-spaces are d-definable in the basic modal language, namely, a topological space $\mathcal{X}$ is a T_0-space if and only if the modal formula (t_0) $p \wedge \Diamond(q \wedge \Diamond p) \to \Diamond p \vee \Diamond(q \wedge \Diamond q)$ is d-valid in $\mathcal{X}$. Furthermore, it is shown in [10] that the modal logic $\mathsf{wK4T_0} = \mathsf{wK4} \oplus \mathrm{t}_0$ is d-logic of all T_0-spaces, and that it is also the d-logic of all spectral spaces.

The separation axiom T_0 says that each pair of different points can be separated by an open set. It is based on T_0 that other point separation axioms are defined. In the present work, we propose the notion of *set separation* and introduce generalized separation axioms. Basically a set $Z \subseteq X$ in a topological space $\langle X, \tau\rangle$ is distinguishable if there exists an open set Y such that $\varnothing \neq Y \cap Z \neq Z$. For each non-zero cardinal κ, we introduce T_0^κ-spaces. For each positive integer n, we introduce a modal formula (t_0^n) which d-defines T_0^n-spaces. We show that $\mathsf{wK4T}_0^n = \mathsf{wK4} \oplus \mathrm{t}_0^n$ is the d-logic of all T_0^n-spaces. However, for each infinite cardinal κ, the class of all T_0^κ-spaces is not d-definable. These generalized separation axioms T_0^κ give new classes of topological spaces. Kripke frames for (t_0^n) are exactly those in which each proper cluster has at most n-irreflexive points. The modal logic $\mathsf{wK4T}_0^n$ is also characterized by frames for (t_0^n).

The structure of this paper is given as follows. Section 2 gives preliminaries on topological spaces and the d-semantics of modal logic. Section 3 gives generalized separation axioms T_0^κ and proves basic properties of T_0^κ-spaces. Section 4 proves that each T_0^n with positive integer n is d-definable in modal logic, and shows the d-completeness of modal logic $\mathsf{wK4T}_0^n$. Section 5 gives some concluding remarks.

2 Preliminaries

Let $\mathbb{N}$ and $\mathbb{Z}$ be sets of natural numbers and integers respectively. Let $\mathbb{Z}^+$ be the set of all positive integers. We use $|X|$ for the cardinality of a set X, and $\mathcal{P}(X)$ for the power set of X. We recall some basic notions of topological spaces from e.g. [24], as well as modal logic of space from e.g. [7–10].

A *topological space* is a pair $\mathcal{X} = \langle X, \tau\rangle$ where $X \neq \varnothing$ and $\tau \subseteq \mathcal{P}(X)$ such that $X, \varnothing \in \tau$ and τ is closed under arbitrary unions and finite intersections. Elements in τ are called *open sets.* A subset $Y \subseteq X$ is *closed* if $X \setminus Y$ is open. The class of all topological spaces is denoted by Topo.

Let $\mathcal{X} = \langle X, \tau\rangle$ be a space and $x \in X$. A subset $Y \subseteq X$ is an *open neighborhood* of x if $x \in Y \in \tau$. Let $N(x)$ be the set of all open neighborhoods of x. For every subset $A \subseteq X$, let $\mathsf{d}(A)$ be the *derived set* of A, i.e.,

$$\mathsf{d}(A) = \{x \in X : \forall U \in N(x)(U \cap (A \setminus \{x\}) \neq \varnothing)\}.$$

A *base* for τ is a $\mathscr{B} \subseteq \tau$ with $\tau = \{\bigcup_{B\in\mathscr{C}} B : \mathscr{C} \subseteq \mathscr{B}\}$. A *subbase* for τ is a $\mathscr{C} \subseteq \tau$ such that all finite intersections of elements in $\mathscr{C}$ form a base for τ.

Point separation in a topological space X are introduced by requiring that X satisfies one of the separation axioms T_0, $T_{\frac{1}{2}}$, T_1 or T_2 etc.

Definition 1. *Let $\mathcal{X} = \langle X, \tau \rangle$ be a topological space.*

(1) $\mathcal{X}$ *is a* T_0-space *if for each pair of different points there exists an open set containing one and not the other.*
(2) $\mathcal{X}$ *is a* $T_{\frac{1}{2}}$-space *if for each $x \in X$ there exists $U \in N(x)$ such that $\{x\}$ is closed in U.*
(3) $\mathcal{X}$ *is a* T_1-space *if for each pair $\langle x, y \rangle$ of different points, there exist open sets $U, V \in \tau$ such that $U \cap \{x, y\} = \{x\}$ and $V \cap \{x, y\} = \{y\}$.*
(4) $\mathcal{X}$ *is a* T_2-space, *or* Hausdorff space *if for each pair of different points x and y there exist disjoint open neighborhoods of x and y.*

Let T_0, $\mathsf{T}_{\frac{1}{2}}$, T_1 and T_2 be classes of all T_0-spaces, $T_{\frac{1}{2}}$-spaces T_1-spaces and T_2-spaces respectively. It is shown in [4] that $T_{\frac{1}{2}}$ is between T_0 and T_1. Then we see that $\mathsf{T}_2 \subsetneq \mathsf{T}_1 \subsetneq \mathsf{T}_{\frac{1}{2}} \subsetneq \mathsf{T}_0$.

Let $\mathcal{X} = \langle X, \tau \rangle$ and $\mathcal{Y} = \langle Y, \sigma \rangle$ be topological spaces. Then $\mathcal{Y}$ is a *subspace* of $\mathcal{X}$ if $Y \subseteq X$ and $\sigma = \{U \cap Y : U \in \tau\}$. Let $f : X \to Y$ be a map.

(1) f is *continuous* if $V \in \sigma$ implies $f^{-1}(V) \in \tau$.
(2) f is *open* if $U \in \tau$ implies $f(U) \in \sigma$.
(3) f is *interior* if f is continuous and open.

We say $\mathcal{Y}$ is an *interior image* of $\mathcal{X}$ if there is an interior map from X onto Y.

Let $\{\mathcal{X}_i = \langle X_i, \tau_i \rangle\}_{i \in I}$ be a family of topological spaces. The *topological sum* of $\{\mathcal{X}_i\}_{i \in I}$ is defined as the topological space $\bigoplus_{i \in I} \mathcal{X}_i = \langle \biguplus_{i \in I} X_i, \tau \rangle$ where $\biguplus_{i \in I} X_i$ is the disjoint union of $\{X_i\}_{i \in I}$ and $\tau = \{U \subseteq X : \forall i \in I(U \cap X_i \in \tau_i)\}$.

The *topological product* of $\{\mathcal{X}_i\}_{i \in I}$ is defined as $\prod_{i \in I} \mathcal{X}_i = \langle \prod_{i \in I} X_i, \tau \rangle$ where $\prod_{i \in I} X_i$ is the product of $\{X_i\}_{i \in I}$ and τ is obtained by taking as a base, sets of the form $\prod_{i \in I} U_i$, where $U_i \in \tau_i$ for all $i \in I$; and $U_i = X_i$ for all but finitely many coordinates. If $I = \{n_1, \cdots, n_k\}$ is finite, we write $\mathcal{X}_{n_1} \times \cdots \times \mathcal{X}_{n_k}$.

Let $\mathcal{X} = \langle X, \tau \rangle$ be a space, Z a set and $g : X \to Z$ a onto mapping. Let $\tau_g = \{H \subseteq Z : g^{-1}(H) \in \tau\}$. The topological space $\mathcal{X}_g = \langle Z, \tau_g \rangle$ is called the *quotient* of $\mathcal{X}$ induced on Z by g. The inducing map g is called a *quotient map.*

A class of topological spaces $\mathcal{K}$ is closed under an operation $\mathcal{O}$ if $\mathcal{O}(\mathcal{X}) \in \mathcal{K}$ for all $\mathcal{X} \in \mathcal{K}$. It is well-known that, for each $i \in \{0, \frac{1}{2}, 1, 2\}$, T_i is closed under subspaces and topological products; T_i is not closed under quotients. Moreover, T_2 is not closed under interior images.

Now we introduce the d-semantics of modal logic in topological spaces, which we refer to [8]. We take the basic modal language with only a diamond $\Diamond$.

Definition 2. *Let $\mathbb{V} = \{p_i : i \in \mathbb{N}\}$ be a set of propositional variables. The set of all modal formulas $\mathcal{L}$ is defined inductively as follows:*

$$\mathcal{L} \ni \varphi ::= p \mid \neg\varphi \mid (\varphi_1 \vee \varphi_2) \mid \Diamond\varphi$$

where $p \in \mathbb{V}$. The connectives $\top$, $\bot$, $\wedge$, $\to$ and $\leftrightarrow$ are defined as usual. We use the abbreviation $\Box\varphi := \neg\Diamond\neg\varphi$.

Definition 3. *A* topological model *is a triple* $\mathcal{M} = \langle X, \tau, \nu \rangle$ *where* $\mathcal{X} = \langle X, \tau \rangle$ *is a topological space and* $\nu : \mathsf{Prop} \to \mathcal{P}(X)$ *is a function which is called a* valuation *in* $\mathcal{X}$. *A valuation* ν *is extended to all modal formulas* $\mathcal{L}$ *as follows:*

$$\nu(\neg\varphi) = X \setminus \nu(\varphi);\ \nu(\varphi \vee \psi) = \nu(\varphi) \cup \nu(\psi);\ \nu(\Diamond\varphi) = \mathsf{d}(\nu(\varphi)).$$

Note that $\nu(\Box\varphi) = \{x \in X : \exists U \in N(x)(U \subseteq \nu(\varphi) \cup \{x\})\}$. *For every formula* φ, *topological model* $\mathcal{M} = \langle X, \tau, \nu \rangle$, $\mathcal{X} \in \mathsf{Topo}$ *and* $\mathcal{K} \subseteq \mathsf{Topo}$,

(1) φ *is* true *at* w *in* $\mathcal{M}$ *(notation:* $\mathcal{M}, w \models \varphi$*) if* $w \in \nu(\varphi)$.
(2) φ *is* true *in* $\mathcal{M}$ *(notation:* $\mathcal{M} \models \varphi$*) if* $w \in \nu(\varphi)$ *for all* $w \in X$.
(3) φ *is* valid *in* $\mathcal{X}$ *(notation:* $\mathcal{X} \models \varphi$*) if* $\mathcal{X}, \nu \models \varphi$ *for every valuation* ν *in* $\mathcal{X}$.
(4) φ *is* valid *in* $\mathcal{K}$ *(notation:* $\mathcal{K} \models \varphi$*) if* $\mathcal{X} \models \varphi$ *for all* $\mathcal{X} \in \mathcal{K}$.

The set of modal formulas $\mathsf{L_d}(\mathcal{K}) = \{\varphi \in \mathcal{L} : \mathcal{K} \models \varphi\}$ *is called the* d-logic *of a class of topological spaces* $\mathcal{K}$. *If* $\mathcal{K} = \{\mathcal{X}\}$, *we write* $\mathsf{L_d}(\mathcal{X})$.

Many d-logics of classes of topological spaces are finitely axiomatizable. Here a *normal modal logic* is defined as a set of modal formulas L such that

(1) all instance of classical propositional tautologies belong to L;
(2) $\Box(p \to q) \to (\Box p \to \Box q) \in L$;
(3) L is closed under (MP) and uniform substitution.

Let K denote the least normal modal logic. For a normal modal logic L and a set of formulas Σ, let $L \oplus \Sigma$ be the least normal modal logic containing $L \cup \Sigma$.

Definition 4. *Let* w4, t_0 *and* 4 *be the following modal formulas:*

$$\Diamond\Diamond p \to p \vee \Diamond p \qquad \text{(w4)}$$
$$p \wedge \Diamond(q \wedge \Diamond p) \to \Diamond p \vee \Diamond(q \wedge \Diamond q) \qquad (\mathrm{t}_0)$$
$$\Diamond\Diamond p \to \Diamond p \qquad (4)$$

Let $\mathsf{wK4} = \mathsf{K} \oplus \mathrm{w4}$, $\mathsf{wK4T_0} = \mathsf{wK4} \oplus \mathsf{t_0}$ *and* $\mathsf{K4} = \mathsf{K} \oplus 4$.

It is known that $\mathsf{wK4} = \mathsf{L_d}(\mathsf{Topo})$ and $\mathsf{K4} = \mathsf{L_d}(\mathsf{T}_{\frac{1}{2}})$ (cf. [15,16]). Moreover, $\mathsf{wK4T_0} = \mathsf{L_d}(\mathsf{T_0})$ (cf. [10]). Kripke semantics is used in the proof of topological completeness and related consequences. Recall some definitions from e.g. [8,10].

Definition 5. *A* frame *is a pair* $\mathfrak{F} = \langle W, R \rangle$ *where* $W \neq \varnothing$ *and* $R \subseteq W \times W$. *For each* $w \in W$ *and* $S \subseteq W$, *we define* $R(w) = \{u \in W : wRu\}$ *and* $R[S] = \bigcup_{w \in S} R(w)$. *A point* $w \in W$ *is* reflexive *if* $w \in R(w)$. *A point* $w \in W$ *is* irreflexive *if* $w \notin R(w)$. *Let* $\Diamond_R : \mathcal{P}(W) \to \mathcal{P}(W)$ *be the function defined as*

$$\Diamond_R Q = \{w \in W : R(w) \cap Q \neq \varnothing\} \text{ for each } Q \in \mathcal{P}(W).$$

A valuation *in a frame* $\mathfrak{F} = \langle W, R \rangle$ *is a function* $\theta : \mathbb{V} \to \mathcal{P}(W)$. *A valuation* θ *is extended to the set of all modal formulas* $\mathcal{L}$ *by the following rule:*

$$\theta(\neg\varphi) = W \setminus \theta(\varphi);\ \theta(\varphi \vee \psi) = \theta(\varphi) \cup \theta(\psi);\ \theta(\Diamond\varphi) = \Diamond_R\theta(\varphi).$$

A Kripke model *is a triple* $\mathfrak{M} = \langle W, R, \theta\rangle$ *where* $\langle W, R\rangle$ *is a frame and* θ *is a valuation. A formula* φ *is* true *at* w *in* $\mathfrak{M} = \langle W, R, \theta\rangle$ *(notation:* $\mathfrak{M}, w \models \varphi$*) if* $w \in \theta(\varphi)$. *A formula* φ *is* valid *at* w *in* $\mathfrak{F} = \langle W, R\rangle$ *(notation:* $\mathfrak{F}, w \models \varphi$*) if* $w \in \theta(\varphi)$ *for each valuation* θ *in* $\mathfrak{F}$. *A formula* φ *is* valid *in* $\mathfrak{F}$ *(notation:* $\mathfrak{F} \models \varphi$*) if* $\mathfrak{F}, w \models \varphi$ *for every* w *in* $\mathfrak{F}$.

For a set of modal formulas Σ, *let* $\mathsf{Fr}(\Sigma)$ *be the class of all frames validating every formula in* Σ. *Let* $\mathsf{Fr}_{<\omega}(\Sigma)$ *be the set of all finite members in* $\mathsf{Fr}(\Sigma)$. *For every class of frames* $\mathcal{C}$, *let* $\mathcal{C} \models \varphi$ *denote that* φ *is valid in* $\mathcal{C}$. *Let* $\mathsf{Th}(\mathcal{C}) = \{\varphi \in \mathcal{L} : \mathcal{C} \models \varphi\}$ *and we call it the* theory *of* $\mathcal{C}$. *If* $\mathcal{C} = \{\mathfrak{F}\}$, *we write* $\mathsf{Th}(\mathfrak{F})$. *A normal modal logic* L *is* Kripke-complete *if* $L = \mathsf{Th}(\mathsf{Fr}(L))$; L *has the* finite model property *(FMP) if* $L = \mathsf{Th}(\mathsf{Fr}_{<\omega}(L))$.

A frame $\mathfrak{F} = \langle W, R\rangle$ is *weakly transitive* if for all $w, u, v \in W$, if wRu and uRv, then $w = v$ or wRv. A frame $\mathfrak{F} \models \mathrm{w4}$ if and only if $\mathfrak{F}$ is weakly transitive. For every weakly transitive frame $\mathfrak{F} = \langle W, R\rangle$, $\mathfrak{F} \models \mathrm{t_0}$ if and only if for all $w, v \in W$, if wRv and vRw, then wRw or vRv. Note that $\mathsf{wK4}$ and $\mathsf{wK4T_0}$ have the FMP (cf. [10,15]).

Definition 6. *Let* $\mathfrak{F} = \langle W, R\rangle$ *be a weakly transitive frame and* $w \in W$. *The* cluster generated by w, *denoted by* $C(w)$, *is defined as follows:*

$$C(w) = \{w\} \cup \{u \in R(w) : w \in R(u)\}.$$

A subset $C \subseteq W$ *is a* cluster *if* $C = C(w)$ *for some* $w \in W$. *Let* C^{ir} *(or* $C^{ir}(w)$*) denote the set of all irreflexive points in* C *(or* $C(w)$*). A cluster* C *is* degenerate *if* $|C| = |C^{ir}| = 1$. *A cluster is* proper *if it is not degenerate.*

Clearly, for every weakly transitive frame $\mathfrak{F} = \langle W, R\rangle$, $\mathfrak{F} \models \mathrm{t_0}$ if and only if every proper cluster in $\mathfrak{F}$ contains at most one irreflexive point.

3 Generalized Separation Axioms

In this section, we generalize point separation to set separation, and introduce generalized separation axioms which define new classes of topological spaces. They are generalized from T_0 by considering set separation.

Definition 7. *Let* $\mathcal{X} = \langle X, \tau\rangle$ *be a topological space. A subset* $Z \subseteq X$ *is* distinguishable *if there exists an open set* $Y \in \tau$ *such that* $\varnothing \neq Y \cap Z \neq Z$. *In this case,* Y *is called a* separator *for* Z, *or we say that* Y separates Z.

For each $n \in \mathbb{N}$, let $\mathbf{n}$ denote the set $\{0, \cdots, n-1\}$. Consider the set $\mathbf{3}$ and topological space $\mathcal{X}_3 = \langle \mathbf{3}, \tau\rangle$ where $\tau = \{\mathbf{3}, \varnothing, \{0,1\}\}$. Clearly the set $\{0,1\}$ is a separator for $\{1,2\}$. However, the set $\{0,1\}$ itself is not distinguishable.

In what follows, we use Card for the class of all cardinals and Card^+ for the class of all non-zero cardinals. For each $\kappa \in \mathsf{Card}^+$, let κ^+ be the successor of κ. We also use cardinal arithmetic lightly.

Definition 8. *Let $\mathcal{X} = \langle X, \tau \rangle \in \mathsf{Topo}$ and $\kappa \in \mathsf{Card}^+$. Then $\mathcal{X}$ is a T_0^κ-space if every $Z \subseteq X$ with $|Z| \geq 1 + \kappa$ is distinguishable. Let T_0^κ be the class of all T_0^κ-spaces.*

For $\kappa \in \mathbb{Z}^+$, $\mathcal{X}$ is a T_0^κ-space if every $Z \subseteq X$ with $|Z| > \kappa$ is distinguishable. For $\kappa \geq \aleph_0$, $\mathcal{X}$ is a T_0^κ-space if every $Z \subseteq X$ with $|Z| \geq \kappa$ is distinguishable. Note that all T_0^1-spaces are exactly T_0-spaces.

Example 1. Partitions of sets yield T_0^κ-spaces in a natural way. Let $X \neq \varnothing$ be a set and π be a partition of X. For every $x \in X$, let $\pi(x)$ be the block of x. Suppose $\max\{|\pi(x)| : x \in X\} = \kappa^+$ with $0 < \kappa < \aleph_0$. Consider the topological space $\mathcal{X} = \langle X, \tau \rangle$ where $\tau = \{\bigcup_{y \in Y} \pi(y) : Y \subseteq X\}$. Now we show $\mathcal{X} \in \mathsf{T}_0^{\kappa^+} \setminus \mathsf{T}_0^\kappa$. To show that $\mathcal{X} \in \mathsf{T}_0^{\kappa^+}$, let $Y \subseteq X$ be a set such that $|Y| \geq 1 + \kappa^+ > \kappa^+$ and $y \in Y$. Then $|\pi(y)| \leq \kappa^+$ and so $\varnothing \neq Y \cap \pi(y) \neq Y$. Then $\pi(y)$ is a separator for Y. Hence $\mathcal{X} \in \mathsf{T}_0^{\kappa^+}$. To show $\mathcal{X} \notin \mathsf{T}_0^\kappa$, take $x \in X$ such that $|\pi(x)| = \kappa^+ \geq 1 + \kappa$. Let $U = \bigcup_{z \in Z} \pi(z) \in \tau$ be an open set. Clearly either $U \cap \pi(x) = \varnothing$ or $\pi(x) \subseteq U$. Then $\pi(x)$ is not distinguishable. Hence $\mathcal{X} \notin \mathsf{T}_0^\kappa$.

Proposition 1. *Let $\kappa \in \mathsf{Card}^+$. The following hold:*

(1) *Subspaces of T_0^κ-spaces are T_0^κ-spaces.*
(2) *Sum of T_0^κ-spaces are T_0^κ-spaces.*

Proof. (1) Let $\mathcal{X} = \langle X, \tau \rangle$ be a T_0^κ-spaces and $\mathcal{Y} = \langle Y, \sigma \rangle$ a subspace of $\mathcal{X}$. If $Z \subseteq Y$ and $|Z| \geq 1 + \kappa$, then $Z \subseteq X$ is distinguishable. Hence $\mathcal{Y}$ is a T_0^κ-space.

(2) Let $\{\mathcal{X}_i\}_{i \in I}$ be a family of T_0^κ-spaces and $\mathcal{X}_i = \langle X_i, \tau_i \rangle$ for each $i \in I$. Let $\mathcal{X} = \langle X, \tau \rangle$ be the sum $\bigoplus_{i \in I} \mathcal{X}_i$. Suppose $S \subseteq X$ is not distinguishable. If $S = \varnothing$, then $|S| < 1 + \kappa$. Suppose $S \neq \varnothing$ and $s \in S$. Then $s \in X_i$ for some $i \in I$. By $X_i \in \tau$, we have $t \in X_i$ for all $t \in S$. Otherwise, X_i is a separator for S. Hence $S \subseteq X_i$. Recall that $\mathcal{X}_i$ is a T_0^κ-space and S is not distinguishable. It follows that $|S| < 1 + \kappa$. □

Theorem 1. *Let $\kappa \in \mathsf{Card}^+$. The following hold:*

(1) $\bigcup_{0 < \lambda < \kappa} \mathsf{T}_0^\lambda \subsetneq \mathsf{T}_0^\kappa$.
(2) *For every $\mathcal{X} \in \mathsf{Topo}$, there exists $\kappa \in \mathsf{Card}^+$ with $\mathcal{X} \in \mathsf{T}_0^\kappa \setminus \bigcup_{0 < \lambda < \kappa} \mathsf{T}_0^\lambda$.*

Proof. (1) Note that $\mathsf{T}_0^\lambda \subseteq \mathsf{T}_0^\kappa$ for all $\lambda < \kappa$. Then $\bigcup_{0<\lambda<\kappa} \mathsf{T}_0^\lambda \subseteq \mathsf{T}_0^\kappa$. Suppose $\kappa = \mu^+$ for some $\mu \in \mathsf{Card}$. Let $X \neq \varnothing$ and $\mathcal{T}_X$ be the trivial topology on X. If $\mu < \aleph_0$ and $|X| = \kappa$, then clearly $\mathcal{T}_X \in \mathsf{T}_0^\kappa \setminus \bigcup_{0<\lambda<\kappa} \mathsf{T}_0^\lambda$. If $\mu \geq \aleph_0$ and $|X| = \mu$, then we can also readily check that $\mathcal{T}_X \in \mathsf{T}_0^\kappa \setminus \bigcup_{0<\lambda<\kappa} \mathsf{T}_0^\lambda$.

Suppose κ is a limit cardinal. Let $\{\mathcal{T}_{X_\lambda} : 0 < \lambda < \kappa\}$ be a family of trivial topological spaces such that $|X_\lambda| = \lambda$ for all $0 < \lambda < \kappa$. Note that $\mathcal{T}_{X_\lambda}$ are T_0^κ-spaces for all $0 < \lambda < \kappa$, by Proposition 1 (2), $\bigoplus_{0<\lambda<\kappa} \mathcal{T}_{X_\lambda}$ is T_0^κ. Let $\lambda < \kappa$. Then we see that $\mathcal{T}_{X_{\lambda^+}} \notin \mathsf{T}_0^\lambda$. Since κ is a limit cardinal, $\lambda^+ < \kappa$ and so $\mathcal{T}_{X_{\lambda^+}}$ is a subspace of $\bigoplus_{0<\lambda<\kappa} \mathcal{T}_{X_\lambda}$. By Proposition 1(1) and the arbitrariness of λ, $\bigoplus_{0<\lambda<\kappa} \mathcal{T}_{X_\lambda}$ is not T_0^λ for any $\lambda < \kappa$, which entails $\bigoplus_{0<\lambda<\kappa} \mathcal{T}_{X_\lambda} \notin \bigcup_{0<\lambda<\kappa} \mathsf{T}_0^\lambda$.

(2) Let $\mathcal{X} = \langle X, \tau \rangle$ be a topological space. Clearly $\mathcal{X} \in \mathsf{T}_0^{|X|^+}$. Then we see that $T = \{\lambda \in \mathsf{Card}^+ : \langle X, \tau \rangle \in \mathsf{T}_0^\lambda\} \neq \varnothing$. Let $\kappa = \min(T)$ and we are done. □

Lemma 1. *Let $\{\mathcal{X}_i : i \in I\}$ be a family of topological spaces and $\prod_{i\in I} \mathcal{X}_i = \langle X, \tau\rangle$. A subset $S \subseteq X$ is distinguishable if and only if $S(i) = \{f(i) : f \in S\}$ is distinguishable in $\mathcal{X}_i$ for some $i \in I$.*

Proof. Let $\mathcal{X}_i = \langle X_i, \tau_i\rangle$ for each $i \in I$. Assume $S(i)$ is distinguishable in $\mathcal{X}_i$ for some $i \in I$. Then there are $f, g \in S$ and $U \in \tau_i$ such that $f(i) \in U$ and $g(i) \notin U$. For each $j \in I$, we define

$$Y_j = \begin{cases} X_j, & \text{if } j \neq i \\ U, & \text{if } j = i. \end{cases}$$

Then $\prod_{j\in I} Y_j \in \tau$ is a separator for S.

Assume S is distinguishable. There are $f, g \in S$ and $U = \prod_{i\in I} U_i \in \tau$ such that $f \in U$ and $g \notin U$. Then there exists $i \in I$ with $g(i) \notin U_i$. Since $f(i) \in U_i$ and $U_i \in \tau_i$, we have $\varnothing \neq S(i) \cap U_i \neq S(i)$. Then $S(i)$ is distinguishable. □

Proposition 2. *Let $\{\kappa_i : i \in I\} \subseteq \mathsf{Card}^+$ and $\{\mathcal{X}_i : i \in I\}$ be a family of topological spaces where $\mathcal{X}_i \in \mathsf{T}_0^{\kappa_i}$ for each $i \in I$. Then $\prod_{i\in I} \mathcal{X}_i \in \mathsf{T}_0^{\kappa}$ where $\kappa = (\prod_{i\in I} \kappa_i)^+$.*

Proof. Let $\mathcal{X}_i = \langle X_i, \tau_i\rangle$ for $i \in I$. Suppose that $S \subseteq \prod_{i\in I} X_i$ is not distinguishable. By Lemma 1, for each $i \in I$, $S(i) = \{f(i) : f \in S\}$ is not distinguishable in $\mathcal{X}_i$. Hence $|S(i)| < 1 + \kappa_i$ and so $|S(i)| \leq \kappa_i$. Note that $S \subseteq \prod_{i\in I} S(i)$. Then $|S| \leq |\prod_{i\in I} S(i)| \leq \prod_{i\in I} \kappa_i < \kappa \leq 1 + \kappa$. It follows that $\prod_{i\in I} \mathcal{X}_i \in \mathsf{T}_0^{\kappa}$. □

When finite non-zero cardinals are concerned, the product of T_0^{κ}-spaces is more elegant. For example, consider the topological space $\mathcal{X}_2 = \langle \mathbf{2}, \{\mathbf{2}, \varnothing\}\rangle$. Clearly $\mathcal{X}_2 \in \mathsf{T}_0^2$. However, $\mathcal{X}_2 \times \mathcal{X}_2 = \langle \mathbf{4}, \{\mathbf{4}, \varnothing\}\rangle$ which is clearly not a T_0^2-space but a T_0^4-space.

Proposition 3. *Let $n, m \in \mathbb{Z}^+$. If $\mathcal{X}_1 \in \mathsf{T}_0^n$ and $\mathcal{X}_2 \in \mathsf{T}_0^m$, then $\mathcal{X}_1 \times \mathcal{X}_2 \in \mathsf{T}_0^{n\times m}$.*

Proof. Let $\mathcal{X}_1 = \langle X_1, \tau_1\rangle \in \mathsf{T}_0^n$ and $\mathcal{X}_2 = \langle X_2, \tau_2\rangle \in \mathsf{T}_0^m$. Suppose $S \subseteq X_1 \times X_2$ is not distinguishable. By Lemma 1, $S(i) = \{f(i) : f \in S\}$ is not distinguishable in X_i for $i \in \{1, 2\}$. Then $|S(1)| < 1 + n$ and $|S(2)| < 1 + m$. Then $|S(1)| \leq n$ and $|S(2)| \leq m$. By $S \subseteq S(1) \times S(2)$, $|S| \leq |S(1) \times S(2)| \leq n \times m < 1 + n \times m$. Hence $\mathcal{X}_1 \times \mathcal{X}_2 \in \mathsf{T}_0^{n\times m}$. □

For the operations of quotient and open continuous image on the class of all topological spaces, we have the following general negative result on T_0^{κ}.

Proposition 4. *Let $\kappa \in \mathsf{Card}^+$. There are spaces $\mathcal{H}_1$ and $\mathcal{H}_2$ such that $\mathcal{H}_1 \in \mathsf{T}_0$, $\mathcal{H}_2 \notin \mathsf{T}_0^{\kappa}$, $\mathcal{H}_2$ is a quotient of $\mathcal{H}_1$, and $\mathcal{H}_2$ is an interior image of $\mathcal{H}_1$.*

Proof. Let $\kappa \in \mathsf{Card}^+$ and $\lambda = \kappa^+ + \aleph_0$. Let X, Y be sets with $|X| = \lambda$ and $|Y| = \kappa^+$. Let $\mathcal{H}_1 = \langle X \times Y, \tau\rangle$ be the topological space where $\tau = \{\varnothing\} \cup \{Z \subseteq X \times Y : |(X \times Y) \setminus Z| < \aleph_0\}$. Let $\mathcal{H}_2$ be the trivial space over Y. Clearly $\mathcal{H}_1 \in \mathsf{T}_0$ and $\mathcal{H}_2 \notin \mathsf{T}_0^{\kappa}$. Let $f : X \times Y \to Y$ be the function defined by $f(\langle x, y\rangle) = y$. Then the quotient of $\mathcal{H}_1$ induced by f is exactly $\mathcal{H}_2$. Note that f is an interior map and onto. Hence $\mathcal{H}_2$ is an interior image of $\mathcal{H}_1$. □

4 Modal Logic $\mathsf{wK4T}_0^n$

Now we introduce modal logics of T_0^κ-spaces under the d-semantics. Recall that T_0-spaces are d-defined by the modal formula t_0, and $\mathsf{wK4T}_0$ is exactly the modal logic of all T_0-spaces. This section generalizes these results for T_0^κ-spaces.

Definition 9. *For each $n \in \mathbb{Z}^+$, let t_0^n denote the modal formula*

$$p_0 \wedge \bigwedge_{1 \leq i \leq n} \Diamond(\neg p_1 \wedge \cdots \wedge \neg p_{i-1} \wedge p_i \wedge \Diamond p_0) \to \Diamond p_0 \vee \bigvee_{1 \leq i \leq n} \Diamond(p_i \wedge \Diamond p_i).$$

Let $\mathsf{wK4T}_0^n$ denote the modal logic $\mathsf{wK4} \oplus \mathrm{t}_0^n$.

Note that t_0^1 is exactly t_0. A frame $\mathfrak{F} = \langle W, R \rangle$ is *irreflexive* if $w \notin R(w)$ for all $w \in W$. The following proposition gives the condition for the validity of t_0^n.

Proposition 5. *Let $\mathfrak{F} = \langle W, R \rangle$ be a weakly transitive frame. For each $n \in \mathbb{Z}^+$, $\mathfrak{F} \models \mathrm{t}_0^n$ if and only if $|C^{ir}(w)| \leq n$ for all $w \in W$.*

Proof. Assume $|C^{ir}(w)| > n$ for some $w \in W$. Then there exist pairwise different irreflexive points $w_0, \cdots, w_n \in C(w)$. Let θ be a valuation in $\mathfrak{F}$ such that $\theta(p_i) = \{w_i\}$ for all $i \leq n$. Then $\mathfrak{F}, \theta, w_i \models \neg p_1 \wedge \cdots \wedge \neg p_{i-1} \wedge p_i \wedge \Diamond p_0$ for all $i \in \{1, \cdots, n\}$, which implies $\mathfrak{F}, \theta, w_0 \models p_0 \wedge \bigwedge_{1 \leq i \leq n} \Diamond(\neg p_1 \wedge \cdots \wedge \neg p_{i-1} \wedge p_i \wedge \Diamond p_0)$. Since $w_0, \cdots, w_n$ are irreflexive, we have $\mathfrak{F}, \theta, w_0 \models \neg\Diamond p_0 \wedge \bigwedge_{1 \leq i \leq n} \Box\neg(p_i \wedge \Diamond p_i)$. Then $\mathfrak{F}, \theta, w_0 \not\models \mathrm{t}_0^n$ and so $\mathfrak{F} \not\models \mathrm{t}_0^n$.

Assume $\mathfrak{F} \not\models \mathrm{t}_0^n$. There exists a point $w \in W$ and a valuation θ in $\mathfrak{F}$ with $\mathfrak{F}, \theta, w \not\models \mathrm{t}_0^n$. Let $\mathfrak{M} = \langle W, R, \theta \rangle$ and $w_0 = w$. Then $\mathfrak{M}, w_0 \models p_0 \wedge \Box\neg p_0$ and for each $1 \leq i \leq n$ there exists $w_i \in R(w_0)$ with $\mathfrak{M}, w_i \models \neg p_1 \wedge \cdots \wedge \neg p_{i-1} \wedge p_i \wedge \Diamond p_0 \wedge \Box\neg p_i$. For all $i \neq j \leq n$, $w_i \neq w_j$ are irreflexive. Then $|\{w_0, \cdots, w_n\}| = n+1$. Let $l \in \{1, \cdots, n\}$. Since $\mathfrak{M}, w_l \models \Diamond p_0$, there exists $u \in R(w_l)$ with $\mathfrak{M}, u \models p_0$. Since $\mathfrak{F}$ is weakly transitive, by $w_0 R w_l$ and $\mathfrak{M}, w_0 \models \Box\neg p_0$, $u = w_0$. Then $w_l R w_0$ for all $l \in \{1, \cdots, n\}$. Hence $i \neq j \leq n$ implies $w_i R w_j$. Then $|C^{ir}(w)| > n$. □

Clearly (t_0^n) is a Sahlqvist-formula and so $\mathsf{wK4T}_0^n$ is complete (cf. e.g. [12]).

Proposition 6. $\bigcap_{n \in \mathbb{Z}^+} \mathsf{wK4T}_0^n = \mathsf{wK4}$.

Proof. Suppose $\varphi \notin \mathsf{wK4}$. Since $\mathsf{wK4}$ has the FMP (cf. [9]), there exists a finite weakly transitive frame $\mathfrak{F} = \langle W, R \rangle$ such that $\mathfrak{F} \not\models \varphi$. Let $|W| = k \in \mathbb{Z}^+$. Obviously $\mathfrak{F} \models \mathsf{wK4T}_0^k$. Then $\mathfrak{F} \models \bigcap_{n \in \mathbb{Z}^+} \mathsf{wK4T}_0^n$. Hence $\varphi \notin \bigcap_{n \in \mathbb{Z}^+} \mathsf{wK4T}_0^n$. □

Proposition 7. *For each $n \in \mathbb{Z}^+$, $\mathsf{wK4T}_0^n$ has the FMP.*

Proof. Note that $\mathsf{Fr}(\mathrm{t}_0^n)$ is closed under taking subframes and so $\mathsf{wK4T}_0^n$ is a subframe logic over $\mathsf{wK4}$. By [11, Theorem 5.9], $\mathsf{wK4T}_0^n$ has the FMP. □

By Proposition 7, for each $n \in \mathbb{Z}^+$, the logic $\mathsf{wK4T}_0^n$ is decidable since it is finitely axiomatized. Now we are ready for proving the d-definability of T_0^n-spaces and further the topological completeness of modal logics $\mathsf{wK4T}_0^n$.

Theorem 2. *Let* $\mathcal{X} \in \mathsf{Topo}$ *and* $n \in \mathbb{Z}^+$. *Then* $\mathcal{X} \models \mathrm{t}_0^n$ *if and only if* $\mathcal{X} \in \mathsf{T}_0^n$.

Proof. Let $\mathcal{X} = \langle X, \tau \rangle$. Suppose $\mathcal{X} \notin \mathsf{T}_0^n$. Then there exist pairwise different points $x_0, \cdots, x_n \in X$ such that $\{x_0, \cdots, x_n\}$ is not distinguishable. Let ν be a valuation in $\mathcal{X}$ with $\nu(p_i) = \{x_i\}$ for all $i \leq n$, and $\mathcal{M} = \langle X, \tau, \nu \rangle$. Let $i \neq j \in \{0, \cdots, n\}$. Since $\{x_0, \cdots, x_n\}$ is not distinguishable, we have $x_i \in U \cap (\{x_i\} \setminus \{x_j\})$ for all $U \in N(x_j)$. Then $x_j \in \mathsf{d}(\{x_i\})$. By $\nu(p_0) = \{x_0\}$, we have $\{x_1, \cdots, x_n\} \subseteq \mathsf{d}(\nu(p_0))$. Then $\{x_m\} = \nu(\neg p_1 \wedge \cdots \wedge \neg p_{m-1} \wedge p_m \wedge \Diamond p_0)$ for all $1 \leq m \leq n$. It follows that $\mathcal{M}, x_0 \models p_0 \wedge \bigwedge_{1 \leq m \leq n} \Diamond(\neg p_1 \wedge \cdots \wedge \neg p_{m-1} \wedge p_m \wedge \Diamond p_0)$. By $X \in N(x_i)$ and $X \cap (\{x_i\} \setminus \{x_i\}) = \varnothing$, we have $x_i \notin \mathsf{d}(\{x_i\})$. By $\nu(p_i) = \{x_i\}$, we have $\nu(p_i \wedge \Diamond p_i) = \nu(p_i) \cap \mathsf{d}(\nu(p_i)) = \varnothing$ and $\nu(\Diamond(p_i \wedge \Diamond p_i)) = \varnothing$. Hence $\mathcal{M}, x_0 \not\models \Diamond p_0 \vee \bigvee_{1 \leq i \leq n} \Diamond(p_i \wedge \Diamond p_i)$. It follows that $\mathcal{X} \not\models \mathrm{t}_0^n$.

Suppose $\mathcal{X} \not\models \mathrm{t}_0^n$. Then there exists a valuation ν on $\mathcal{X}$ and $w_0 \in X$ such that $\mathcal{M}, w_0 \not\models \mathrm{t}_0^n$ where $\mathcal{M} = \langle X, \tau, \nu \rangle$. Then $\mathcal{M}, w_0 \models \Box \neg p_0$. Hence there exists $U_0 \in N(w_0)$ such that $U_0 \subseteq \nu(\neg p_0) \cup \{w_0\}$. Let $k \in \{1, \cdots, n\}$. Then $\mathcal{M}, w_0 \models \Box(p_k \to \Box \neg p_k)$. Then there exists $U_k' \in N(w_0)$ such that $U_k' \subseteq \nu(p_k \to \Box \neg p_k) \cup \{w_0\}$. By $w_0 \in U_0 \in N(w_0)$, we have $U_k = U_k' \cap U_0 \in N(w_0)$. Note that $\mathcal{M}, w_0 \models \Diamond(\neg p_1 \wedge \cdots \wedge \neg p_{k-1} \wedge p_k \wedge \Diamond p_0)$. Then there exists $w_k \in U_k \cap \nu(\neg p_1 \wedge \cdots \wedge \neg p_{k-1} \wedge p_k \wedge \Diamond p_0)$ such that $w_k \neq w_0$. By $\mathcal{M}, w_k \models p_k \wedge (p_k \to \Box \neg p_k)$, we have $\mathcal{M}, w_k \models \Box \neg p_k$. Then there exists $V_k' \in N(w_k)$ such that $V_k' \subseteq \nu(\neg p_k) \cup \{w_k\}$. By $w_k \in U_k \in \tau$, we have $V_k = U_k \cap V_k' \in N(w_k)$. By $\mathcal{M}, w_k \models \Diamond p_0$, we have $V_k \cap (\nu(p_0) \setminus \{w_k\}) \neq \varnothing$. Note that $V_k \subseteq U_k \subseteq U_0 \subseteq \nu(\neg p_0) \cup \{w_0\}$. Then $w_0 \in V_k$. Otherwise, $V_k \cap \nu(p_0) = \varnothing$. Hence $V_k \in N(w_0)$. By $k \in \{1, \cdots, n\}$, we define such V_k for all $k \in \{1, \cdots, n\}$.

Now we show that $W = \{w_k : k \leq n\}$ is not distinguishable. For a contradiction, suppose that $U \in \tau$ is a separator for W. We have two cases:

(1) $w_0 \in U$. Then there exists $m \in \{1, \cdots, n\}$ with $w_m \notin U$. Since $U \cap V_m \in N(w_0)$, by $\mathcal{M}, w_0 \models \Diamond p_m$, we have $(U \cap V_m \cap \nu(p_m)) \neq \varnothing$. Note that $V_m \subseteq V_m' \subseteq \nu(\neg p_m) \cup \{w_m\}$ and $w_m \notin U$. Then $U \cap V_m \subseteq \nu(\neg p_m)$ and $(U \cap V_m \cap \nu(p_m)) = \varnothing$ which contradict the assumption.
(2) $w_0 \notin U$. Then there exists $m \in \{1, \cdots, n\}$ with $w_m \in U$. Since $U \cap V_m \in N(w_m)$, by $\mathcal{M}, w_m \models \Diamond p_0$, we have $(U \cap V_m \cap \nu(p_0)) \neq \varnothing$. Note that $V_m \subseteq U_m \subseteq U_0 \subseteq \nu(\neg p_0) \cup \{w_0\}$ and $w_0 \notin U$. Then $U \cap V_m \subseteq \nu(\neg p_0)$ and $(U \cap V_m \cap \nu(p_0)) = \varnothing$ which contradict the assumption.

Note that $|W| = n+1 > n$ and W is not distinguishable. It follows that $\mathcal{X} \notin \mathsf{T}_0^n$. $\Box$

In what follows, we shall prove the d-completeness of $\mathsf{wK4T}_0^n$ for each $n \in \mathbb{Z}^+$. Let $\mathfrak{F} = \langle W, R \rangle$ be a weakly transitive frame. Then we get a frame $\mathfrak{F}^* = (W, R^*)$ where R^* is the reflexive closure of R. Let $\mathcal{X}_{\mathfrak{F}} = \langle W, \tau_R \rangle$ be the *Alexandroff space induced by* $\mathfrak{F}^*$, namely, $\tau_R = \{R^*(Y) : Y \subseteq W\}$.

Definition 10. *Let* $\mathcal{X} = \langle X, \tau \rangle$ *be a topological space,* $\mathfrak{F} = \langle W, R \rangle$ *a weakly transitive frame and* $f : X \to W$ *a mapping. We say that* (1) f *is* irreflexively discrete *(i-discrete) if* $f^{-1}(w)$ *is a discrete subspace of* X *for all irreflexive point*

$w \in W$; *and* (2) f *is* reflexively dense *(r-dense) if* $f^{-1}(w) \subseteq \mathsf{d}(f^{-1}(w))$ *for all reflexive point* $w \in W$. *We call* f *a* d-morphism *if* f *is i-discrete, r-dense and* $f : \mathcal{X} \to \mathcal{X}_{\mathfrak{F}}$ *is interior.*

Let $\mathfrak{F} = \langle W, R\rangle$ be weakly transitive and $\mathcal{X} = \langle X, \tau\rangle$ a topological space. By [8, Corollary 2.8], if $\mathfrak{F}$ is finite, then $f : X \to W$ is a d-morphism if and only if $\mathsf{d}(f^{-1}(w)) = f^{-1}(R^{-1}(w))$ for all $w \in W$. By [8, Corollary 2.9], if $f : X \to W$ is a surjective d-morphism, then $\mathsf{L_d}(\mathcal{X}) \subseteq \mathsf{Th}(\mathfrak{F})$.

Definition 11. *Let* $\mathfrak{F} = \langle W, R\rangle$ *be a frame and* $\{\mathcal{X}_w\}_{w\in W}$ *a family of topological spaces where* $\mathcal{X}_w = \langle X_w, \tau_w\rangle$. *Let* $X_\oplus$ *be the disjoint union of* $\{X_w\}_{w\in W}$, *i.e.,* $X_\oplus = \{\langle x, w\rangle : x \in X_w \text{ and } w \in W\}$. *For each* $A \subseteq X_\oplus$ *and* $w \in W$, *let* $A_w = A \cap X_w$. *The* $\mathfrak{F}$-sum *of* $\{\mathcal{X}_w\}_{w\in W}$ *is defined as* $\bigoplus_{\mathfrak{F}} \mathcal{X}_w = \langle X_\oplus, \tau_\oplus\rangle$ *where* $U \in \tau_\oplus$ *if and only if for all* $w, v \in W$, *(1)* $U_w \in \tau_w$; *and (2) if* wRv, $w \neq v$ *and* $U_w \neq \varnothing$, *then* $U_v = X_v$.

Note that the $\mathfrak{F}$-sum $\bigoplus_{\mathfrak{F}} X_w$ in Definition 11 is indeed a topological space. For each $n \in \mathbb{Z}^+$, let n^* denote the set $\{i^* : i < n\}$ and $\mathbb{N}^*$ the set $\{i^* : i \in \mathbb{N}\}$. Now we define more topological spaces.

Definition 12. *The topological space* $\mathcal{X}_n = \langle X_n, \tau_n\rangle$ *is defined as follows:*

(1) $X_n = \mathbb{N} \cup n^*$.
(2) $\tau_n = \{\varnothing\} \cup \{Y \subseteq X_n : n^* \subseteq Y \text{ and } |X_n \setminus Y| < \aleph_0\}$.

Let $\mathfrak{F} = \langle W, R\rangle$ *be finite weakly transitive. We define* $\mathfrak{F}_s = (W_s, R_s)$ *as*

(1) $W_s = \{C(w) : w \in W\}$.
(2) $R_s = \{\langle C(w), C(v)\rangle : C(w) \neq C(v) \text{ and } wRv\}$.

For each $C \in W_s$, *the topological space* $\mathcal{X}_C = \langle X_C, \tau_C\rangle$ *is defined as follows:*

$$\mathcal{X}_C = \begin{cases} \langle |C|^*, \{|C|^*, \varnothing\}\rangle, & \text{if } |C| = |C^{ir}|. \\ \langle X_{|C^{ir}|}, \tau_{|C^{ir}|}\rangle, & \text{if } |C| \neq |C^{ir}|. \end{cases}$$

Let $\mathcal{X}^\circ_{\mathfrak{F}} = \langle X^\circ_{\mathfrak{F}}, \tau^\circ_{\mathfrak{F}}\rangle = \bigoplus_{\mathfrak{F}_s} \mathcal{X}_C$.

Note that the frame $\mathfrak{F}_s$ in Definition 12 is transitive and irreflexive.

Lemma 2. *Let* $\mathfrak{F}$ *be finite, weakly transitive and* $\mathfrak{F} \models \mathrm{t}^n_0$. *Then* $\mathcal{X}^\circ_{\mathfrak{F}}$ *is a* T^n_0*-space.*

Proof. Let $\mathfrak{F} = \langle W, R\rangle$. We show $|Y| \leq n$ for all $Y \subseteq X^\circ_{\mathfrak{F}}$ which are not distinguishable. Suppose not. Let $Y \subseteq X^\circ_{\mathfrak{F}}$ be not distinguishable and $|Y| > n$. Every element in Y is of the form $\langle x, C\rangle$ where $x \in \mathbb{N} \cup \mathbb{N}^*$ and $C \in W_s$. We have the following claims:

(1) if $\langle x, C\rangle, \langle y, D\rangle \in Y$, then $C = D$. Suppose $\langle x, C\rangle, \langle y, D\rangle \in Y$ and $C \neq D$. Since R_s is transitive and irreflexive, either $\langle C, D\rangle \notin R_s$ or $\langle D, C\rangle \notin R_s$. If $\langle C, D\rangle \notin R_s$, then $U = \{\langle z, E\rangle \in X^\circ_{\mathfrak{F}} : E \in R_s(C) \cup \{C\}\}$ is an open set which separates Y. The case $\langle D, C\rangle \notin R_s$ is similar. This contradicts the assumption.

(2) if $\langle x, C\rangle \in Y$, then $x \notin \mathbb{N}$. Suppose $\langle x, C\rangle \in Y$ and $x \in \mathbb{N}$. By $|Y| > n \geq 1$, there exists $\langle y, C\rangle \in Y$ with $y \neq x$. Let $V = \{\langle z, E\rangle \in X^\circ_{\mathfrak{F}} : E \in R_s(C) \cup \{C\}\} \setminus \{\langle x, C\rangle\}$. Then we see that V is an open set of $\mathcal{X}^\circ_{\mathfrak{F}}$. Note that $\langle x, C\rangle \notin V$ and $\langle y, C\rangle \in V$. Then V is a separator for Y. This contradicts the assumption.

By (1) and (2), $Y \subseteq \{\langle z, C\rangle : z \in |C^{ir}|^*\}$ for some $C \in W_s$. Note that $\mathfrak{F} \models \mathrm{t}^n_0$, we see $|C^{ir}| \leq n$ and so $|Y| \leq |C^{ir}| \leq n$ which contradicts the assumption. Hence $|Y| \leq n$ for all $Y \subseteq X^\circ_{\mathfrak{F}}$ which is not distinguishable. Hence $\mathcal{X}^\circ_{\mathfrak{F}}$ is a T^n_0-space. □

Lemma 3. *Let $\mathfrak{F} = \langle W, R\rangle$ be finite weakly transitive. Then $\mathsf{L_d}(\mathcal{X}^\circ_{\mathfrak{F}}) \subseteq \mathsf{Th}(\mathfrak{F})$.*

Proof. By [8, Corollary 2.9], it suffices to show there is a d-morphism from $\mathcal{X}^\circ_{\mathfrak{F}}$ to $\mathfrak{F}$. For each $C \in W_s$, we define $f_C : X_C \to C$ as follows:

(1) $|C| = |C^{ir}|$. Then $|X_C| = |C|$. Let f_C be a bijection from X_c to c.
(2) $|C| \neq |C^{ir}|$. Since $\mathfrak{F}$ is finite, there are $k+1 = |C| - |C^{ir}|$ reflexive points $v_0, \cdots, v_k \in C$, and $l = |C^{ir}|$ irreflexive points $u_1, \cdots, u_l \in C$. The function f_C is defined as follows:

$$f_C(\langle x, C\rangle) = \begin{cases} v_i, & \text{if } x \in \mathbb{N} \text{ and } x \equiv i \pmod{k+1}. \\ u_j, & \text{if } x = j^*. \end{cases}$$

Let $f = \bigcup_{C \in W_s} f_C$. Clearly f is onto. By [8, Corollary 2.8], it suffices to show $\mathsf{d}(f^{-1}(w)) = f^{-1}(R^{-1}(w))$ for all $w \in W$. Let $w \in W$.

(1) $\mathsf{d}(f^{-1}(w)) \subseteq f^{-1}(R^{-1}(w))$. Let $\langle y, D\rangle \notin f^{-1}(R^{-1}(w))$. Then $f(\langle y, D\rangle) \notin R^{-1}(w)$. Let $U = \{\langle z, E\rangle \in X^\circ_{\mathfrak{F}} : E \in R_s(D) \cup \{D\}\}$. We show $(f^{-1}(w) \cap U) \setminus \{\langle y, D\rangle\} = \varnothing$. Clearly $f(\langle y, D\rangle) \in D$. Then $\langle D, C(w)\rangle \notin R_s$. We have two cases:
  (1.1) $D \neq C(w)$. Then $C(w) \notin R_s(D) \cup \{D\}$. If $\langle z, E\rangle \in f^{-1}(w)$, then $E = C(w)$ for all $\langle z, E\rangle \in X^\circ_{\mathfrak{F}}$. Then $U \cap f^{-1}(w) = \varnothing$. So $(f^{-1}(w) \cap U) \setminus \{\langle y, D\rangle\} = \varnothing$.
  (1.2) $D = C(w)$. By $f(\langle y, D\rangle) \notin R^{-1}(w)$, $f(\langle y, D\rangle) = w$ is irreflexive. Then $f^{-1}(w) = \{\langle y, D\rangle\}$ and so $f^{-1}(w) \setminus \{\langle y, D\rangle\} = \varnothing$. So $(f^{-1}(w) \cap U) \setminus \{\langle y, D\rangle\} = \varnothing$.
  Note that U is an open set of $\mathcal{X}^\circ_{\mathfrak{F}}$ and $\langle y, D\rangle \in U$. Then $\langle y, D\rangle \notin \mathsf{d}(f^{-1}(w))$.
(2) $f^{-1}(R^{-1}(w)) \subseteq \mathsf{d}(f^{-1}(w))$. Assume $\langle y, D\rangle \in f^{-1}(R^{-1}(w))$ and $\langle y, D\rangle \notin \mathsf{d}(f^{-1}(w))$. Then $f(\langle y, D\rangle) \in R^{-1}(w)$ and so $C(w) \in R_s(D) \cup \{D\}$.
  (2.1) $C(w) \in R_s(D)$. Since R_s is irreflexive, we have $C(w) \neq D$. Then for all open neighborhood V of $\langle y, D\rangle$, $V \cap X_{C(w)} = X_{C(w)}$ and so $(f^{-1}(w) \cap V) \setminus \{\langle y, D\rangle\} = f^{-1}(w) \neq \varnothing$. Then $\langle y, D\rangle \in \mathsf{d}(f^{-1}(w))$ which is a contradiction.
  (2.2) $C(w) = D$. Note that $\langle y, D\rangle \notin \mathsf{d}(f^{-1}(w))$. Then there exists an open neighborhood V of $\langle y, D\rangle$ such that $(V \cap f^{-1}(w)) \setminus \{\langle y, D\rangle\} = \varnothing$. Now we show that w is irreflexive. For a contradiction, suppose that

w is reflexive. Then $|f^{-1}(w)| = \aleph_0$. Since $V \cap X_{C(w)}$ is an open set of $\mathcal{X}_{C(w)}$, we have $|X_{C(w)} \setminus V| < \aleph_0$. Note that $f^{-1}(w) \subseteq X_{C(w)}$. Then $|V \cap f^{-1}(w)| = \aleph_0$ which contradicts $(V \cap f^{-1}(w)) \setminus \{\langle y, D\rangle\} = \varnothing$. Hence w is irreflexive and so $f^{-1}(w) = \{\langle m^*, D\rangle\}$ for some $m \in \mathbb{N}$. Note that $f(\langle y, D\rangle) \in R^{-1}(w)$. Then $f(\langle y, D\rangle) \neq w$ which yields $y \neq m^*$. Clearly, for each open neighborhood V of $\langle y, D\rangle$, we have $\langle m^*, D\rangle \in V \cap f^{-1}(w)$. Hence $\langle m^*, D\rangle \in (V \cap f^{-1}(w)) \setminus \{\langle y, D\rangle\}$ which is a contradiction.

By (1) and (2), $\mathsf{d}(f^{-1}(w)) = f^{-1}(R^{-1}(w))$ for all $w \in W$. Then f is a d-morphism from $\mathcal{X}_{\mathfrak{F}}^{\circ}$ to $\mathfrak{F}$ and hence $\mathsf{L_d}(\mathcal{X}_{\mathfrak{F}}^{\circ}) \subseteq \mathsf{Th}(\mathfrak{F})$. □

Theorem 3. *For each $n \in \mathbb{Z}^+$, $\mathsf{wK4T}_0^n$ is the* d*-logic of all T_0^n-spaces.*

Proof. Assume $\varphi \notin \mathsf{wK4T}_0^n$. Since $\mathsf{wK4T}_0^n$ has the FMP, there exists a finite frame $\mathfrak{F}$ for $\mathsf{wK4T}_0^n$ such that $\mathfrak{F} \not\models \varphi$. By Lemma 3, $\mathcal{X}_{\mathfrak{F}}^{\circ} \not\models \varphi$. By Lemma 2 and Theorem 2, $\mathcal{X}_{\mathfrak{F}}^{\circ} \models \mathsf{wK4T}_0^n$. Hence $\mathsf{wK4T}_0^n = \mathsf{L_d}(\mathsf{T}_0^n)$. □

By Theorem 2, for each $n \in \mathbb{Z}^+$, t_0^n defines the class of all T_0^n-spaces. By Theorem 3, $\mathsf{wK4T}_0^n$ is the d-logic of the class of T_0^n-spaces for each $n \in \mathbb{Z}^+$. However, for an infinite cardinal κ, this is not the case.

Proposition 8. *For each cardinal $\kappa \geq \aleph_0$, the class T_0^κ is not* d*-definable.*

Proof. Let $\kappa \geq \aleph_0$ be a cardinal. By Theorem 1, we have $\mathsf{Topo} \supsetneq \mathsf{T}_0^\kappa \supseteq \mathsf{T}_0^\omega \supsetneq \bigcup_{1 \leq n < \omega} \mathsf{T}_0^n$. By Theorem 3, $\bigcap_{n \in \mathbb{Z}^+} \mathsf{wK4T}_0^n$ is the d-logic of $\bigcap_{1 \leq n < \omega} \mathsf{T}_0^n$. By Proposition 6, $\mathsf{L_d}(\mathsf{T}_0^\kappa) = \mathsf{wK4} = \mathsf{L_d}(\mathsf{Topo})$. Hence T_0^κ is not d-definable. □

5 Concluding Remarks

This work proposes generalized separation axioms for topological spaces and explores their modal logics. Point separation in the traditional study of topological spaces is generalized to set separation. For each non-zero cardinal κ, a separation axiom T_0^κ is given. We show that the d-logic of T_0^n for each $n \in \mathbb{Z}^+$ is axiomatized by the modal logic $\mathsf{wK4T}_0^n = \mathsf{wK4} \oplus \mathsf{t}_0^n$ where t_0^n is a new formula proposed in this work. We also show that T_0^κ for each infinite cardinal κ is not d-definable. These contributions make a progress in the logical study of separation axioms. For further exploration, we can define more topological spaces based on these axioms T_0^κ. We can also explore topological duality of modal algebras $\mathsf{wK4T}_0^n$.

References

1. Adhikari, A., Adhikari, M.R.: Separation axioms. In: Adhikari, A., Adhikari, M.R. (eds.) Basic Topology 1: Metric Spaces and General Topology, pp. 233–267. Springer, Singapore (2022). https://doi.org/10.1007/978-0-8176-8126-5_5
2. Aiello, M., van Benthem, J.: A modal walk through space. J. Appl. Non-Classical Logics **12**(3–4), 319–364 (2002)

3. Aiello, M., van Benthem, J., Bezhanishvili, G.: Reasoning about space: the modal way. J. Log. Comput. **13**(6), 889–920 (2003)
4. Aull, C.E., Thron, W.J.: Separation axioms between T_0 and T_1. Indag. Math. **24**, 26–37 (1962)
5. Banerjee, A.K., Pal, J.: New separation axioms in generalized bitopological spaces. Math. Sci. **14**(2), 185–192 (2020). https://doi.org/10.1007/s40096-020-00330-z
6. van Benthem, J., Bezhanishvili, G., ten Cate, B., Sarenac, D.: Multimodal logics of products of topologies. Stud. Logica **84**(3), 369–392 (2006)
7. van Benthem, J., Bezhanishvili, G.: Modal logic of space. In: Aiello, M., Pratt-Hartmann, I., van Benthem, J. (eds.) Handbook of Spatial Logics, pp. 217–298. Springer, Dordrecht (2007). https://doi.org/10.1007/978-1-4020-5587-4_5
8. Bezhanishvili, G., Esakia, L., Gabelaia, D.: Some results on modal axiomatization and definability for topological spaces. Stud. Logica **81**, 325–355 (2004)
9. Bezhanishvili, G., Esakia, L., Gabelaia, D.: The modal logic of stone spaces: diamond as derivative. Rev. Symbolic Logic **3**(1), 26–40 (2010)
10. Bezhanishvili, G., Esakia, L., Gabelaia, D.: Spectral and T_0-spaces in d-semantics. In: Bezhanishvili, N., Löbner, S., Schwabe, K., Spada, L. (eds.) TbiLLC 2009. LNCS (LNAI), vol. 6618, pp. 16–29. Springer, Heidelberg (2011). https://doi.org/10.1007/978-3-642-22303-7_2
11. Bezhanishvili, G., Ghilardi, S., Jibladze, M.: An algebraic approach to subframe logics. Modal case. Notre Dame J. Formal Logic **52**(2), 187–202 (2011)
12. Blackburn, P., de Rijke, M., Venema, Y.: Modal Logic. Cambridge University Press, Cambridge (2001)
13. Borzooei, R.A., Rezaei, G.R., Kouhestani, N.: Separation axioms in (semi)topological quotient BL-algebras. Soft. Comput. **16**, 1219–1227 (2012)
14. Cramer, T.: A definition of separation axiom. Can. Math. Bull. **17**(4), 485–491 (1974)
15. Esakia, L.: Weak transitivity - a restitution. Logical Invest. **8**, 244–255 (2001). (in Russian)
16. Esakia, L.: Intuitionistic logic and modality via topology. Ann. Pure Appl. Logic **127**(1–3), 155–170 (2004)
17. Ghanim, M.H., Kerre, E.E., Mashhour, A.S.: Separation axioms, subspaces and sums in fuzzy topology. J. Math. Anal. Appl. **102**(1), 189–202 (1984)
18. McKinsey, J.C.C., Tarski, A.: The algebra of topology. Ann. Math. **45**, 141–191 (1944)
19. Nelson, E.D.: Separation axioms in topology. Master thesis, University of Montana (1966)
20. Parikh, R., Moss, L.S., Steinsvold, C.: Topology and epistemic logic. In: Aiello, M., Pratt-Hartmann, I., van Benthem, J. (eds.) Handbook of Spatial Logics, pp. 299–341. Springer, Dordrecht (2007). https://doi.org/10.1007/978-1-4020-5587-4_6
21. Pultr, A., Tozzi, A.: Separation axioms and frame representation of some topological facts. Appl. Categ. Struct. **2**, 107–118 (1994)
22. Sarsak, M.S.: New separation axioms in generalized topological spaces. Acta Math. Hungar. **132**, 244–252 (2011)
23. Sustretov, D.: Hybrid logics of separation axioms. J. Logic Lang. Inform. **18**, 541–558 (2009)
24. Willard, S.: General Topology. Dover Publications Inc., New York (2004)

Multiple-Valued Semantics for Metric Temporal Logic

Fan He[1,2](✉)

[1] Institute of Logic and Cognition, Sun Yat-sen University, Guangzhou, China
hephan@gmail.com

[2] Department of Philosophy, Sun Yat-sen University, Guangzhou, China

Abstract. Metric temporal frames are introduced based on multiple-valued semantics. The intended metric temporal language is interpreted in models based on metric temporal frames. Normal metric temporal logics are introduced, and some completeness results are naturally given by adjusting the canonical model method. The finite model property to the minimal normal metric temporal logic is established.

Keywords: Metric temporal logic · Multiple-valued semantics · Completeness · Finite model property

1 Introduction

Temporal logic is a branch of logic that studies the reasoning about time. In basic temporal language, we have F and P as its additional operators. We interpret the formula '$F\varphi$' as 'at some moment in the future, φ is true', and '$P\varphi$' is to be read as 'at some moment in the past, φ was true'. But this language is too limited in expressivity for many applications. For example, the following two statements:

The construction of the new railway will be finished in two years.
The construction of the new railway will be finished in six years.

which present some possible occasion in the future and whose only difference is their temporal distances from now. This difference plays an important role in the reasoning about time. Beyond the expressivity, this difference cannot be described and analyzed in the basic temporal language. Therefore, A.N. Prior considered the use of what he called 'metric tense logic' (e.g., [8–10]). This logic is the tense logic in which the future and past operators have an index representing a temporal distance. This logic have been extensively studied by computer scientists under the name 'metric temporal logic' (e.g., [6,7] and others).

We have introduced multiple-valued semantics for multimodal logics in [5]. From the semantic perspective, a Kripke frame for a monomodal language is a pair (W, R) where W is a non-empty set of states, and R is a binary relation on W. Each modal formula $\Box\varphi$ is true at a state w if and only if φ is true at

M. Banerjee and A. V. Sreejith (Eds.): ICLA 2023, LNCS 13963, pp. 105–116, 2023.
https://doi.org/10.1007/978-3-031-26689-8_8

all R-accessible states of w (e.g., [1,2]). An accessibility relation R is indeed a *bivalent* function $R : W \times W \to \{0, 1\}$. We generalize Kripke frames by changing the set $\{0, 1\}$ into a set of values Q and obtain *multiple-valued* frames. This leads to a general framework for the investigation of multimodal logics. One can impose additional structure on Q, and study the modal logics of these special class of frames. With some additional restrictions, the multiple-valued frames can be used to interpret the metric temporal operators. Thus we have a new perspective of the metric temporal logic.

In this paper, we do some primary work on the multiple-valued semantics for the metric temporal logic, and the article is structured as follows. Section 2 gives the modal language and semantics where metric temporal frames are introduced. Section 3 introduces normal metric temporal logics and proves the completeness of the minimal normal metric temporal logic. Section 4 proves the minimal normal metric temporal logic has the finite model property with respect to the class of all temporal metric models. Section 5 gives concluding remarks.

2 Language and Semantics

Let $\mathfrak{N} = (\mathbb{N}, 0, \leqslant^{\mathfrak{N}}, +^{\mathfrak{N}})$ be the standard model of arithmetic. An *initial finite segment* of $\mathbb{N}$ is a set $\{x \in \mathbb{N} : x \leqslant m\}$ for some $m \in \mathbb{N}$. An *inversely well-ordered set* ('i.w.o set' for short) is a pair $(Q, \leqslant)$ such that $\geqslant$ well-orders the nonempty set Q. Obviously, the set $\{x \in \mathbb{N} : x \leqslant m\}$ is inversely well-ordered by $\geqslant^{\mathfrak{N}}$, i.e., the converse of the relation $\leqslant^{\mathfrak{N}}$ in $\mathfrak{N}$. Therefore, every subset X of $\{x \in \mathbb{N} : x \leqslant m\}$ with $X \neq \varnothing$ has a maximal element $\bigvee X$. If Q is an ordered set ordered by $\leqslant$, a *downset* in Q is a subset $X \subseteq Q$ such that $a \leqslant b \in X$ implies $a \in X$. An *upset* in Q is a subset $X \subseteq Q$ such that $a \in X$ and $a \leqslant b$ imply $b \in X$. Let $\downarrow X$ and $\uparrow X$ be the downset and upset in Q generated by X respectively. The cardinal of a set X is denoted by $|X|$.

Definition 1 (Temporal Metric). *A* temporal metric $\mathfrak{M} = (\mathbb{M}, 0, \leqslant, +)$ *is a structure where* $\mathbb{M}$ *is an initial finite segment of* $\mathbb{N}$, $\leqslant$ *is the binary relation which is the restriction of* $\leqslant^{\mathfrak{N}}$ *to* $\mathbb{M}$, *and* $+$ *is the binary partial function such that:*

$$a + b = \begin{cases} a +^{\mathfrak{N}} b & \textit{if } a, b, a +^{\mathfrak{N}} b \in \mathbb{M}. \\ \textit{undefined} & \textit{otherwise.} \end{cases}$$

Temporal metric will be used to measure the temporal distance between different moments. For technical reasons, we don't choose $\mathfrak{N}$, the standard model of arithmetic, as our temporal metric. But in most cases, we could find a sufficient large natural number m such that the set $\{x \in \mathbb{N} : x \leqslant m\}$ is suitable for applications.

Definition 2 (Metric Temporal Language). *Let* $\mathfrak{M} = (\mathbb{M}, 0, \leqslant, +)$ *be a temporal metric. The* metric temporal language $\mathcal{L}_{MT}(\mathfrak{M})$ *consists of a denumerable set of propositional variables* $\mathbb{P} = \{p_i : i < \omega\}$, *connectives* $\bot$ *and* $\to$, *and unary* *modal operators* $\{G_a : a \in \mathbb{M}\}$ *and* $\{H_a : a \in \mathbb{M}\}$*. The set of formulas* $Fm(\mathfrak{M})$ *is defined inductively as follows:*

$$Fm(\mathfrak{M}) \ni \varphi ::= p \mid \bot \mid (\varphi_1 \to \varphi_2) \mid G_a\varphi \mid H_a\varphi$$

where $p \in \mathbb{P}$ *and* $a \in \mathbb{M}$*. Connectives* $\top, \neg, \wedge, \vee$ *and* $\leftrightarrow$ *are defined as usual. For every* $a \in \mathbb{M}$*, one defines* $F_a\varphi := \neg G_a\neg\varphi$ *and* $P_a\varphi := \neg H_a\neg\varphi$*. The* complexity *of a formula* $\varphi \in Fm(\mathfrak{M})$*, denoted by* $\delta(\varphi)$*, is defined inductively as follows:*

$$\delta(p) = \delta(\bot) = 0$$

$$\delta(\varphi \to \psi) = max\{\delta(\varphi), \delta(\psi)\} + 1$$

$$\delta(G_a\varphi) = \delta(H_a\varphi) = \delta(\varphi) + 1$$

A substitution *is a function* $s : \mathbb{P} \to Fm(\mathfrak{M})$*. For every formula* $\varphi \in Fm(\mathfrak{M})$*, let* φ^s *be obtained from* φ *by the substitution* s*.*

Definition 3 (Metric Temporal Frame). *Let* $\mathfrak{M} = (\mathbb{M}, 0, \leqslant, +)$ *be a temporal metric. A* metric temporal frame *over* $\mathfrak{M}$ *('*$\mathfrak{M}$*-frame' for short) is a pair* $\mathcal{F} = (W, \sigma)$ *where* $W \neq \varnothing$ *is a set of states, and* $\sigma : W \times W \to \mathbb{M}$ *is a partial function from* $W \times W$ *to* $\mathbb{M}$ *satisfying the following conditions:*

(1) for all $w \in W$*,* $\sigma(w, w)! \geqslant 0$*;*
(2) for all $w, u, v \in W$ *and* $a, b, a + b \in \mathbb{M}$*, if* $\sigma(w, u)! \geqslant a$ *and* $\sigma(u, v)! \geqslant b$*, then* $\sigma(w, v)! \geqslant a + b$*.*

where the notation $\sigma(w, u)!$ *means that* $\sigma(w, u)$ *exists in* $\mathbb{M}$*, and* $\sigma(w, u)! \geqslant a$ *means* $\sigma(w, u)!$ *and* $\sigma(w, u) \geqslant a$*. For every* $a \in \mathbb{M}$*, the binary relation* R^σ_a *on* W *is defined as follows:*

$$wR^\sigma_a u \text{ if and only if } \sigma(w, u)! \geqslant a.$$

Let $R^\sigma_a(w) = \{u \in W : \sigma(w, u)! \geqslant a\}$*. Let* $\mathscr{F}_{\mathfrak{M}}$ *be the class of all* $\mathfrak{M}$*-frames.*

A valuation *in a* $\mathfrak{M}$*-frame* $\mathcal{F} = (W, \sigma)$ *is a function* $V : \mathbb{P} \to \mathcal{P}(W)$ *from* $\mathbb{P}$ *to the powerset of* W*. A* $\mathfrak{M}$-model *is a triple* $\mathcal{M} = (W, \sigma, V)$ *where* (W, σ) *is a* $\mathfrak{M}$*-frame and* V *is a valuation in* (W, σ)*. Let* $\mathscr{M}_{\mathfrak{M}}$ *be the class of all* $\mathfrak{M}$*-models.*

Intuitively, in a frame $\mathcal{F}$, states of $\mathcal{F}$ represent all possible moments in consideration. And the *measure function* σ assigns (or not) a *temporal distance* for a pair of states in $\mathcal{F}$. As the temporal structures we are interested in may have some kind of branching, we do not assume our measure function to be total.

Moreover, suppose w, u, v are any moments in our consideration, clearly we have $\sigma(w, w) = 0$, which means the temporal distance between a moment w and itself is 0. Furthermore, if $\sigma(w, u) = a$ and $\sigma(u, v) = b$, then $\sigma(w, v) = a + b$, which represents the fact that if u is in the future of w with temporal distance a, and v is in the future of u with temporal distance b, then v is in the future of w with temporal distance $a + b$. The corresponding conditions in our definition are weakened mainly because of technical reasons.

The usual semantics of metric temporal logic uses two-sorted structure $(T, \Delta, <, d, +, 0)$ where T is a nonempty set of 'moments', $<$ is a binary total order on T, and Δ is a nonempty set called metric domain. The *temporal distance function* $d : T \times T \to \Delta$ is surjective and satisfies usual topological conditions apart from the replacement of the triangular inequality by a conditional equality, and the structure $(\Delta, +, 0)$ satisfies some arithmetical laws (cf. [6]).

Definition 4 (Satisfaction Relation). *Let $\mathcal{F} = (W, \sigma)$ be a $\mathfrak{M}$-frame, $\mathcal{M} = (W, \sigma, V)$ a $\mathfrak{M}$-model and $w \in W$. For every $\varphi \in Fm(\mathfrak{M})$, the* satisfaction relation *$\mathcal{M}, w \models \varphi$ is defined inductively as follows:*

(1) $\mathcal{M}, w \models p$ *if and only if* $w \in V(p)$.
(2) $\mathcal{M}, w \not\models \bot$.
(3) $\mathcal{M}, w \models \varphi \to \psi$ *if and only if* $\mathcal{M}, w \not\models \varphi$ *or* $\mathcal{M}, w \models \psi$.
(4) $\mathcal{M}, w \models G_a\varphi$ *if and only if* $\mathcal{M}, u \models \varphi$ *for all* u *such that* $\sigma(w, u)! \geqslant a$.
(5) $\mathcal{M}, w \models H_a\varphi$ *if and only if* $\mathcal{M}, u \models \varphi$ *for all* u *such that* $\sigma(u, w)! \geqslant a$.

Let $V(\varphi) = \{w \in W : \mathcal{M}, w \models \varphi\}$. A formula φ is true *in $\mathcal{M}$, notation $\mathcal{M} \models \varphi$, if $V(\varphi) = W$. A formula φ is* valid *at w in $\mathcal{F} = (W, \sigma)$, notation $\mathcal{F}, w \models \varphi$, if $\mathcal{F}, V, w \models \varphi$ for every valuation V in $\mathcal{F}$. A formula φ is* valid *in $\mathcal{F}$, notation $\mathcal{F} \models \varphi$, if $\mathcal{F}, w \models \varphi$ for every $w \in W$. A formula φ is* valid *in a class of $\mathfrak{M}$-frames $\mathscr{K}$, notation $\mathscr{K} \models \varphi$, if $\mathcal{F} \models \varphi$ for every $\mathcal{F} \in \mathscr{K}$.*

Let $\Gamma \subseteq Fm(\mathfrak{M})$ be a set of formulas. Suppose $\mathbb{S}$ is a metric temporal structure (model or frame) or a class of metric temporal structures, let $\mathbb{S} \models \Gamma$ stand for that $\mathbb{S} \models \varphi$ for all $\varphi \in \Gamma$. The class of all $\mathfrak{M}$-frames defined by Γ is denoted by $\mathsf{Fr}_{\mathfrak{M}}(\Gamma) = \{\mathcal{F} : \mathcal{F} \models \Gamma\}$. If $\Gamma = \{\varphi\}$, one writes $\mathsf{Fr}_{\mathfrak{M}}(\varphi)$. The modal theory *of a class of $\mathfrak{M}$-frames $\mathscr{K}$ is defined as the set $\mathsf{Th}(\mathscr{K}) = \{\varphi \in Fm(\mathfrak{M}) : \mathscr{K} \models \varphi\}$. We say that $\mathscr{K}$ is* modally $\mathfrak{M}$-definable, *if $\mathscr{K} = \mathsf{Fr}_{\mathfrak{M}}(\mathsf{Th}(\mathscr{K}))$.*

3 Normal $\mathfrak{M}$-Modal Logics

In this section, we introduce normal metric temporal logics over $\mathfrak{M}$, or normal $\mathfrak{M}$-modal logics. As expected, the canonical method is applied in showing the completeness of the minimal normal $\mathfrak{M}$-modal logic.

Definition 5 (Normal Metric Temporal Logic). *A* normal $\mathfrak{M}$-modal logic *is a set of formulas $L \subseteq Fm(\mathfrak{M})$ such that L contains the following formulas:*

(Tau) *All instances of classical propositional tautologies.*
(K) $G_a(p \to q) \to (G_ap \to G_aq)$ *and* $H_a(p \to q) \to (H_ap \to H_aq)$.
(R) $p \to G_aP_ap$ *and* $p \to H_aF_ap$.
(T_0) $G_0p \to p$ *and* $H_0p \to p$.
(C) $G_ap \to G_bp$ *and* $H_ap \to H_bp$, *where* $a \leqslant b$ *in* $\mathfrak{M}$.
(A) $G_{a+b}p \to G_aG_bp$ *and* $H_{a+b}p \to H_aH_bp$, *where* $a, b, a + b \in \mathbb{M}$.

and L is closed under the following rules:

(MP) *if $\varphi, \varphi \to \psi \in L$, then $\psi \in L$.*
(Gen) *if $\varphi \in L$, then $G_a\varphi \in L$ and $H_a\varphi \in L$.*
(Sub) *if $\varphi \in L$, then $\varphi^s \in L$ for every substitution s.*

A formula φ is a theorem *of L, notation $\vdash_L \varphi$, if $\varphi \in L$.*

For every family of normal $\mathfrak{M}$-modal logics $\{L_i : i \in I\}$, $\bigcap_{i\in I} L_i$ is a normal $\mathfrak{M}$-modal logic. The *minimal* normal $\mathfrak{M}$-modal logic is denoted by $\mathsf{K}_{\mathfrak{M}}$. Let $\bigoplus_{i\in I} L_i$ be the smallest normal $\mathfrak{M}$-modal logic containing $\bigcup_{i\in I} L_i$. For every set of formulas Σ, let $\mathsf{K}_{\mathfrak{M}} \oplus \Sigma = \bigcap\{L : \Sigma \subseteq L\}$, the minimal normal $\mathfrak{M}$-modal logic containing Σ. If $\Sigma = \{\varphi\}$, we write $\mathsf{K}_{\mathfrak{M}} \oplus \varphi$ instead of $\mathsf{K}_{\mathfrak{M}} \oplus \{\varphi\}$. For every normal $\mathfrak{M}$-modal logic L, let $\mathsf{NExt}(L)$ be the set of all normal $\mathfrak{M}$-modal logics containing L.

Remark 1. Suppose $[a] \in \{G_a, H_a\}$ and $\langle a\rangle \in \{F_a, P_a\}$, the following hold for every normal $\mathfrak{M}$-modal logic L:

(1) $[a]\top \leftrightarrow \top \in L$ and $\langle a\rangle\bot \leftrightarrow \bot \in L$.
(2) $[a](\varphi_1 \wedge \ldots \wedge \varphi_n) \leftrightarrow ([a]\varphi_1 \wedge \ldots [a]\varphi_n) \in L$.
(3) $\langle a\rangle(\varphi_1 \vee \ldots \vee \varphi_n) \leftrightarrow (\langle a\rangle\varphi_1 \vee \ldots \langle a\rangle\varphi_n) \in L$.
(4) $G_a\varphi \wedge F_a\psi \to F_a(\varphi \wedge \psi) \in L$ and $H_a\varphi \wedge P_a\psi \to P_a(\varphi \wedge \psi) \in L$.
(5) if $\varphi \to \psi \in L$, then $[a]\varphi \to [a]\psi \in L$ and $\langle a\rangle\varphi \to \langle a\rangle\psi \in L$.

Let L be a normal $\mathfrak{M}$-modal logic. A formula φ is a *L-consequence* of a set of formulas Γ, notation $\Gamma \vdash_L \varphi$, if $\varphi \in L$ or there exist $\psi_1, \ldots, \psi_n \in \Gamma$ with $\psi_1 \wedge \ldots \wedge \psi_n \to \varphi \in L$. A set of formulas Γ is *L-consistent*, if $\Gamma \nvdash_L \bot$; and Γ is *maximal L-consistent*, if Γ is L-consistent and $\subseteq$-maximal. One obtains the deduction theorem and Lindenbaum-Tarski lemma for L: (i) $\Gamma, \varphi \vdash_L \psi$ if and only if $\Gamma \vdash_L \varphi \to \psi$; (ii) if Γ is L-consistent, there is a maximal L-consistent set Σ with $\Gamma \subseteq \Sigma$.

A normal $\mathfrak{M}$-modal logic L is *complete*, if $L = \mathsf{Th}(\mathsf{Fr}_{\mathfrak{M}}(L))$. One can obtain some completeness results using the canonical method.

Definition 6 (Canonical Model). *Let W^L be the set of all maximal L-consistent sets of formulas. For every $a \in \mathbb{M}$, one defines $R_a^L \subseteq W^L \times W^L$ as follows:*

$$\Sigma R_a^L \Theta \text{ if and only if } \varphi \in \Theta \text{for all} G_a\varphi \in \Sigma.$$

For every pair $\langle \Sigma, \Theta\rangle \in W^L \times W^L$, one defines $X_{\mathbb{M}}^L(\Sigma, \Theta) = \{a \in \mathbb{M} : \Sigma R_a^L \Theta\}$. The canonical $\mathfrak{M}$-model *for L is defined as $\mathcal{M}^L = (W^L, \sigma^L, V^L)$ where*

$$\sigma^L(\Sigma, \Theta) = \begin{cases} \bigvee X_{\mathbb{M}}^L(\Sigma, \Theta) & \text{if } X_{\mathbb{M}}^L(\Sigma, \Theta) \neq \varnothing. \\ \text{undefined} & \text{otherwise.} \end{cases}$$

and $V^L(p) = \{\Sigma \in W^L : p \in \Sigma\}$ for every $p \in \mathbb{P}$. The canonical $\mathfrak{M}$-frame *for L is defined as $\mathcal{F}^L = (W^L, \sigma^L)$.*

Lemma 1. *For every $\Sigma, \Theta \in W^L$ and $a \in \mathbb{M}$, the following hold:*

(1) $X^L_{\mathbb{M}}(\Sigma, \Theta)$ *is a downset in* $\mathfrak{M}$.
(2) *if* $X^L_{\mathbb{M}}(\Sigma, \Theta) \neq \varnothing$, *then* $X^L_{\mathbb{M}}(\Sigma, \Theta) = \downarrow \bigvee X_{\mathbb{M}}(\Sigma, \Theta)$.
(3) $a \in X^L_{\mathbb{M}}(\Sigma, \Theta)$ *if and only if* $\sigma^L(\Sigma, \Theta)! \geqslant a$.

Proof.(1) Assume $a \leqslant b$ and $b \in X^L_{\mathbb{M}}(\Sigma, \Theta)$. Then $\Sigma R^L_b \Theta$. Suppose $G_a\varphi \in \Sigma$. By (C), $G_a\varphi \to G_b\varphi \in L$. Hence $G_b\varphi \in \Sigma$. By $\Sigma R^L_b \Theta$, one obtains $\varphi \in \Theta$. It follows that $a \in X^L_{\mathbb{M}}(\Sigma, \Theta)$.
(2) Assume $X^L_{\mathbb{M}}(\Sigma, \Theta) \neq \varnothing$. Then $\bigvee X^L_{\mathbb{M}}(\Sigma, \Theta)$ is the maximal element of $X^L_{\mathbb{M}}(\Sigma, \Theta)$. By (1), $X^L_{\mathbb{M}}(\Sigma, \Theta) = \downarrow \bigvee X^L_{\mathbb{M}}(\Sigma, \Theta)$.
(3) Assume $a \in X^L_{\mathbb{M}}(\Sigma, \Theta)$. Then $\sigma^L(\Sigma, \Theta)! = \bigvee X^L_{\mathbb{M}}(\Sigma, \Theta) \geqslant a$. Assume $\sigma^L(\Sigma, \Theta)! \geqslant a$. Then $a \leqslant \bigvee X^L_{\mathbb{M}}(\Sigma, \Theta)$. By (2), $a \in X^L_{\mathbb{M}}(\Sigma, \Theta)$.

Lemma 2. *For every $\Sigma \in W^L$:*

(1) if $G_a\varphi \notin \Sigma$, there exists $\Theta \in W^L$ with $a \in X^L_{\mathbb{M}}(\Sigma, \Theta)$ and $\varphi \notin \Theta$.
(2) if $H_a\varphi \notin \Sigma$, there exists $\Theta \in W^L$ with $a \in X^L_{\mathbb{M}}(\Theta, \Sigma)$ and $\varphi \notin \Theta$.

Proof.(1) $G_a\varphi \notin \Sigma$. Let $\Gamma = \{\psi : G_a\psi \in \Sigma\} \cup \{\neg\varphi\}$. Assume that Γ is not L-consistent. Then $\Gamma \vdash_L \bot$. There exist $\psi_1, \ldots, \psi_n \in \Gamma$ with $(\psi_1 \wedge \ldots \wedge \psi_n) \to \varphi \in L$. By (Gen), (K) and (MP), $G_a(\psi_1 \wedge \ldots \wedge \psi_n) \to G_a\varphi \in L$. By $G_a(\psi_1 \wedge \ldots \wedge \psi_n) \leftrightarrow (G_a\psi_1 \wedge \ldots G_a\psi_n) \in L$, one obtains $G_a\varphi \in L$. Then $G_a\varphi \in \Sigma$, which contradicts the assumption. Hence Γ is L-consistent. Let $\Gamma \subseteq \Theta \in W^L$. Then $a \in X^L_{\mathbb{M}}(\Sigma, \Theta)$ and $\varphi \notin \Theta$.
(2) If $H_a\varphi \notin \Sigma$, let $\Gamma = \{\psi : H_a\psi \in \Sigma\} \cup \{\neg\varphi\}$. We can prove Γ is L-consistent as above. Let $\Gamma \subseteq \Theta \in W^L$, then we have $\varphi \notin \Theta$. Suppose $G_a\varphi \in \Theta$ and $\varphi \notin \Sigma$, then $\neg\varphi \in \Sigma$. By (R), $H_aF_a\neg\varphi \in \Sigma$ and $F_a\neg\varphi \in \Gamma \subseteq \Theta$, which contradicts the assumption. Hence, if $G_a\varphi \in \Theta$ then $\varphi \in \Sigma$, we have $a \in X^L_{\mathbb{M}}(\Theta, \Sigma)$.

Lemma 3. *Suppose L is a normal $\mathfrak{M}$-modal logic, then the canonical $\mathfrak{M}$-frame for L is a metric temporal frame.*

Proof. We check the canonical $\mathfrak{M}$-frame for L satisfies the conditions in the definition of metric temporal frame.

(1) Suppose $\Sigma \in W^L$, by (T_0), we have $0 \in X^L_{\mathbb{M}}(\Sigma, \Sigma)$, hence $\sigma^L(\Sigma, \Sigma)! \geqslant 0$.
(2) Suppose $\Sigma, \Theta, \Gamma \in W^L$ and $a, b, a+b \in \mathbb{M}$. If $\sigma^L(\Sigma, \Theta)! \geqslant a$ and $\sigma^L(\Theta, \Gamma)! \geqslant b$, we show $\sigma^L(\Sigma, \Gamma)! \geqslant a + b$. Assume $G_{a+b}\varphi \in \Sigma$, by (A), $G_aG_b\varphi \in \Sigma$, then $G_b\varphi \in \Theta$ and $\varphi \in \Gamma$ by Lemma 1. Hence $a + b \in X^L_{\mathbb{M}}(\Sigma, \Gamma)$, and $\sigma^L(\Sigma, \Gamma)! \geqslant a + b$.

Lemma 4. *For every $\Sigma \in W^L$, $\mathcal{M}^L, \Sigma \models \varphi$ if and only if $\varphi \in \Sigma$.*

Proof. The proof proceeds by induction on the complexity $\delta(\varphi)$. The atomic and Boolean cases are obvious. We check the modal cases:

(1) Let $\varphi = G_a\psi$. Assume $G_a\psi \in \Sigma$. Suppose $\sigma^L(\Sigma, \Theta)! \geqslant a$. By Lemma 1, $a \in X^L_{\mathbb{M}}(\Sigma, \Theta)$. Then $\Sigma R^L_a \Theta$. Hence $\psi \in \Theta$. By induction hypothesis, $\mathcal{M}^L, \Theta \models \psi$. Hence $\mathcal{M}^L, \Sigma \models G_a\psi$. Assume $G_a\psi \notin \Sigma$. By Lemma 2, there exists $\Theta \in W^L$ with $a \in X^L_{\mathbb{M}}(\Sigma, \Theta)$ and $\psi \notin \Theta$. Then $X^L_{\mathbb{M}}(\Sigma, \Theta) \neq \varnothing$ and $\sigma^L(\Sigma, \Theta)! = \bigvee X^L_{\mathbb{M}}(\Sigma, \Theta) \geqslant a$. By induction hypothesis, $\mathcal{M}^L, \Theta \not\models \psi$. Hence $\mathcal{M}^L, \Sigma \not\models G_a\psi$.

(2) Let $\varphi = H_a\psi$. Assume $H_a\psi \in \Sigma$. Suppose $\sigma^L(\Theta, \Sigma)! \geqslant a$. By Lemma 1, $a \in X^L_{\mathbb{M}}(\Theta, \Sigma)$. If $\psi \notin \Theta$, then $\neg\psi \in \Theta$, and by (R) we have $G_aP_a\neg\psi \in \Theta$. By $a \in X^L_{\mathbb{M}}(\Theta, \Sigma)$, $P_a\neg\psi \in \Sigma$, a contradiction. Hence $\psi \in \Theta$. By induction hypothesis, $\mathcal{M}^L, \Theta \models \psi$. Hence $\mathcal{M}^L, \Sigma \models G_a\psi$. Assume $H_a\psi \notin \Sigma$. By Lemma 2, there exists $\Theta \in W^L$ with $a \in X^L_{\mathbb{M}}(\Theta, \Sigma)$ and $\psi \notin \Theta$. Then $X^L_{\mathbb{M}}(\Theta, \Sigma) \neq \varnothing$ and $\sigma^L(\Theta, \Sigma)! = \bigvee X^L_{\mathbb{M}}(\Theta, \Sigma) \geqslant a$. By induction hypothesis, $\mathcal{M}^L, \Theta \not\models \psi$. Hence $\mathcal{M}^L, \Sigma \not\models H_a\psi$.

Theorem 1. $\mathsf{K}_{\mathfrak{M}}$ *is complete.*

Proof. Clearly $\mathsf{Fr}(\mathsf{K}_{\mathfrak{M}}) = \mathscr{F}_{\mathfrak{M}}$. Obviously $\mathsf{K}_{\mathfrak{M}} \subseteq \mathsf{Th}(\mathscr{F}_{\mathfrak{M}})$. Assume $\varphi \notin \mathsf{K}_{\mathfrak{M}}$. Then $\{\neg\varphi\}$ is $\mathsf{K}_{\mathfrak{M}}$-consistent. Let $\Sigma \in W^{\mathsf{K}_{\mathfrak{M}}}$ with $\neg\varphi \in \Sigma$. By Lemma 4, $\mathcal{M}^{\mathsf{K}_{\mathfrak{M}}}, \Sigma \not\models \varphi$. Hence $\varphi \notin \mathsf{Th}(\mathscr{F}_{\mathfrak{M}})$.

In the rest of this section, we make some observations on some extensions of $\mathsf{K}_{\mathfrak{M}}$. Consider the following formulas:

($\mathrm{T_a}$) $G_ap \to p$ and $H_ap \to p$.
($\mathrm{B_a}$) $p \to G_aF_ap$ and $p \to H_aP_ap$.
($\mathrm{4_a}$) $G_ap \to G_aG_ap$ and $H_ap \to H_aH_ap$.

We have $T_0 \in \mathsf{K}_{\mathfrak{M}}$. Suppose $\mathcal{F} = (W, \sigma)$ is a $\mathfrak{M}$-frame, $a \in \mathbb{M}$ and $a \neq 0$, clearly $\mathcal{F} \models T_a$ if and only if for any $w \in W$, $\sigma(w, w) \geqslant a$. According to our intended interpretation of measure function, this means the temporal distance of a moment and itself is not 0, which is not so consistent with our intuition of metric temporal systems. Moreover, if $a \in \mathbb{M}$ and $a \neq 0$, then $\mathcal{F} \models p \to G_aF_ap$ if and only if for any $w, u \in W$, $\sigma(w, u) \geqslant a$ implies $\sigma(u, w) \geqslant a$. This indicates that our metric temporal frame has a kind of temporal loop structure. Since $\mathsf{K}_{\mathfrak{M}}$ has $G_ap \to G_{a+a}p$ and $G_{a+a}p \to G_aG_ap$ as its axioms, (4_a) is a theorem of $\mathsf{K}_{\mathfrak{M}}$.

Recall our definition of metric temporal frames, if $\mathcal{F} = (W, \sigma)$ is a metric temporal frame, we demand that the measure function σ is a partial function. Let $\mathscr{F}^t_{\mathfrak{M}}$ be the class of all metric temporal frames over $\mathfrak{M}$ with a total measure function, i.e., if $\mathcal{F} = (W, \sigma) \in \mathscr{F}^t_{\mathfrak{M}}$, then for every pair (w, u) in $W \times W$, $\sigma(w, u)$ is defined. We will prove that this class is characterized by $\mathsf{K}_{\mathfrak{M}} \oplus B_0$.

Theorem 2. $\mathsf{K}_{\mathfrak{M}} \oplus B_0 = \mathsf{Th}(\mathscr{F}^t_{\mathfrak{M}})$.

Proof. Suppose $\mathcal{F} = (W, \sigma) \in \mathsf{Th}(\mathscr{F}^t_{\mathfrak{M}})$, then $\sigma(w, u) \geqslant 0$ for all $w, u \in W$. It follows that $\mathcal{F} \models p \to G_0F_0p$ and $\mathcal{F} \models p \to H_0P_0p$. Therefore $\mathsf{K}_{\mathfrak{M}} \oplus B_0 \subseteq \mathsf{Th}(\mathscr{F}^t_{\mathfrak{M}})$. For the completeness, let $\mathcal{M} = (W, \sigma, V)$ be the canonical model for $\mathsf{K}_{\mathfrak{M}} \oplus B_0$. Note that $0 \in X_{\mathbb{M}}(\Sigma, \Theta)$ and so $\sigma(\Sigma, \Theta) = \bigvee X_{\mathbb{M}}(\Sigma, \Theta)$ which is a total function. If $\varphi \notin \mathsf{K}_{\mathfrak{M}} \oplus B_0$, then $\mathcal{M} \not\models \varphi$. Hence $\mathsf{Th}(\mathscr{F}^t_{\mathfrak{M}}) \subseteq \mathsf{K}_{\mathfrak{M}} \oplus B_0$.

4 Finite Model Property

In this section, we prove for a temporal metric $\mathfrak{M}$, the minimal normal $\mathfrak{M}$-modal logic $\mathsf{K}_{\mathfrak{M}}$ has the finite model property with respect to $\mathscr{M}_{\mathfrak{M}}$, the class of all temporal metric models over $\mathfrak{M}$.

Definition 7 (Metric Temporal Subformula Closed). *A set of $\mathcal{L}_{MT}(\mathfrak{M})$ formulas Σ is* closed under metric temporal subformulas *('*subformula closed*' for short) if for all formulas φ and ψ and $a, b, c, a+b \in \mathbb{M}$:*

(1) if $(\varphi \to \psi) \in \Sigma$, then so are φ and ψ.
(2) if $G_a\varphi \in \Sigma$ or $H_a\varphi \in \Sigma$, then so is φ.
(3) if $G_a\varphi \in \Sigma$ or $H_a\varphi \in \Sigma$, and $a \leqslant c$, then so is $G_c\varphi$ or $H_c\varphi$ respectively.
(4) if $G_{a+b}\varphi \in \Sigma$ or $H_{a+b}\varphi \in \Sigma$, then $G_aG_b\varphi, G_b\varphi \in \Sigma$ or $H_aH_b\varphi, H_b\varphi \in \Sigma$ respectively.

We say φ and ψ are subformulas of $(\varphi \to \psi)$, φ is a subformula of $G_a\varphi$ or $H_a\varphi$. And (1) if $a \leqslant c$, $G_c\varphi$ and $H_c\varphi$ are metric temporal subformulas *(or simply 'subformulas') of $G_a\varphi$ and $H_a\varphi$ respectively. (2) $G_aG_b\varphi, G_b\varphi$ and $H_aH_b\varphi, H_b\varphi$ are* metric temporal subformulas *(or simply 'subformulas') of $G_{a+b}\varphi$ and $H_{a+b}\varphi$ respectively.*

Definition 8 (Filtration). *Let $\mathcal{M} = (W, \sigma, V)$ be a $\mathfrak{M}$-model and Σ a subformula closed set of $\mathcal{L}_{MT}(\mathfrak{M})$ formulas. Let $\leftrightsquigarrow_\Sigma$ be the equivalence relation on the states of $\mathcal{M}$ defined by: $w \leftrightsquigarrow_\Sigma u$ if and only if for all $\varphi \in \Sigma$, $\mathcal{M}, w \models \varphi$ if and only if $\mathcal{M}, u \models \varphi$. We denote the equivalence class of a state w of $\mathcal{M}$ with respect to $\leftrightsquigarrow_\Sigma$ by $[w]_\Sigma$, or simply by $[w]$ if no confusion will arise. Then a* filtration of $\mathcal{M}$ through Σ *is any $\mathfrak{M}$-model $\mathcal{M}^f_\Sigma = (W^f, \sigma^f, V^f)$ such that:*

(1) $W^f = \{[w] : w \in W\}$.
(2) if $\sigma(w, u) \geqslant a$, then $\sigma^f([w], [u]) \geqslant a$.
(3) if $\sigma^f([w], [u]) \geqslant a$, then for all $G_a\varphi, H_a\varphi \in \Sigma$: if $\mathcal{M}, w \models G_a\varphi$ then $\mathcal{M}, u \models \varphi$, and if $\mathcal{M}, u \models H_a\varphi$ then $\mathcal{M}, w \models \varphi$.
(4) $V^f(p) = \{[w] : \mathcal{M}, w \models p\}$, for all propositional variables $p \in \Sigma$.

We will often write $\mathcal{M}^f$ instead of $\mathcal{M}^f_\Sigma$ if there is no confusion.

Theorem 3. *Let $\mathcal{M} = (W, \sigma, V)$ be a $\mathfrak{M}$-model and Σ a subformula closed set of $\mathcal{L}_{MT}(\mathfrak{M})$ formulas. Suppose $\mathcal{M}^f = (W^f, \sigma^f, V^f)$ is a filtration of $\mathcal{M}$ through Σ, then for all formulas $\varphi \in \Sigma$ and all states $w \in W$, $\mathcal{M}, w \models \varphi$ if and only if $\mathcal{M}^f, [w] \models \varphi$.*

Proof. The proof proceeds by induction on φ. The base case is from the definition of V^f. The boolean cases follows from the fact that Σ is closed under subfomulas, this allows us to apply the inductive hypothesis.

(1) Let $\varphi = G_a\psi \in \Sigma$. Assume $\mathcal{M}, w \models G_a\psi$ and $\sigma^f([w], [u]) \geqslant a$, by the definition of filtration, we have $\mathcal{M}, u \models \psi$. As Σ is subformula closed, $\psi \in \Sigma$, therefore by the inductive hypothesis, $\mathcal{M}^f, [u] \models \psi$. Hence $\mathcal{M}^f, [w] \models G_a\psi$.

Conversely, suppose $\mathcal{M}^f, [w] \models G_a\psi$ and $\sigma(w,u) \geqslant a$, by the definition of filtration, we have $\sigma^f([w],[u]) \geqslant a$. Thus $\mathcal{M}^f, [u] \models \psi$. As Σ is subformula closed, $\psi \in \Sigma$, therefore by the inductive hypothesis, $\mathcal{M}, u \models \psi$. Hence $\mathcal{M}, w \models G_a\psi$.

(2) Let $\varphi = H_a\psi \in \Sigma$. Assume $\mathcal{M}, w \models H_a\psi$ and $\sigma^f([u],[w]) \geqslant a$, by the definition of filtration, we have $\mathcal{M}, u \models \psi$. As Σ is subformula closed, $\psi \in \Sigma$, therefore by the inductive hypothesis, $\mathcal{M}^f, [u] \models \psi$. Hence $\mathcal{M}^f, [w] \models H_a\psi$. Conversely, suppose $\mathcal{M}^f, [w] \models H_a\psi$ and $\sigma(u,w) \geqslant a$, by the definition of filtration, we have $\sigma^f([u],[w]) \geqslant a$. Thus $\mathcal{M}^f, [u] \models \psi$. As Σ is subformula closed, $\psi \in \Sigma$, therefore by the inductive hypothesis, $\mathcal{M}, u \models \psi$. Hence $\mathcal{M}, w \models H_a\psi$.

The next lemma is clear, we just state it below and the proof is omitted.

Lemma 5. *Let Σ be a finite subformula closed set of $\mathcal{L}_{MT}(\mathfrak{M})$ formulas. For any $\mathfrak{M}$-model $\mathcal{M} = (W, \sigma, V)$, if $\mathcal{M}^f = (W^f, \sigma^f, V^f)$ is a filtration of $\mathcal{M}$ through Σ, then $|W^f|$ is finite.*

Definition 9 (Metric Temporal Filtration). *Let $\mathcal{M} = (W, \sigma, V)$ be a $\mathfrak{M}$-model and Σ a subformula closed set of $\mathcal{L}_{MT}(\mathfrak{M})$ formulas. Let the equivalence relation $\leftrightsquigarrow_\Sigma$ and its equivalence classes be defined as before. Suppose $w, u \in W$, we define:*

$$X^G_{\mathbb{M}}(w,u) = \{a \in \mathbb{M} : \textit{for all } G_a\varphi \in \Sigma,\ \textit{if } \mathcal{M}, w \models G_a\varphi \textit{ then } \mathcal{M}, u \models \varphi\}$$

$$X^H_{\mathbb{M}}(w,u) = \{a \in \mathbb{M} : \textit{for all } H_a\varphi \in \Sigma,\ \textit{if } \mathcal{M}, u \models H_a\varphi \textit{ then } \mathcal{M}, w \models \varphi\}$$

Let $X^\Sigma_{\mathbb{M}}(w,u) = X^G_{\mathbb{M}}(w,u) \cap X^H_{\mathbb{M}}(w,u)$. The metric temporal filtration of $\mathcal{M}$ *through Σ is the structure $\mathcal{M}^t_\Sigma = (W^f, \sigma^t, V^f)$ where W^f and V^f are defined as before, and*

$$\sigma^t([w],[u]) = \begin{cases} \bigvee X^\Sigma_{\mathbb{M}}(w,u) & \textit{if } X^\Sigma_{\mathbb{M}}(w,u) \neq \varnothing. \\ \textit{undefined} & \textit{otherwise.} \end{cases}$$

We will often write $\mathcal{M}^t$ instead of $\mathcal{M}^t_\Sigma$ if there is no confusion.

Remark 2. We need to check this definition is well defined. Suppose $w \leftrightsquigarrow_\Sigma w'$ and $u \leftrightsquigarrow_\Sigma u'$, we prove $X^\Sigma_{\mathbb{M}}(w,u) = X^\Sigma_{\mathbb{M}}(w',u')$. If $a \in X^\Sigma_{\mathbb{M}}(w,u)$, then for all $G_a\varphi \in \Sigma$ and $H_a\varphi \in \Sigma$, we have if $\mathcal{M}, w \models G_a\varphi$ then $\mathcal{M}, u \models \varphi$ and if $\mathcal{M}, u \models H_a\varphi$ then $\mathcal{M}, w \models \varphi$. By the definition of $\leftrightsquigarrow_\Sigma$, and $\varphi \in \Sigma$ as Σ is closed under subformulas, we have if $\mathcal{M}, w' \models G_a\varphi$ then $\mathcal{M}, u' \models \varphi$ and if $\mathcal{M}, u' \models H_a\varphi$ then $\mathcal{M}, w' \models \varphi$. Therefore $a \in X^\Sigma_{\mathbb{M}}(w',u')$. Conversely, if $a \in X^\Sigma_{\mathbb{M}}(w',u')$, then for all $G_a\varphi \in \Sigma$ and $H_a\varphi \in \Sigma$, we have if $\mathcal{M}, w' \models G_a\varphi$ then $\mathcal{M}, u' \models \varphi$ and if $\mathcal{M}, u' \models H_a\varphi$ then $\mathcal{M}, w' \models \varphi$. By the definition of $\leftrightsquigarrow_\Sigma$, and $\varphi \in \Sigma$ as Σ is closed under subformulas, we have if $\mathcal{M}, w \models G_a\varphi$ then $\mathcal{M}, u \models \varphi$ and if $\mathcal{M}, u \models H_a\varphi$ then $\mathcal{M}, w \models \varphi$. Hence $a \in X^\Sigma_{\mathbb{M}}(w,u)$.

Lemma 6. *Let $\mathcal{M} = (W, \sigma, V)$ be a $\mathfrak{M}$-model and Σ a subformula closed set of $\mathcal{L}_{MT}(\mathfrak{M})$ formulas. For every $w, u \in W$ and $a \in \mathbb{M}$, the following hold:*

(1) *$X^{\Sigma}_{\mathbb{M}}(w, u)$ is a downset in $\mathfrak{M}$.*
(2) *if $X^{\Sigma}_{\mathbb{M}}(w, u) \neq \varnothing$, then $X^{\Sigma}_{\mathbb{M}}(w, u) = \downarrow \bigvee X^{\Sigma}_{\mathbb{M}}(w, u)$.*
(3) *$a \in X^{\Sigma}_{\mathbb{M}}(w, u)$ if and only if $\sigma^t([w], [u])! \geqslant a$.*

Proof.(1) Assume $a \leqslant b$ and $b \in X^{\Sigma}_{\mathbb{M}}(w, u)$. Then for all $G_b\varphi \in \Sigma$ and $H_b\varphi \in \Sigma$, we have if $\mathcal{M}, w \models G_b\varphi$ then $\mathcal{M}, u \models \varphi$ and if $\mathcal{M}, u \models H_b\varphi$ then $\mathcal{M}, w \models \varphi$. Suppose $G_a\varphi \in \Sigma$ and $\mathcal{M}, w \models G_a\varphi$, then we have $G_b\varphi \in \Sigma$ as Σ is metric temporal subformula closed and $a \leqslant b$. And by axiom (C) $G_a\varphi \to G_b\varphi$, we have $\mathcal{M}, w \models G_b\varphi$, and $\mathcal{M}, u \models \varphi$ as $b \in X^{\Sigma}_{\mathbb{M}}(w, u)$ and $G_b\varphi \in \Sigma$. Hence $a \in X^{G}_{\mathbb{M}}(w, u)$. We can prove $a \in X^{H}_{\mathbb{M}}(w, u)$ in the same way. It follows that $a \in X^{\Sigma}_{\mathbb{M}}(w, u)$.

(2) Assume $X^{\Sigma}_{\mathbb{M}}(w, u) \neq \varnothing$. Then $\bigvee X^{\Sigma}_{\mathbb{M}}(w, u)$ is the maximal element of $X^{\Sigma}_{\mathbb{M}}(w, u)$. By (1), $X^{\Sigma}_{\mathbb{M}}(w, u) = \downarrow \bigvee X^{\Sigma}_{\mathbb{M}}(w, u)$.

(3) Assume $a \in X^{\Sigma}_{\mathbb{M}}(w, u)$. Then $\sigma^t([w], [u])! = \bigvee X^{\Sigma}_{\mathbb{M}}(w, u) \geqslant a$. Assume $\sigma^t([w], [u])! \geqslant a$. Then $a \leqslant \bigvee X^{\Sigma}_{\mathbb{M}}(w, u)$. By (2), $a \in X^{\Sigma}_{\mathbb{M}}(w, u)$.

Lemma 7. *Let $\mathcal{M} = (W, \sigma, V)$ be a $\mathfrak{M}$-model and Σ a subformula closed set of $\mathcal{L}_{MT}(\mathfrak{M})$ formulas. Then $\mathcal{M}^t = (W^f, \sigma^t, V^f)$, the metric temporal filtration of $\mathcal{M}$ through Σ is a filtration and a metric temporal model over $\mathfrak{M}$.*

Proof. First we prove $\mathcal{M}^t$ is a filtration:

(1) Suppose $\sigma(w, u) \geqslant a$, we prove $a \in X^{\Sigma}_{\mathbb{M}}(w, u)$, by Lemma 6 this implies $\sigma^t([w], [u]) \geqslant a$. Assume $G_a\varphi \in \Sigma$ and $\mathcal{M}, w \models G_a\varphi$, then $\mathcal{M}, u \models \varphi$ as $\sigma(w, u) \geqslant a$. Hence $a \in X^{G}_{\mathbb{M}}(w, u)$. Suppose $H_a\varphi \in \Sigma$ and $\mathcal{M}, u \models H_a\varphi$, then $\mathcal{M}, w \models \varphi$ as $\sigma(w, u) \geqslant a$. Hence $a \in X^{H}_{\mathbb{M}}(w, u)$. It follows that $a \in X^{\Sigma}_{\mathbb{M}}(w, u)$.

(2) Suppose $\sigma^t([w], [u]) \geqslant a$, by Lemma 6, we have $a \in X^{\Sigma}_{\mathbb{M}}(w, u)$. Thus for all $G_a\varphi, H_a\varphi \in \Sigma$: if $\mathcal{M}, w \models G_a\varphi$ then $\mathcal{M}, u \models \varphi$, and if $\mathcal{M}, u \models H_a\varphi$ then $\mathcal{M}, w \models \varphi$.

Next we show $\mathcal{M}^t$ is a metric temporal model over $\mathfrak{M}$:

(1) Suppose $w \in W$ and $G_0\varphi \in \Sigma$. If $\mathcal{M}, w \models G_0\varphi$, then by axiom (T_0) $G_0\varphi \to \varphi$, we have $\mathcal{M}, w \models \varphi$. Hence $0 \in X^{G}_{\mathbb{M}}(w, w)$. Assume $H_0\varphi \in \Sigma$. If $\mathcal{M}, w \models H_0\varphi$, then by axiom (T_0), we have $\mathcal{M}, w \models \varphi$. Hence $0 \in X^{H}_{\mathbb{M}}(w, w)$. It follows that $0 \in X^{\Sigma}_{\mathbb{M}}(w, w)$. By Lemma 6, we have $\sigma^t([w], [w])! \geqslant 0$.

(2) Suppose $w, u, v \in W$, $\sigma^t([w], [u])! \geqslant a$ and $\sigma^t([u], [v])! \geqslant b$. By Lemma 6 we have $a \in X^{\Sigma}_{\mathbb{M}}(w, u)$ and $b \in X^{\Sigma}_{\mathbb{M}}(u, v)$. Assume $G_{a+b}\varphi \in \Sigma$ and $\mathcal{M}, w \models G_{a+b}\varphi$, then $G_aG_b\varphi, G_b\varphi \in \Sigma$ as Σ is metric temporal subformula closed. And by axiom (A) $G_{a+b}\varphi \to G_aG_b\varphi$, we have $\mathcal{M}, w \models G_aG_b\varphi$. By $a \in X^{\Sigma}_{\mathbb{M}}(w, u)$ we have for all $G_a\varphi \in \Sigma$, if $\mathcal{M}, w \models G_a\varphi$ then $\mathcal{M}, u \models \varphi$, thus $\mathcal{M}, u \models G_b\varphi$. By $b \in X^{\Sigma}_{\mathbb{M}}(u, v)$ we have for all $G_b\varphi \in \Sigma$, if $\mathcal{M}, u \models G_b\varphi$ then $\mathcal{M}, v \models \varphi$, thus $\mathcal{M}, v \models \varphi$. Hence $a + b \in X^{G}_{\mathbb{M}}(w, v)$. We can prove $a + b \in X^{H}_{\mathbb{M}}(w, v)$ in the same way. It follows that $a + b \in X^{\Sigma}_{\mathbb{M}}(w, v)$. By Lemma 6, we have $\sigma^t([w], [v])! \geqslant a + b$.

Theorem 4. *Suppose $\mathfrak{M}$ is a temporal metric, then $\mathsf{K}_{\mathfrak{M}}$, the minimal normal $\mathfrak{M}$-modal logic has the finite model property with respect to $\mathscr{M}_{\mathfrak{M}}$, the class of all temporal metric models over $\mathfrak{M}$.*

Proof. Suppose φ is a $\mathcal{L}_{MT}(\mathfrak{M})$ formula, take Σ to be the set of metric temporal subformulas of φ. Suppose $\mathcal{M}$ is a $\mathfrak{M}$-model, let $\mathcal{M}^t$ be the metric temporal filtration of $\mathcal{M}$ through Σ. By Lemma 7, $\mathcal{M}^t$ is a filtration and a $\mathfrak{M}$-model. Since $\mathbb{M}$ is finite, clearly Σ is finite. By Lemma 5, $\mathcal{M}^t$ is a finite $\mathfrak{M}$-model. And by Theorem 3, we have $\mathcal{M}, w \models \varphi$ if and only if $\mathcal{M}^t, [w] \models \varphi$.

Corollary 1. $\mathsf{K}_{\mathfrak{M}}$ *is decidable.*

5 Concluding Remarks

In the present work, we contribute multiple-valued semantics for metric temporal logic. We have introduced metric temporal frames based on multiple-valued frames, and take them as the semantic ontology to interpret the metric temporal operators. In the study of normal metric temporal logics, we adjust the canonical model method and obtain some completeness results. We also obtain some finite model property results for normal metric temporal logics. There are many problems that are interesting for further exploration. Here we mention two of them:

(1) Recall our definition of satisfaction relation, we have:

$\mathcal{M}, w \models G_a\varphi$ if and only if $\mathcal{M}, u \models \varphi$ for all u such that $\sigma(w, u)! \geqslant a$.
$\mathcal{M}, w \models H_a\varphi$ if and only if $\mathcal{M}, u \models \varphi$ for all u such that $\sigma(u, w)! \geqslant a$.

There are other kinds of metric temporal operators, let's denote them by $G_a^=, H_a^=$ and $G_a^\leqslant, H_a^\leqslant$. Their satisfaction relation could be defined by:

$\mathcal{M}, w \models G_a^=\varphi$ if and only if $\mathcal{M}, u \models \varphi$ for all u such that $\sigma(w, u)! = a$.
$\mathcal{M}, w \models H_a^=\varphi$ if and only if $\mathcal{M}, u \models \varphi$ for all u such that $\sigma(u, w)! = a$.
$\mathcal{M}, w \models G_a^\leqslant\varphi$ if and only if $\mathcal{M}, u \models \varphi$ for all u such that $\sigma(w, u)! \leqslant a$.
$\mathcal{M}, w \models H_a^\leqslant\varphi$ if and only if $\mathcal{M}, u \models \varphi$ for all u such that $\sigma(u, w)! \leqslant a$.

One could study the systems of these kinds of metric temporal operators and the system combine these operators together.

(2) In our definition of metric temporal frames, the conditions to be satisfied by the measure function are weakened for technical reasons. One could study the metric temporal frames with unweakened conditions, i.e., the frames with measure function σ satisfies:

(a) $\sigma(w, w) = 0$ for every state w in the frame;
(b) if $\sigma(w, u) = a$ and $\sigma(u, v) = b$, then $\sigma(w, v) = a + b$.

We wish these problems should be solved in further investigations, and a fully developed account of metric temporal frames based on multiple-valued semantics may lead to fruitful theories about metric temporal reasoning, which may find applications in philosophical analysis of some problems involving metric temporal operators.

References

1. Blackburn, P., van Benthem, J.: Modal logic: a semantic perspective. In: Blvbackburn, P., van Benthem, J., Wolter, F. (eds.) Handbook of Modal Logic, pp. 1–84. Elsevier, Amsterdam (2007)
2. Blackburn, P., Rijke, M.D., Venema, Y.: Modal Logic. Cambridge University Press, Cambridge (2001)
3. Burgess, J.P.: Basic tense logic. In: Gabbay, D.M., Guenthner, F. (eds.) Handbook of Philosophical Logic, vol. 7, pp. 1–42. Springer, Dordrecht (2002). https://doi.org/10.1007/978-94-017-0462-5_1
4. Chagrov, A., Zakharyaschev, M.: Modal Logic. Clarendon Press, Oxford (1997)
5. He, F.: Modal logic of multivalued frames over inversely well-ordered sets. Stud. Log. **15**(3), 52–72 (2022)
6. Koymans, R.: Specifying real-time properties with metric temporal logic. Real-Time Syst. **2**(4), 255–299 (1990). https://doi.org/10.1007/BF01995674
7. Ouaknine, J., Worrell, J.: Some recent results in metric temporal logic. In: Cassez, F., Jard, C. (eds.) FORMATS 2008. LNCS, vol. 5215, pp. 1–13. Springer, Heidelberg (2008). https://doi.org/10.1007/978-3-540-85778-5_1
8. Prior, A.N.: Time and Modality. Clarendon Press, Oxford (1957)
9. Prior, A.N.: Past, Present and Future. Clarendon Press, Oxford (1967)
10. Prior, A.N., Hasle, P.F.: Papers on Time and Tense. Clarendon Press, Oxford (2003)

Segment Transit Function of the Induced Path Function of Graphs and Its First-Order Definability

Jeny Jacob and Manoj Changat(✉)

Department of Futures Studies, University of Kerala, Trivandrum 695581, India
jenyjacobktr@gmail.com, mchangat@keralauniversity.ac.in

Abstract. The first-order definability of the shortest path function (I) as well as the first-order non-definability of the induced path function (J) of graphs is established by Nebeský. Inspired by these results, we try to investigate the first-order logic axiomatisation of the segment transit functions associated with the induced path ($\hat{J}$) in graphs and obtained that $\hat{J}$ does not possess first-order axiomatisation.

Keywords: Segment transit function · Induced path function · First order definability

1 Introduction

First-order logic is a natural object of study, it is semantically complete and is adequate to the axiomatisation of all ordinary mathematics. Further Lindström's theorem shows that it is the maximal logic satisfying the compactness and Löwenheim-Skolem properties [12]. So it is not surprising that first-order logic has long been regarded as the "right" logic for investigations into the foundations of mathematics.

In this paper, we consider only a connected finite and simple graph, denoted as $G = (V, E)$ with V (denoted some times as $V(G)$) being the vertex set and E (denoted some times as $(E(G))$, the edge set of G. A u, v*-path* is a sequence of distinct vertices $u = u_1, u_2, \cdots, u_n = v$ in G where $u_i u_j$ is an edge of G whenever $|i-j| = 1$. If P is a path then length of P is the number of edges in P. A u, v-path is called a u, v*-shortest path*, if P is a path of minimum length. The *distance* between two vertices u and v of a graph G is the length of a shortest u, v-path and is denoted by $d_G(u, v)$ or $d(u, v)$.

The *interval function* $I : V \times V \to 2^V$ of a graph G is defined as: $I(u, v) = \{z \in V : z$ lies on some shortest u, v-path in $G\}$, is an important tool in studying distance properties in graphs and is a part of folklore in metric graph theory [14]. In [16,17], Ladislav Nebeský gave an interesting turn in the axiomatic study in graphs by characterising the interval function of a connected graph $G = (V, E)$, using a set of simple first-order (FO) axioms defined on an arbitrary function R defined on V. This function is later termed as a *transit function* (denoted as R)

M. Banerjee and A. V. Sreejith (Eds.): ICLA 2023, LNCS 13963, pp. 117–129, 2023.
https://doi.org/10.1007/978-3-031-26689-8_9

by Mulder in [13], defined as the function $R : V \times V \to 2^V$ satisfying the three transit axioms: $(t1)$ $\forall x \, \forall y \, (x \in R(x, y))$; $(t2)$ $\forall x \, \forall y \, (R(x, y) = R(y, x))$; and $(t3)$ $\forall x \, (R(x, x) = \{x\})$. This first order axiomatisation of the function I is further refined in [15] and extended to arbitrary graphs in [6].

Moreover, the transit function is used to generalise the notion of betweenness, intervals and convex sets, present in several areas in mathematics. It is not difficult to see that, corresponding to any function $R : V \times V \to 2^V$ there is a ternary relation $T \subseteq V \times V \times V$ and vice versa. Thus any axiom defined on R can be associated with a corresponding axiom in the ternary relation T and hence the axioms (t1), (t2) and (t3) can be translated into corresponding axioms on T and vice versa.

Motivated by the study of the interval function, other functions defined by natural path properties and betweenness is studied in graphs. An immediate generalisation of a shortest path is the *induced path*; a u, v-path, say $P := u = v_1, v_2, ..., v_k = v$ in G is an induced u, v-path if there is no edge in G joining non-consecutive vertices of P; that is, $v_i v_j$ is not an edge in G with $|j - i| > 1$. The corresponding *induced path function* J *(or monophonic interval)* is defined as $J(u, v) = \{z \in V : z \text{ lies on an induced } u, v\text{-path}\}$. Similarly, if we consider all paths between u and v instead of shortest or induced paths, we get the so called *all-paths transit function*, defined as $A(u, v) = \{z \in V : z \text{ lies on a } u, v\text{-path}\}$. The functions I, J, A and the associated betweenness and convexities, known as respectively, the shortest path betwenness and the geodesic convexity, the induced path betwenness and induced path convexity(monophonic convexity), all-paths betweenness and all paths convexity, are well studied in graphs, for e.g., see [4,5,19], also the survey [7], and references therein.

Given a transit function R on V. A subset X of V is *R-convex*, if $R(x, y) \subseteq X$, for all $x, y \in X$. The family of R-convex sets in V is called the *R-convexity* on V. The *R-closure* $R(X)$ of a subset X of V is given by $R(X) = \bigcup_{u,v \in X} R(u, v)$. The smallest R-convex set containing X is denoted by $\langle X \rangle_R$ and is called the *R-convex hull* of X. $\langle X \rangle_R$ is also defined recursively as follows: $R^1(X) = R(R(X))$ and $R^k(X) = R(R^{k-1}(X))$. For the least k satisfying $R^k(X) = R^{k+1}(X)$, we say that $\langle X \rangle_R = R^k(X)$.

Given a transit function R, one can define another transit function, named as the *segment transit function* associated with R, as $\hat{R} : V \times V \to 2^V$ defined by $\hat{R}(u, v) = \langle \{u, v\} \rangle_R$, for $u, v \in V$.

In 2000, Nebeský [18] gave another interesting result that the induced path function of a connected graph is not FO definable. In 2010, Changat et.al [5] gave special cases involving forbidden induced subgraphs, in which J can be characterised by transit axioms. It may be noted that the all paths transit function $A(u, v)$ also possess a first order axiomatic characterisation, see [3]. It is also interesting to observe that for the transit function A, $A(u, v)$ is always A-convex and hence both the transit function A and the segment transit function $\hat{A}$ are the same which implies that the function $\hat{A}$ is also first order axiomatisable.

In this paper, we consider the induced path function J of a connected graph and its corresponding segment transit function. In 1984, Duchet [9] and in 2010,

Dourado et all [8] gave elegant characterisations of $\hat{J}$. It is proved in [8] that the time complexity of determining the J-convex hull of a set of vertices of connected graph is polynomial ($O(n^2m)$), while that of computing the interval $J(u, v)$ is NP-complete. Motivated from these results and the work of Nebeský on the non-FO definability of the function J, we attempt to study the feasibility of first order axiomatic characterisation of $\hat{J}$.

2 Preliminaries

In this section, we fix the preliminary concepts, notations and terminologies that we follow in this paper. Given a transit function R, one can define a sequence of transit functions R^1, R^2, $\cdots$ so that for the least k satisfying $R^k(u, v) = R^{k+1}(u, v)$, we get $\hat{R}(u, v) = R^k(u, v)$. We define $R^0 = R$ and $R^k(u, v) = \{z : z \in R(x, y), \text{ where } x, y \in R^{k-1}(u, v)\}$.

If for a transit function R, all sets $R(x, y)$ are convex, then it can be easily seen that $R = \hat{R}$. We can define a simple first order axiom to specify the convexity of $R(x, y)$ known as *monotone axiom* on R denoted as (m). A transit function R satisfies axiom (m), if

$$x, y \in R(u, v) \Rightarrow R(x, y) \subseteq R(u, v), \text{ for every } x, y, u, v \in V.$$

Thus for a transit function R satisfying the axiom (m), R and $\hat{R}$ coincides. In other words, a graph with convex intervals satisfies the axiom (m). Generally, it might be noted that the convexity defined by R and $\hat{R}$ are the same since convex hull of a convex set is the set itself and the convex hull of a set is basically a convex set.

Two natural betweenness axioms of a transit function R are the following.

$(b1)$ $x \in R(u, v), x \neq v \Rightarrow v \notin R(u, x)$
$(b2)$ $x \in R(u, v) \Rightarrow R(u, x) \subseteq R(u, v)$.

It was shown in [2] that the function J of a connected graph G need not satisfy these axioms and further proved that J satisfies the axioms $(b1)$ and $(b2)$ if and only if G is HHD-free (that is, G has no house, hole or domino as induced subgraphs). It may be observed that axiom (m) is a stronger axiom than $(b2)$ as the axiom (m) is a special case of axiom $(b2)$. If we compare the function J and $\hat{J}$ of an arbitrary connected graph G with respect to the betweeness properties, we observe that both J and $\hat{J}$ satisfies the simple betweenness axiom $(j2)$, defined on a transit function R as $R(u, x) = \{u, x\}, R(x, v) = \{x, v\}, u \neq v$, and $R(u, v) \neq \{u, v\}$ implies $x \in R(u, v)$. It is already observed that $\hat{J}$ satisfies the monotone axiom (m), but J need not satisfy (m) and as noted above, J even need not satisfy the weaker axiom $(b2)$. Thus the function $\hat{J}$ possesses stronger betweenness axioms than the function J. We prove that, still the function $\hat{J}$ doesn't possess a first order axiomatisation.

We use the method of Ehrenfeucht Fraïssé game [10,12] (EF game) to show the inexpressibility of $\hat{J}$ whereas Nebeský uses the back and forth condition in [18]. EF game is one of the main tools in proving that a query is not definable in first order logic. This game is called "Ehernfeucht Fraïssé Game", in honor of their developers Ehrenfeucht and Fraïssé. Some of the other tools used to check the inexpressibility of queries in first-order logic are zero-one law, Hanf-locality, Gaifman-locality, etc. [12]. The following is a brief outline of the method of Ehrenfeucht Fraïssé game:

The tuple $\mathbf{X} = (X, \mathcal{S})$ is called a *structure* when X is a nonempty set called *universe* and $\mathcal{S}$ is a finite set of function symbols, relation symbols and constant symbols called *signature*. Here we assume that the signature contains only relation symbols. The *quantifier rank* of a formula ϕ is its depth of quantifier nesting and is denoted by $qr(\phi)$. Let $\mathbf{A}$ and $\mathbf{B}$ be two structures with same signatures. A map q is said to be a *partial isomorphism* from $\mathbf{A}$ to $\mathbf{B}$ if and only if $dom(q) \subset A$, $rg(q) \subset B$, q is injective and for any n-ary relation R in the signature and a_0, $\ldots$, $a_{l-1} \in dom(q)$, $R^{\mathcal{A}}(a_0, \ldots, a_{l-1})$ iff $R^{\mathcal{B}}(q(a_0), \ldots, q(a_{l-1}))$.

Let r be a positive integer. The *r-move Ehrenfeucht-Fraisse Game* on $\mathbf{A}$ and $\mathbf{B}$ is played between 2 players called the *Spoiler* and the *Duplicator*, according to the following rules:

Each run of the game has r moves. In each move, the Spoiler plays first and picks an element from the universe A of the structure $\mathbf{A}$ or from the universe B of the structure $\mathbf{B}$; the Duplicator then responds by picking an element from the universe of the other structure (that is if the Spoiler has picked an element from B, then the duplicator picks an element from A and vice versa).

Let $a_i \in A$ and $b_i \in B$ be the two elements picked by the Spoiler and the Duplicator in their ith move, $1 \leq i \leq r$. The Duplicator wins the run $(a_1, b_1), ..., (a_r, b_r)$ if the mapping $a_i \rightarrow b_i$, where $1 \leq i \leq r$ is a partial isomorphism from the structure $\mathbf{A}$ to $\mathbf{B}$. Otherwise the Spoiler wins the run $(a_1, b_1), ..., (a_r, b_r)$.

The *Duplicator wins the r-move EF-game on $\mathbf{A}$ and $\mathbf{B}$* or *the Duplicator has a winning strategy for the EF-game on $\mathbf{A}$ and $\mathbf{B}$* if the duplicator can win every run of the game; no matter how the spoiler plays. Otherwise, the *Spoiler wins the r-move EF-game on $\mathbf{A}$ and $\mathbf{B}$*. For more details on first order logic and game concepts refer [10,12]. The following theorem is our main tool in proving the inexpressibility results.

Theorem 1. *[12] Let $\mathbf{A}$ and $\mathbf{B}$ be two structures in a relational vocabulary. Then the following are equivalent.*

1. *$\mathbf{A}$ and $\mathbf{B}$ satisfy the same sentence σ with $qr(\sigma) \leq n$.*
2. *The Duplicator has an n- round winning strategy in the EF game on $\mathbf{A}$ and $\mathbf{B}$.*

The study of graphs using different signatures other than I, J, and A can be seen in [11] and [1].

3 Segment Function Corresponding to the Induced Path Function

Here we show that it is not possible to give a characterisation of $\hat{J}$ using a set of first-order axioms defined on R.

By a *ternary structure* we mean an ordered pair (X, T) where X is a finite nonempty set and T is a ternary relation on X. By the *underlying graph* of a ternary structure (X, T) we mean the graph G with the properties that X is its vertex set and distinct vertices u and v of G are adjacent if and only if

$$\{x \in X; T(u, x, v)\} \cup \{x \in X; T(v, x, u)\} = \{u, v\}.$$

We call a ternary structure (X, T), 'the B^n- *structure* of a graph G', if X is the vertex set of G and T is the ternary relation corresponding to J^n. A ternary structure (X, T), is 'the C- *structure* of a graph G', if X is the vertex set of G and T is the ternary relation corresponding to $\hat{J}$. We say that the ternary structure (X, T) is a C-structure (or B^k-structure), if there exists a connected graph G such that (X, T) is the C- structure (or B^k-structure) of G. From the definition of underlying graph of a ternary structure, it is easy to see that if (X, T) is a C-structure (or B^k-structure), then it is the C-structure (or B^k-structure) of exactly one connected graph, namely the underlying graph of (X, T). Remember that, corresponding to any ternary relation $T \subseteq X \times X \times X$, there is a function $F : X \times X \to 2^X$ defined as $F(x, y) = \{u \in X : T(x, u, y)\}$. So, for any ternary structure (X, T), we can associate the function F corresponding to T. We say that (X, T) is *scant* if the function F corresponding to the ternary relation T, satisfies the condition: (s) $\forall x\ \forall y\ (F(x, y) \neq \{x, y\},\ x \neq y \Rightarrow F(x, y) = X)$; along with the axioms $(t1)$, $(t2)$, and $(t3)$ or in other words F is a transit function satisfying the axiom (s).

For each $k \geq 1$, we present two graphs G_d and G'_d such that the B^k-structure of one of which is scant and the other is not. Clearly, if the B^k-structure of G_d is scant then the C-structure of G_d will also be scant. Moreover, the proof will get settled, once we prove that the Duplicator wins the EF game on G_d and G'_d.

For $d \geq 2$ let G_d be a graph with vertices $\{u_1, u_2, \ldots, u_{4d}, v_1, v_2, \ldots, v_{4d}, w_1, w_2, x_1, x_2\}$, and edges

$$\begin{aligned} E(G_d) = \{&u_1u_2, u_2u_3, \ldots, u_{4d-1}u_{4d}, u_{4d}u_1, \\ &v_1v_2, v_2v_3, \ldots, v_{4d-1}v_{4d}, v_{4d}v_1, \\ &u_1v_2, u_2v_3, \ldots, u_{4d-1}v_{4d}, u_{4d}v_1, \\ &v_1u_2, v_2u_3, \ldots, v_{4d-1}u_{4d}, v_{4d}u_1, \\ &u_1x_1, v_1x_1, u_1w_1, v_1w_1, w_1x_1 \\ &u_{2d+1}x_2, v_{2d+1}x_2, u_{2d+1}w_2, v_{2d+1}w_2, w_2x_2 \\ &x_1x_2, x_1w_2, w_1x_2, w_1w_2\} \end{aligned}$$

For $d \geq 2$ let G'_d be a graph with vertices $\{u'_1, u'_2, \ldots, u'_{4d}, v'_1, v'_2, \ldots, v'_{4d}, w'_1, w'_2, x'_1, x'_2\}$, and edges

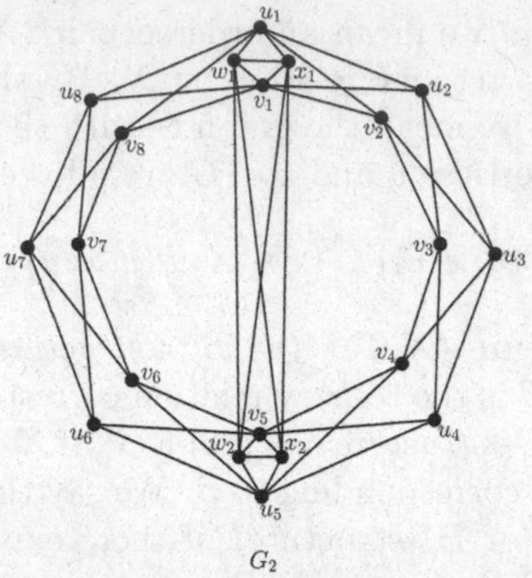

Fig. 1. Diagram for G_2

$$
\begin{aligned}
E(G'_d) = \{&u'_1u'_2, u'_2u'_3, \ldots, u'_{2d-1}u'_{2d}, u'_{2d}u'_1,\\
&u'_{2d+1}u'_{2d+2}, u'_{2d+2}u'_{2d+3}, \ldots, u'_{4d-1}u'_{4d}, u'_{4d}u'_{2d+1},\\
&v'_1v'_2, v'_2v'_3, \ldots, v'_{2d-1}v'_{2d}, v'_{2d}v'_1,\\
&v'_{2d+1}v'_{2d+2}, v'_{2d+2}v'_{2d+3}, \ldots, v'_{4d-1}v'_{4d}, v'_{4d}u'_{2d+1},\\
&u'_1v'_2, u'_2v'_3, \ldots, u'_{2d-1}v'_{2d}, u'_{2d}v'_1,\\
&u'_{2d+1}v'_{2d+2}, u'_{2d+2}v'_{2d+3}, \ldots, u'_{4d-1}v'_{4d}, u'_{4d}v'_{2d+1},\\
&v'_1u'_2, v'_2u'_3, \ldots, v'_{2d-1}u'_{2d}, v'_{2d}u'_1,\\
&v'_{2d+1}u'_{2d+2}, v'_{2d+2}u'_{2d+3}, \ldots, v'_{4d-1}u'_{4d}, v'_{4d}u'_{2d+1},\\
&u'_1x'_1, v'_1x'_1, u'_1w'_1, v'_1w'_1, w'_1x'_1\\
&u'_{2d+1}x'_2, v'_{2d+1}x'_2, u'_{2d+1}w'_2, v'_{2d+1}w'_2, w'_2x'_2\\
&x'_1x'_2, x'_1w'_2, w'_1x'_2, w'_1w'_2\}
\end{aligned}
$$

The graphs G_2 and G'_2 are shown in Fig. 1 and 2 respectively. Before going into the next lemma, observe that the subgraph K' induced by the vertices x'_1, x'_2, w'_1, and w'_2 in G'_2 and the subgraph K induced by the vertices x_1, x_2, w_1, and w_2 in G_2 are complete graphs. Due to the presence of this K' in G'_2 we can see that, $J^1(u'_3, v'_3) = \{u'_3, v'_3, u'_1, v'_1, u'_2, v'_2, u'_4, v'_4\} \neq V(G'_2)$. Hence, the B^1-structure of G'_2 is not scant. On the other hand, in G_2, even in the presence of the induced subgraph K, we have $J^1(u_3, v_3) = V(G_2)$. A further checking on each other pair of vertices in $V(G_2)$ will yield that B^1-structure of G_2 is scant. We generalize this observation for G_d and G'_d with $d \geq 2$, in the following lemma.

Lemma 1. *Let $d \geq 2$.*

i. The B^1-structure of G_d is scant.
ii. The B^1-structure of G'_d is not scant.

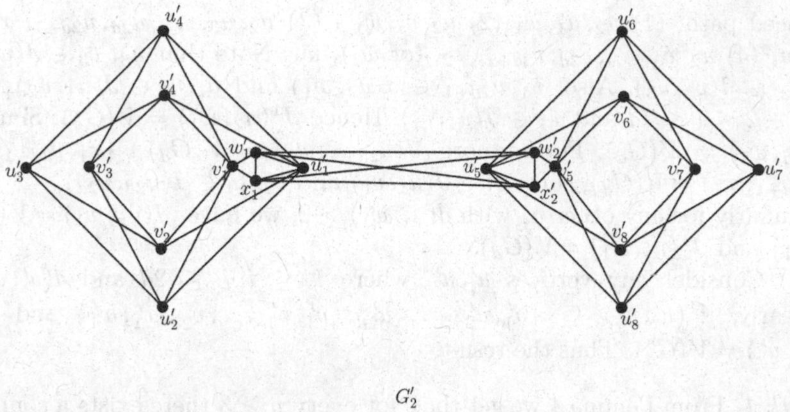

Fig. 2. Diagram for G'_2

Proof. (*i*) Let $V(G_d) = U \cup V \cup W \cup X$, where $U = \{u_1, u_2, \ldots, u_{4d}\}$, $V = \{v_1, v_2, \ldots, v_{4d}\}$, $W = \{w_1, w_2\}$, and $X = \{x_1, x_2\}$. Also, we write $u_0 = u_{4d}$, $v_0 = v_{4d}$, $u_{4d+1} = u_1$ and $v_{4d+1} = v_1$. Let $u, v \in V(G_d)$ with $d(u,v) \geq 2$. Clearly, u and v will be some elements in U, V, W, or X. We have to show that $J^1(u,v) = V(G_d)$.

When $u, v \in W$ (or X), or when $u \in X$ (or W) and $v \in W$ (or X) then $d(u,v) = 1$ and for this particular u and v, the axiom (s) holds trivially. So we consider the following cases:

Case 1: $u, v \in U$ (or V)

First we consider the case when $u = u_2$ and $v = u_{4d}$ (the case when $u = u_{2d}$ and $v = u_{2d+2}$ can be obtained similarly). Now, $J(u_2, u_{4d}) = V(G_d) \setminus \{v_2, v_{4d}, x_1, x_2, w_1, w_2\}$. But $u_1, v_1, u_{2d+1}, v_{2d+1} \in J(u_2, u_{4d})$ and since $v_2, v_{4d}, x_1, w_1 \in J(u_1, v_1)$ and $x_2, w_2 \in J(u_{2d+1}, v_{2d+1})$, we get $J^1(u_2, u_{4d}) = V(G_d)$. Similarly $J^1(u_{2d}, u_{2d+2}) = V(G_d)$.

Furthermore, for any other pair of vertices $u = u_i$ and $v = u_j$ with $d(u_i, u_j) \geq 2$, we have $J(u_i, u_j) = V(G_d) \setminus \{v_i, v_j\}$. Since $u_{i+1}, v_{i+1}, u_{j+1}, v_{j+1} \in J(u_i, u_j)$ we obtain $v_i \in J(u_{i+1}, v_{i+1})$ and $v_j \in J(u_{j+1}, v_{j+1})$. Hence $J^1(u_i, u_j) = V(G_d)$.

Case 2: $u \in U$ and $v \in V$

Let $u = u_i$ and $v = v_j$. Suppose $i = j$. When $i = 1$, $J(u_i, v_i) = \{u_i, v_i, u_{i+1}, v_{i+1}, u_{i-1}, v_{i-1}, x_1, w_1\}$, when $i = 2d+1$, $J(u_i, v_i) = \{u_i, v_i, u_{i+1}, v_{i+1}, u_{i-1}, v_{i-1}, x_2, w_2\}$ and otherwise, that is when $i \neq 1, 2d+1$, we have $J(u_i, v_i) = \{u_i, v_i, u_{i+1}, v_{i+1}, u_{i-1}, v_{i-1}\}$ and $x_1, x_2, w_1, w_2 \in J(u_{i+1}, u_{i-1})$ (for example consider the induced path $u_{i+1}, u_{i+2}, \ldots, u_{2d+1}, x_2$(or w_2), x_1(or w_1), $u_1, u_2, \ldots, u_{i-1}$). Clearly, $U \subset J(u_{i+1}, u_{i-1})$ and $V \subset J(v_{i+1}, v_{i-1})$. Thus, when $i = j$, $J^1(u_i, v_j) = V(G_d)$. If $i \neq j$, then follow Case 1 with $v = v_{4d}$ and $v = v_j$ instead of $v = u_{4d}$ and $v = u_j$ to obtain $J^1(u_i, v_j) = V(G_d)$.

Case 3: $u \in U$ (or V) and $v \in W$ (or X)

Let $u \in U$ and $v \in W$. With out loss of generality we take $v = w_1$. If $d(u, w_1) = 2$ then, $u = u_2$, u_{4d} or u_{2d+1}. The following is a list of u_2, w_1

- induced path: (1) u_2, u_1, w_1 (2) u_2, v_1, w_1 , (3) $u_2, u_3, \ldots, u_{2d}, u_{2d+1}, x_2$, (or w_2), w_1 (4) $u_2, v_3, \ldots, v_{2d}, v_{2d+1}, x_2$, (or w_2), w_1. Note that $u_3, v_3 \in J(u_2, w_1)$ and $v_2 \in J(v_3, u_3)$. Also, $u_1, u_{2d+1} \in J(u_2, w_1)$ and $u_i, v_i \in J(u_1, u_{2d+1})$ for all $2d+2 \leq i \leq 4d$ and $x_1 \in J(u_1, v_1)$. Hence $J^1(u_2, w_1) = V(G_d)$. Similarly, $J^1(u_{4d}, w_1) = V(G_d)$. Furthermore, $J(u_{2d+1}, w_1) = V(G_d) \setminus \{x_1, v_{2d+1}\}$ and $J^1(u_{2d+1}, w_1) = V(G_d)$, since $x_1 \in J(u_1, v_1)$ and $v_{2d+1} \in J(u_{2d}, v_{2d})$.

Similarly for any other u_j with $d(u_j, w_1) > 2$, we have, $J(u_j, w_1) = V(G_d) \setminus \{v_j, x_1\}$ and $J^1(u_j, w_1) = V(G_d)$.

(*ii*) Consider any vertices u'_i, u'_j, where $1 \leq i, j \leq 2d$ and $d(u'_i, u'_j) \geq 2$. Clearly, $J^1(u'_i, u'_j) \subseteq \{u'_1, u'_2, \ldots, u'_{2d}, v'_1, v'_2, \ldots, v'_{2d}, x'_1, w'_1\}$ and hence $J^1(u'_i, u'_j) \neq V(G'_d)$. Thus the result. □

Remark 1. From Lemma 1 we get that, for every $n \geq 3$ there exists a connected graph G of diameter n such that the C-structure of G is scant.

Lemma 2. *Let $n \geq 1$ and $d > 2^{n+1}$. Assume that (X_1, T_1) and (X_2, T_2) are scant ternary structures such that the underlying graph of (X_1, T_1) is G_d and the underlying graph of (X_2, T_2) is G'_d. Then (X_1, T_1) and (X_2, T_2) satisfy the same sentence η with $qr(\eta) \leq n$.*

Proof. Let $X_1 = \{u_1, u_2, \ldots, u_{4d}, v_1, v_2, \ldots, v_{4d}, w_1, w_2, x_1, x_2\}$ and let $X_2 = \{u'_1, u'_2, \ldots, u'_{4d}, v'_1, v'_2, \ldots, v'_{4d}, w'_1, w'_2, x'_1, x'_2\}$. Let $U = \{u_1, u_2, \ldots, u_{4d}\}$, $V = \{v_1, v_2, \ldots, v_{4d}\}$, $W = \{w_1, w_2\}$, and $X = \{x_1, x_2\}$. Also, let $U' = \{u'_1, u'_2, \ldots, u'_{4d}\}$, $V' = \{v'_1, v'_2, \ldots, v'_{4d}\}$, $W' = \{w'_1, w'_2\}$, and $X' = \{x'_1, x'_2\}$. Clearly $X_1 = U \cup V \cup W \cup X$ and $X_2 = U' \cup V' \cup W' \cup X'$. Let d^* and d' denote the distance function of G_d and G'_d respectively.

We will show that the Duplicator wins the n-move EF-game on G_d and G'_d using induction on n. In the ith move we respectively use p_i and q_i to denote points chosen from G_d and G'_d in the n-move game. Clearly, p_i will be some elements in X_1 and q_i will be some element in X_2. Note that, during the game, the elements of U (respectively, V, W and X) will be mapped to element of U' (respectively, V', W' and X').

Let H_1 be the subgraph induced by the vertices $u_1, u_2, \ldots, u_{4d}$ of G_d and H'_1 be the subgraph induced by the vertices $u'_1, u'_2, \ldots, u'_{4d}$ of G'_d. Since (X_1, T_1) and (X_2, T_2) are scant ternary structures, to win the game the Duplicator must preserve the edges in G_d and G'_d.

For, we claim that, for $1 \leq j, l \leq i$, the duplicator can play in G_d and G'_d, in a way that ensures the following conditions after each round $i \leq n$.

$$\text{(1) If } d^*(p_j, p_l) \leq 2^{n-i}, \text{ then } d'(q_j, q_l) = d^*(p_j, p_l).$$

$$\text{(2) If } d^*(p_j, p_l) > 2^{n-i}, \text{ then } d'(q_j, q_l) > 2^{n-i}.$$

Obviously, to win the game, the following correspondence must be preserved by the Duplicator:

$u_1 \mapsto u'_1$, $v_1 \mapsto v'_1$, $w_1 \mapsto w'_1$, $x_1 \mapsto x'_1$, $u_{2d+1} \mapsto u'_{2d+1}$, $v_{2d+1} \mapsto v'_{2d+1}$, $w_2 \mapsto w'_2$, $x_2 \mapsto x'_2$.

For $i = 1$, (1) and (2) trivially hold. Suppose they hold after i moves and that the Spoiler makes his $i+1^{th}$ move. Let the Spoiler picks $p_{i+1} \in X_1$ (the case of $q_{i+1} \in X_2$ is symmetric). If $p_{i+1} = p_j$ for some $j \leq i$, then $q_{i+1} = q_j$ and the conditions (1) and (2) are preserved. If p_{i+1} is some u_r (or v_r) where the vertex v_r (or u_r) is already chosen. With out loss of generality, let $p_{i+1} = u_r$ and let $v_r \mapsto v'_s$. Then set $q_{i+1} = u'_s$. Clearly conditions (1) and (2) will hold, since, for any previously chosen vertex p_j, $d^*(p_j, u_r) = d^*(p_j, v_r)$.

Otherwise, find two previously chosen vertices p_j and p_ℓ closest to p_{i+1} so that there are no other previously chosen vertices on the p_j, p_ℓ-path of G_d which passes via p_{i+1}.

Case 1: p_j, p_ℓ, $p_{i+1} \in U$

First we consider the case when $d^*(p_j, p_\ell) = d_{H_1}(p_j, p_\ell)$. There are two possibilities depending on the value of $d^*(p_j, p_\ell)$. If $d^*(p_j, p_\ell) \leq 2^{n-i}$, then by induction assumption there will be vertices q_j and q_ℓ in G'_d with $d'(q_j, q_\ell) \leq 2^{n-i}$. Then the Duplicator can choose q_{i+1} so that $d^*(p_j, p_{i+1}) = d'(q_j, q_{i+1})$ and $d^*(p_{i+1}, p_\ell) = d'(q_{i+1}, q_\ell)$. Clearly, the conditions (1) and (2) will hold. If $d^*(p_j, p_\ell) > 2^{n-i}$, then by induction assumption $d'(q_j, q_\ell) > 2^{n-i}$. There are two cases:

- Case i. $d^*(p_j, p_{i+1}) \leq 2^{n-(i+1)}$ or $d^*(p_{i+1}, p_\ell) < 2^{n-(i+1)}$, say the first. Then $d^*(p_{i+1}, p_\ell) > 2^{n-(i+1)}$. So the Duplicator can choose q_{i+1} with $d'(q_j, q_{i+1}) = d^*(p_j, p_{i+1})$ and $d'(q_{i+1}, q_\ell) > 2^{n-(i+1)}$.
- Case ii. $d^*(p_j, p_{i+1}) > 2^{n-(i+1)}$ and $d^*(p_{i+1}, p_\ell) > 2^{n-(i+1)}$. Therefore, the Spoiler plays at a distance greater than $2^{n-(i+1)}$ from all previously played vertices. However, since we have chosen d sufficiently large, we can be sure that, if fewer than n-rounds of the game have been played, in G'_d there is a point at distance larger than $2^{n-(i+1)}$ from all the previously played vertices.

Now, suppose $d^*(p_j, p_\ell) \neq d_{H_1}(p_j, p_\ell)$. This case occurs when p_j, p_ℓ-shortest path contains the vertices u_1, u_{2d+1}, v_1, v_{2d+1}, x_1, x_2, w_1 and w_2. Here find out $min\{d^*(p_j, p_{i+1}), d^*(p_\ell, p_{i+1})\}$. Suppose $d^*(p_j, p_{i+1}) \leq d^*(p_\ell, p_{i+1})$. Then choose q_{i+1} so that $d^*(p_j, p_{i+1}) = d'(q_j, q_{i+1})$.

Case 2: p_j, $p_\ell \in V$, $p_{i+1} \in V$

Let $p_j = v_r$, $p_\ell = v_s$, $p_{i+1} = v_t$. Then find the elements u_r, u_s and u_t in U and use case 1 to find the response of Duplictor when Spoiler chooses u_t. Let $u_t \mapsto u'_z$. Then choose $q_{i+1} = v'_z$.

Similarly for other cases (when $p_j \in U($ or $V)$, $p_\ell \in V($ or U$)$, $p_{i+1} \in V($ or $U))$ we can make all the vertices lying in U as in case 2 and is possible to find a response of the Duplicator. Evidently, in all the cases, the conditions (1) and (2) hold.

Furthermore, after n rounds of the game, the Duplicator can preserve the partial isomorphism. For, suppose n rounds have been played and let $\{p_1, p_2, \ldots, p_n\} \in X_1$ and $\{q_1, q_2, \ldots, q_n\} \in X_2$ be the vertices chosen in the n-move EF-game. For, $1 \leq j, \ell \leq n$, let $p_j p_\ell \in E(G_d)$. That is, $d^*(p_j, p_\ell) = 1$, and by (1) $d'(q_j, q_\ell) = 1$. Therefore, $q_j q_\ell \in E(G'_d)$. Conversely, let $q_j q_\ell \in E(G'_d)$. If $p_j p_\ell \notin E(G_d)$, then $d^*(p_j, p_\ell) > 1$ and by (2) we get $d'(q_j, q_\ell) > 1$, contrary to our assumption that $q_j q_\ell \in E(G'_d)$. Thus the Duplicator wins the n-move EF-game on G_d and G'_d. Hence by Theorem 1 we obtain the result. □

In general to show that J^k, $k \geq 2$ is not first order axiomatisable, we need a graph G such that for any a, b in $V(G)$, $\hat{J}(a,b) = J^\ell(a,b)$, with $\ell \leq k$. Consider the following graphs H_d (depicted in Fig. 3) and H'_d (depicted in Fig. 4):

For $d \geq 2$, let H_d be a graph with vertices $V(H_d) = V(G_d) \cup \{a_1, a_2, \ldots, a_k, b_1, b_2, \ldots, b_k\}$, and edges

$$\begin{aligned} E(H_d) = & E(G_d) \cup \{u_1a_1, u_1b_1, v_1a_1, v_1b_1, \\ & a_1a_2, a_2a_3, \ldots, a_{k-1}a_k, \\ & b_1b_2, b_2b_3, \ldots, b_{k-1}b_k, \\ & a_1b_2, a_2b_3, \ldots, a_{k-1}b_k, \\ & b_1a_2, b_2a_3, \ldots, b_{k-1}a_k\} \end{aligned}$$

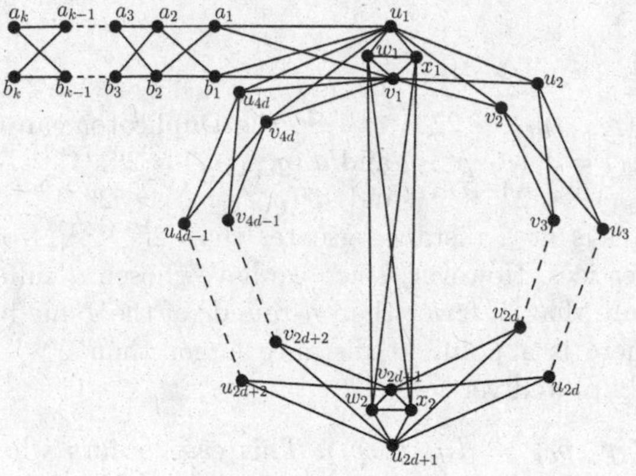

Fig. 3. Diagram for H_d

Let H'_d be a graph with vertices $V(H'_d) = V(G'_d) \cup \{a'_1, a'_2, \ldots, a'_k, b'_1, b'_2, \ldots, b'_k\}$, and edges

$$\begin{aligned} E(H'_d) = E(G'_d) \cup \{ & u'_1a'_1, u'_1b'_1, v'_1a'_1, v'_1b'_1, \\ & a'_1a'_2, a'_2a'_3, \ldots, a'_{k-1}a'_k, \\ & b'_1b'_2, b'_2b'_3, \ldots, b'_{k-1}b'_k, \\ & a'_1b'_2, a'_2b'_3, \ldots, a'_{k-1}b'_k, \\ & b'_1a'_2, b'_2a'_3, \ldots, b'_{k-1}a'_k\} \end{aligned}$$

Notice that for any p, q in $V(H_d)$, $\hat{J}(p,q) = J^\ell(p,q)$, with $\ell \leq k+1$. Clearly, the B^{k+1}-structure of H_d is scant and that of H'_d is not scant. Also analogous

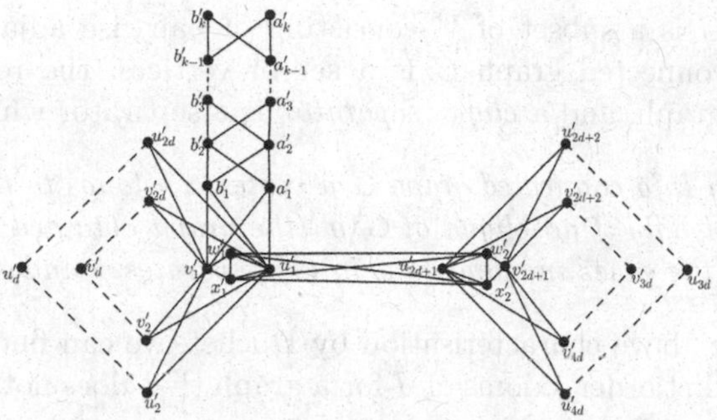

Fig. 4. Diagram for H'_d

of Lemma 2 can be obtained, since the Duplicator can win the game on H_d and H'_d by corresponding a_i to a'_i and b_i to b'_i for each $1 \leq i \leq k$.

By Lemma 2, we get that scant ternary structures with underlying graph G_d and G'_d satisfy the same first-order sentences (that is they cannot be distinguished using any FO sentences). And using Lemma 1, we obtain that the B^1-structure of G_d and G'_d are different. In other words, G_d and G'_d differ in J^1. Thus J^1 is not FO definable. Similarly, using analogous of Lemma 1 and Lemma 2 for the graphs H_d and H'_d, we get, J^k, $k \geq 2$ is not first-order definable. Hence, we obtain the following theorem for every $k \geq 1$.

Theorem 2. *There exists no sentence σ of the first order logic of vocabulary $\{T\}$ such that a connected ternary structure is a B^k-structure if and only if it satisfies σ.*

By Theorem 2 we can conclude that for any $k \geq 1$, the transit function J^k does not posses a first order axiomatic characterisation. In addition, observe that the C-structure of the graph G_d, as well as H_d is scant and C-structure of G'_d and H'_d are not scant. This observations along with Lemma 2 for both the pairs of graphs G_d and G'_d, and, H_d and H'_d, we can conclude the following result.

Theorem 3. *There exists no finite set S of sentences of the first order logic of vocabulary $\{T\}$ such that a connected ternary structure is a C- structure if and only if it satisfies each sentence in S.*

4 Characterisation of Graph Classes Using $\hat{J}$

In this section, we observe that even though the segment function $\hat{J}$ of arbitrary connected graphs are not first order axiomatisable, there are non-trivial graph classes whose $\hat{J}$ possesses first order axiomatic characterisation. The particular problem seems challenging, because of the iterative nature of $\hat{J}$. We prove that $\hat{J}$ possesses a first order axiomatic characterisation in graphs having no clique separator.

A *clique* of G is a subset of V consisting of pairwise adjacent vertices. A *separator* in a connected graph G is a set of vertices, the removal of which disconnects the graph, and a *clique separator* is a separator which is a clique.

Theorem 4. *[9] In a connected graph G a vertex a belongs to the J-convex hull of a set A if and only if no clique of $G\backslash a$ (the graph obtained by removing the vertex a and all the edges incident to a in G) separates a and A*

Adapting the above characterisation by Duchet, we can find a characterisation in terms of first order axioms of $\hat{J}$ for a graph that does not contain a clique separator.

Theorem 5. *Let G be a connected graph which is not a complete graph. Then $\hat{J}$ of G satisfies axiom (s) if and only if G does not contain a clique separator.*

(s) $\forall u\ \forall v\ \forall x\ (R(u,v) \neq \{u,v\} \Rightarrow x \in R(u,v))$

Proof. Prior to beginning the proof, note that if G is a complete graph, then $\hat{J}$ of G holds axiom (s) trivially. Suppose G contains a clique separator C. We will show that $\hat{J}$ of G does not satisfy the axiom (s). Without loss of generality assume that C separates G into precisely two parts say, H_1 and H_2. Let u, v $\in V(H_1)$ which does not frame an edge and let $x \in V(H_2)$. Then, by Theorem 4, we get $x \notin \hat{J}(u,v)$.

Conversely, suppose $\hat{J}$ of G does not satisfy axiom (s). Then there exists vertices u, v, and x with $\hat{J}(u,v) \neq \{u,v\}$ and $x \notin \hat{J}(u,v)$. Now by Theorem 4, $x \notin \hat{J}(u,v)$ (that is, x does not belong to the J-convex hull of the set $\{u,v\}$) means there exists a clique of $G\backslash x$ that separates x and $\{u,v\}$. That is, G contains a clique separator.

5 Conclusion

In this work, we proved that the segment transit function of the induced path function of graphs does not have a first order axiomatic characterisation. Also, due to the iterative nature of the functions the characterisation of graphs using these functions will have less expressibility compared to FOLB (the first order logic with betweenness). It will be interesting to check the FO axiomatisation of segment transit function associated with the interval function.

Acknowledgments. M.C. acknowledges the financial support from SERB, Department of Science & Technology, Govt. of India (research project under MATRICS scheme No. MTR/2017/000238). J.J acknowledges the financial support from University of Kerala, India, for providing University JRF (No: 445/2020/UOK, 391/2021/UOK, 3093/2022/UOK).

References

1. Brandenburg, F.J.: A first order logic definition of beyond-planar graphs. J. Graph Algorithms Appl. **22**(1), 51–66 (2017)
2. Morgana, M.A., Mulder, H.M.: The induced path convexity, betweenness and svelte graphs. Discret. Math. **254**, 349–370 (2002)
3. Changat, M., Klavzar, S., Mulder, H.M.: The all-paths transit function of a graph. Czechoslov. Math. J. **51**(2), 439–448 (2001)
4. Changat, M., Mulder, H.M., Sierksma, G.: Convexities related to path properties on graphs. Discret. Math. **290**(2–3), 117–131 (2005)
5. Changat, M., Mathew, J., Mulder, H.M.: The induced path function, monotonicity and betweenness. Discret. Appl. Math. **158**(5), 426–433 (2010)
6. Changat, M., Nezhad, F.H., Mulder, H.M., Narayanan, N.: A note on the interval function of a disconnected graph. Discuss. Math. Graph Theory **38**(1), 39–48 (2018)
7. Changat, M., Narasimha-Shenoi, P.G., Seethakuttyamma, G.: Betweenness in graphs: a short survey on shortest and induced path betweenness. AKCE Int. J. Graphs Comb. **16**(1), 96–109 (2018)
8. Dourado, M.C., Protti, F., Szwarcfiter, J.L.: Complexity results related to monophonic convexity. Discret. Appl. Math. **158**(12), 1268–1274 (2010)
9. Duchet, P.: Convex sets in graphs, II. Minimal path convexity. J. Comb. Theory Series B **44**(3), 307–316 (1988)
10. Ehrenfeucht, A.: An application of games to the completeness problem for formalized theories. Fund. Math. **49**, 129–141 (1961)
11. Hellings, J., Kuijpers, B., Van den Bussche, J., Zhang, X.: Walk logic as a framework for path query languages on graph databases. In: Proceedings of the 16th International Conference on Database Theory, pp. 117–128. ACM (2013)
12. Libkin, L.: Elements of Finite Model Theory. Springer, Heidelberg (2013)
13. Mulder, H.M.: Transit functions on graphs (and posets). In: Changat, M., Klavžar, S., Mulder, H.M., Vijayakumar, A. (eds.) Convexity in Discrete Structures. Lecture Notes Series, vol. 5, pp. 117–130. Ramanujan Mathematical Society (2008)
14. Mulder, H.M.: The interval function of a graph. MC Tracts (1980)
15. Mulder, H.M., Nebeský, L.: Axiomatic characterization of the interval function of a graph. Eur. J. Comb. **30**(5), 1172–1185 (2009)
16. Nebeský, L.: A characterization of the interval function of a connected graph. Czechoslovak Math. J. **44**(1), 173–178 (1994)
17. Nebeský, L.: Characterizing the interval function of a connected graph. Math. Bohem. **123**(2), 137–144 (1998)
18. Nebeský, L.: The induced paths in a connected graph and a ternary relation determined by them. Math. Bohem. **127**(3), 397–408 (2002)
19. van De Vel, M.L.J.: Theory of Convex Structures. Elsevier, Amsterdam (1993)

Fuzzy Free Logic with Dual Domain Semantics

Bornali Paul[1(✉)] and Sandip Paul[2(✉)]

[1] Department of Philosophy, Rabindra Bharati University, Kolkata, West Bengal, India
talking2bornali@gmail.com
[2] ECSU, Indian Statistical Institute, Kolkata, West Bengal, India
talking2sandip@gmail.com

Abstract. In this work we propose a fuzzy free logic, that is a generalized system which can give proper interpretation to sentences containing vagueness of non-referring singular terms. This logic is an amalgamation of classical positive free logic system with predicate rational Pavelka logic. The semantics given to the fuzzy free logic is based on dual-domain semantics. A graded similarity measure is introduced in the system which allows comparing two objects based on some properties and assign a degree of similarity. Soundness of the proposed system is proved.

Keywords: Free logic · Positive free logic · Dual-domain semantics · Rational Pavelka logic · Graded similarity measure · Empty-names

1 Introduction

Most of our daily communications use ordinary language and a good part of our thinking is done in it. In our daily lives, we often talk about sentences with referentless names, i.e., empty names, like, Dark matter, Santa Claus, Vulcan, Pegasus, Byomkesh Bakshi etc. These names may come from fictional stories (like story of Christmas, Roman Mythology, Greek Mythology, stories of Satyanweshi), from metaphysics ('nothingness') or from scientific discussions about existence of unobserved hypothetical entities (like dark matter). Also, in the ordinary language, we often use words whose meanings are vague, e.g., 'tall', 'beautiful' etc.

Though the referentless names are unavoidable in ordinary discourse, classical first order logic is not well-equipped to reason with such names. Because, if we attempt to assign arbitrary truth values to such sentences by assuming some kind of interpretation of the empty names in the domain of discourse, i.e., both existent and non-existent objects are placed in a single domain without any distinction, then some counter-intuitive results, like "Pegasus exists", "Round square exists", "something exists that does not exist" etc. can be derived. Furthermore, some logical rules/principles, i.e., principle of Existential Generalization (EG) and principle of Universal Instantiation (UI) are invalidated [8,9]. Also classical logic is inadequate to deal with graded valuation arising from vague terms.

So, to cope with the problem of empty names, free logic emerges [10,12], where all classical principles are valid in case of sentences containing empty names. But free

M. Banerjee and A. V. Sreejith (Eds.): ICLA 2023, LNCS 13963, pp. 130–142, 2023.
https://doi.org/10.1007/978-3-031-26689-8_10

logic is unable to deal with vague properties and fuzzy reasoning. To deal with vague properties of objects and for approximate reasoning, fuzzy logic is used, which allows graded valuation according to the degree to which an object satisfies a particular property. But fuzzy logic also has its drawbacks. A sentence, like, "A Pegasus is more similar to a horse than a Hyppogriff", which involves non-referring names and the fuzzy notion of similarity, cannot simultaneously be dealt within fuzzy predicate logic.

In this paper, we propose a new fuzzy logical system that can simultaneously deal with vague properties and also empty names. The system is named as **Fuzzy Free Logic with Dual Domain Semantics** (FFDS). The proposed system is new in the sense that it amalgamates dual domain positive free logic with predicate rational Pavelka logic [5], thus giving a many-valued variation of free logic. Also in the system we introduce a graded similarity measure that broadens the scope to comparing different objects having different degrees of similarity. Not only that, the similarity measure is defined in such a way so that we can express different degrees of similarity between two objects based on different sets of properties. In this respect the proposed system is more generalized than systems containing only strict identity, because strict identity cannot capture different grades of similarity. The similarity relation plays significant role in fuzzy approximate reasoning [14] and also to build information systems [6]. Also the proposed system extends the scope of dealing with sentences that are used in ordinary language.

To the best of our knowledge this type of amalgamation of fuzzy logic and free logic has not been studied yet. In [2] a basic outline of a fuzzy free logic was proposed. But their approach was quite different from the system presented here. Firstly, in [2] the authors allowed truth-value gaps in the semantics; while here the proposed semantics is based on total valuation. Secondly, no detailed mathematical analysis and axiomatization was presented in [2]. Thirdly, we have incorporated a notion of graded similarity in our system.

2 Fuzzy Free Logic with Dual Domain Semantics (FFDS)

In this work, we propose a fuzzy free logic, which is an amalgamation of predicate rational Pavelka logic ($RPL\forall$) and positive free logic with dual domain semantics. This logic resolves the problem of empty names of fuzzy predicate logic by dispensing with the tacit existential assumption of singular terms by means of an existence predicate $E!$. Simultaneously, the proposed system is capable of dealing with vague properties and graded truth values of sentences.

There are three major semantics for free logic [12], namely positive [7, 10], negative [11] and neutral [3]. Positive semantics allows some empty-termed atomic formulas, not of the form '$E!t$' (where, t is a term), to be true. On the other hand, negative free logic semantics require all empty-termed atomic formulas to be false. In neutral semantics all empty-termed atomic formulas, not of the form '$E!t$', are truthvalueless. Here, in positive free logic the sentence $t = t$ is true, even if t is empty; whereas in negative free logic the sentence $t = t$ is false if t is empty.

The semantics of the proposed system is based on the dual domain semantics of positive free logic. Dual domain positive semantics, uses two domains of interpretation [13], namely *inner* and *outer* domain; where, the inner domain captures the class of existing objects and the outer domain includes non-existent objects. Hence, in a dual-domain semantics the interpretation function is total and empty names are not non-referring as they refer to non-existent objects from the outer domain. The quantifiers range over the inner domain only. The domain of interpretation can be given as:

$$D = D_i \cup D_o,$$

where, D_i and D_o are *inner* and *outer domain* respectively, and D_i and D_o are disjoint sets.

Single and dual-domain semantics differ in the ways of assigning truth values to atomic formulas containing empty names. In dual-domain semantics the extension of a predicate ranges over the union of inner and outer domain, and truth value of any atomic formula is evaluated in the Tarskian fashion; i.e., an atomic formula is true if and only if the ordered n-tuple of objects referred to by its singular terms belongs to the extension of the predicate. Such evaluation is not possible in case of single domain semantics as empty names have no reference; hence truth values of such formulas are fixed either to be true or to be false depending on the semantics, which may lead to counter-intuitive results. For instance, the atomic proposition "*Santa Claus is the President of United States*" is true in single-domain positive free semantics and "*Harry Potter is a fictitious character*" is false in negative free semantics. This issue is resolved by considering two domains of interpretation.

Predicate rational Pavelka logic ($RPL\forall$) is a conservative extension of Łukasiewicz's infinite valued logic [5], which extends Łukasiewicz logic with truth constants $\bar{r}$ for each rational $r \in [0, 1]$, where $\bar{r}$ is treated as an atomic formula whose truth value is r under each evaluation. This allows to derive partially true conclusions from partially true statements.

2.1 The Role of Similarity

One essential feature of the proposed system is that, here, along with strict identity (=) a graded similarity measure is introduced, so that two objects can be compared based on how much they have in common.

Strict identity follows Leibnizian principle, which in a modified form, can be stated as [1]:

$$x = y \text{ iff } x \text{ has } p \text{ implies } y \text{ has } p \text{ and conversely;}$$

where, p represents a property belonging to a specified collection of properties or attributes. Say, the specified collection of properties is denoted by A.

However, graded similarity offers a broader scope for comparing two objects, which can't be captured by strict identity. This is illustrated with the help of following two cases.

Firstly, in case of vague properties, e.g., tall, beautiful etc., the concern is not only whether an object has those properties but also to what extents the properties are present in the object. Hence for any object $x \in D$ and any vague property $p \in A$, it is not enough, only to specify that whether x is p or not, but to what extent x has the property of being p is to be specified. For this, $RPL\forall$ is appropriate to assign a graded valuation from $[0, 1]$

to the statement "x is p". In such a scenario, two objects, x and y, may both have some property p, but with different degrees. Thus x and y are not identical, but have certain similarity with respect to the vague property p. The same scenario can arise when a set of vague properties is considered instead of a single property p. This aspect cannot be handled using strict identity.

Secondly, even if we are not considering vague properties, the notion of graded similarity may come into the picture [6]. Instead of the strict identity between two objects (or entities), finding some common properties becomes important in reasoning with ordinary discourse. Bivalent identity does not capture these aspects. Suppose two objects $x, y \in D$ have certain (crisp) properties $p_1, .., p_n \in A$. Whereas, x and y mismatch over the rest of the properties in A. Then clearly x and y are not identical; but they are *similar*. But this similarity is not fuzzy, two objects can either be similar or not.

In the proposed logic, a graded similarity measure is introduced along with the concept of identity of classical free logic, that can deal with both of the aforementioned cases.

In order to define an appropriate similarity measure it must be kept in mind that similarity between two objects (or entities) should be judged with respect to many parameters, not just one. For instance, a horse is similar to a Pegasus with respect to their *appearance*; Santa Claus is similar to Pegasus because both are *non-existent*; Hippogriff is similar to Harry Potter since both are characters from *same fantasy story*, whereas they have very different appearance. Hence, a single similarity measure is not capable of dealing with these various *parameters of comparison*, i.e., sets of properties, based on which similarity of two objects are to be evaluated.

2.2 Representation of Properties and Similarity

Any unary predicate stands for a property or attribute of the corresponding argument. For instance, if $Intelligent$ is considered to be a unary predicate then, $Intelligent(x)$ depicts a property of x. In case of binary predicates there are two arguments. Thus a binary predicate depicts a property of one of its arguments relative to the other one. For instance, in case of the binary predicate *Greater_than*; $Greater\_than(x,x_1)$ says *x is greater than* x_1; i.e. the property of x being greater than x_1. This binary predicate can be considered to be an attribute for one of its arguments if the other argument is not a free variable. If the second argument is some constant, say 2, then $Greater\_than(x,2)$, having only one free variable, corresponds to the attribute *being greater than* 2. The position of the free variable is important in case of non-symmetric non-unary predicates. $Greater\_than(x,2)$ does not depict the same property as $Greater\_than(2,x)$. Same holds true for higher arity predicates, where only one argument is kept for the free variable.

In a general way it can be said that any well formed formula with a single free variable can be associated with a property. For instance, the property of *being tall and blonde* can be represented with $Tall(x) \wedge Blonde(x)$. The property of *being tall or not blonde* can be depicted as $Tall(x) \vee \neg Blonde(x)$.

Hence, corresponding to each property, depicted by any wff with a single free variable, a similarity measure can be induced over any two objects.

2.3 Language of *FFDS*

In the language, L_F, of the proposed system, there are: a list of variables $(x, x_1, x_2, x_3, ..)$, a list of individual constants $(c, c_1, c_2, ...)$, a list of predicate symbols, P_i^n, where, n stands for the arity and i for identifier, logical connectives $\neg$, &, $\underline{\vee}$, $\rightarrow$, $\wedge$, $\vee$, the equality predicate $=$, truth constants $\overline{r}$ for each $r \in [0,1]$, universal and existential quantifiers $\forall$, $\exists$, an *existence* predicate $E!$, a special binary predicate symbol $Sim\_meas_{\varphi,s}$. Here φ and s are parameters.

The corresponding truth functions of t-norms & and $\wedge$ are taken to be Lukaseiwicz t-norm ($\odot$) and *min* respectively; truth functions of t-conorms $\underline{\vee}$ and $\vee$ are taken to be Lukaseiwicz t-conorm ($\oplus$) and *max* respectively; truth function of implication ($\rightarrow$) is taken to be Lukaseiwicz residuated implicator ($\Rightarrow$) [4]. The *existence* predicate $E!$ is taken from free logic.

In the language L_F, a *term* is a variable or an individual constant. For any n-ary predicate P_i^n, $P_i^n(t_1, .., t_n)$ is an atomic formula, where $t_1, ..t_n$ are terms. For any wff φ and terms s, t, t_1, $Sim\_meas_{\varphi,s}(t, t_1)$ and $E!s$ are atomic formulas. For each rational $r \in [0,1]$, $\overline{r}$ is a formula. If φ, ψ, χ are formulas, x is a variable and s is a term, then $\neg\varphi$, $\varphi \rightarrow \psi$, $\varphi \& \psi$, $\varphi \wedge \psi$, $\varphi \vee \psi$, $\varphi \underline{\vee} \psi$, $(\forall x)\psi$, $(\exists x)\varphi$ are formulas. Each formula that results from atomic formulas and using the deduction rule, are also formulas in the system.

The similarity measure between two terms t and t_1 with respect to a wff φ and a term s, denoted as $Sim\_meas_{\varphi,s}(t, t_1)$, is used to represent the similarity between two terms t and t_1 with respect to the property depicted by the wff φ, where the term s serves to denote the arguments to be substituted to capture the property depicted by φ.

For instance, consider φ is $IQ(t_1) \wedge Pass\_Exm(t_1) \wedge Good\_univ(t_2) \wedge Admit(t_1, t_2)$, which considers a student's IQ, whether he/she passed the exam and whether he/she has taken admission to a good university. Then $Sim\_Meas_{\varphi,t_1}(s, t)$ stands for the similarity of two students based on φ, where, s and t are to be substituted in place of t_1.

2.4 Semantics of FFDS

The semantics of the system, *FFDS*, is based on dual-domain positive semantics and semantics of *RPL*$\forall$. In the dual domains, inner domain represents the domain of existent entities and the outer domain represents the domain of non-existent entities. So, in the semantics, the interpretation function is total.

The model structure is specified as follows;

$$M = \langle D_i, D_o, \langle d_c \rangle_c, \langle r_P \rangle_P \rangle .$$

The truth values range over $[0,1]$ and the domain of interpretation $D_i \bigcup D_o$ is non-empty.

In the model structure,

- D_i, D_o are two disjoint sets, called *inner* and *outer* domain respectively.
- For each object constant c, d_c is an element of $D_i \bigcup D_o$.
- For an n-ary predicate P_i^n, $r_{P_i^n} : (D_i \bigcup D_o)^n \rightarrow [0,1]$ associates to each tuple $(d_{c_1}, ..., d_{c_n}) \in (D_i \bigcup D_o)^n$ the membership value of $(d_{c_1}, ..., d_{c_n})$ to the fuzzy relation P_i^n, which is $r_{P_i^n}(d_{c_1}, ..., d_{c_n}) \in [0,1]$.

Definition 1. *For the language L_F and model structure M for L_F, an M-evaluation of variables is a mapping v, that assigns an element from $D_i \bigcup D_o$ to each object variable x. Let, v, v' be two assignments, then $v \equiv_x v'$ means $v(x_1) = v'(x_1)$ for each variable x_1 distinct from x.*

The interpretation of a term by M, v, is given as; $\|x\|_{M,v} = v(x)$ and $\|c\|_{M,v} = d_c$. The truth value of a sentence φ, denoted as $\|\varphi\|_{M,v}$, is obtained under the following conditions:

1. $\|P_i^n(t_1,...,t_n)\|_{M,v} = r_{P_i^n}(\|t_1\|_{M,v},...,\|t_n\|_{M,v})$;
2. $\|\varphi \to \psi\|_{M,v} = \|\varphi\|_{M,v} \Rightarrow \|\psi\|_{M,v}$;
3. (a) $\|\varphi \& \psi\|_{M,v} = \|\varphi\|_{M,v} \odot \|\psi\|_{M,v}$;
   (b) $\|\varphi \veebar \psi\|_{M,v} = \|\varphi\|_{M,v} \oplus \|\psi\|_{M,v}$;
   (c) $\|\varphi \wedge \psi\|_{M,v} = min(\|\varphi\|_{M,v}, \|\psi\|_{M,v})$;
   (d) $\|\varphi \vee \psi\|_{M,v} = max(\|\varphi\|_{M,v}, \|\psi\|_{M,v})$;
   (e) $\|\neg\varphi\|_{M,v} = 1 - \|\varphi\|_{M,v}$
4. $\|\bar{r}\|_{M,v} = r, r \in \mathbb{Q} \cap [0,1]$;
5. $\|\forall x\varphi\|_{M,v} = inf\{\|\varphi\|_{M,v'} \mid v \equiv_x v' \text{ and } v'(x) \in D_i\}$;
6. $\|\exists x\varphi\|_{M,v} = sup\{\|\varphi\|_{M,v'} \mid v \equiv_x v' \text{ and } v'(x) \in D_i\}$;
7. $\|t = t_1\|_{M,v} = \begin{cases} 1, & \|t\|_{M,v} = \|t_1\|_{M,v} \\ 0, & \text{otherwise} \end{cases}$
8. $\|E!t\|_{M,v} = \begin{cases} 1, & \|t\|_{M,v} \in D_i \\ 0, & \text{otherwise} \end{cases}$;
9. $\left\|Sim\_meas_{\varphi,s}(t,t_1)\right\|_{M,v} = 1 - \left| \|\varphi(t//s)\|_{M,v} - \|\varphi(t_1//s)\|_{M,v} \right|$
   where, $\varphi(t//s)$ means replacing zero or more (but not necessarily all) occurrences of s in φ with t. Also a restriction is imposed that, in $\varphi(t//s)$ and $\varphi(t_1//s)$, t and t_1 are substituted in the same occurrences of s.

It can be seen that in the proposed semantics the valuation is *total*, because each and every well formed formula is given some valuation and no truth-value gap is allowed. Here, one thing is to be noted that predicates' extensions range over both the inner and outer domain. Semantic condition 1 ensures that the truth values of atomic sentences with empty names, that are not of the form $E!t$, are being assigned in the Tarskian fashion, rather that being rigidly true or false. This would result in positive dual-domain semantics if the valuation would have been restricted to $\{0,1\}$ and also this clearly points out the distinction of the proposed system from negative and neutral semantics. The valuation can be constructed in such a way to assign 0 to the sentence "*Santa Claus is the President of the Unites States*" and to assign 1 to the sentence "*Harry Potter is a fictitious character*"; whereas negative semantics would assign false to both of them and neutral semantics would assign truth-value gap to both of them.

In condition 9, the similarity between two terms t and t_1 with respect to any property depicted by φ is measured by the extents to which t and t_1 satisfies φ. If both $\|\varphi(t//s)\|_{M,v}$ and $\|\varphi(t_1//s)\|_{M,v}$ are close to each other then both t and t_1 are similar with respect to φ and $\left\|Sim\_Meas_{\varphi,s}(t,t_1)\right\|_{M,v}$ is close to 1. On the other hand if t and

t_1 are dissimilar with respect to φ, then one of $\|\varphi(t//s)\|_{M,v}$ and $\|\varphi(t_1//s)\|_{M,v}$ will be close to 1 and the other one is close to 0; thus resulting $\left\|Sim\_Meas_{\varphi,s}(t,t_1)\right\|_{M,v}$ to be close to 0. The similarity defined here, is reflexive, symmetric and transitive.

Lemma 1: $\|\varphi(t_1//s)\|_{M,v} \geq \left\|Sim\_Meas_{\varphi,s}(t,t_1)\right\|_{M,v} \odot \|\varphi(t//s)\|_{M,v}$.

Proof: $R.H.S. = \left\|Sim\_Meas_{\varphi,s}(t,t_1)\right\|_{M,v} \odot \|\varphi(t//s)\|_{M,v}$

$$= \left(1 - \left| \|\varphi(t//s)\|_{M,v} - \|\varphi(t_1//s)\|_{M,v} \right|\right) \odot \|\varphi(t//s)\|_{M,v}$$
$$= max\left\{0, 1 - \left| \|\varphi(t//s)\|_{M,v} - \|\varphi(t_1//s)\|_{M,v} \right| + \|\varphi(t//s)\|_{M,v} - 1\right\}$$
$$= max\left\{0, \|\varphi(t//s)\|_{M,v} - \left| \|\varphi(t//s)\|_{M,v} - \|\varphi(t_1//s)\|_{M,v} \right|\right\}$$

Case 1: If $\|\varphi(t//s)\|_{M,v} \leq \|\varphi(t_1//s)\|_{M,v}$

$$R.H.S. = max\left\{0, 2\|\varphi(t//s)\|_{M,v} - \|\varphi(t_1//s)\|_{M,v}\right\}$$
$$= max\left\{0, 2\left(\|\varphi(t//s)\|_{M,v} - \|\varphi(t_1//s)\|_{M,v}\right) + \|\varphi(t_1//s)\|_{M,v}\right\} \leq \|\varphi(t_1//s)\|_{M,v}$$

Case 2: If $\|\varphi(t_1//s)\|_{M,v} \leq \|\varphi(t//s)\|_{M,v}$

$$R.H.S. = max\left\{0, \|\varphi(t//s)\|_{M,v} - \|\varphi(t//s)\|_{M,v} + \|\varphi(t_1//s)\|_{M,v}\right\}$$
$$= max\left\{0, \|\varphi(t_1//s)\|_{M,v}\right\} = \|\varphi(t_1//s)\|_{M,v}$$

2.5 Axiom Schema of FFDS

The axiomatization of the proposed system is developed by converging free logic with predicate rational Pavelka logic ($RPL\forall$), with introduction of a similarity measure.

Let φ, ψ, χ be well formed formulas, x is a variable and $s, t, t_1, t_2, \ldots t_{i-1}$ be terms, then the axioms are as follows:

1. $\varphi \rightarrow (\psi \rightarrow \varphi)$;
2. $(\varphi \rightarrow \psi) \rightarrow ((\psi \rightarrow \chi) \rightarrow (\varphi \rightarrow \chi))$;
3. $(\neg\varphi \rightarrow \neg\psi) \rightarrow (\psi \rightarrow \varphi)$;
4. $((\varphi \rightarrow \psi) \rightarrow \psi) \rightarrow ((\psi \rightarrow \varphi) \rightarrow \varphi)$;
5. $(\varphi \rightarrow \forall x\varphi)$, x is not free in φ (Blanket Axiom);
6. $(\forall x(\varphi \rightarrow \psi)) \rightarrow (\forall x\varphi \rightarrow \forall x\psi)$ (Dist);
7. $\forall x\varphi \rightarrow (E!t \rightarrow \varphi(t/x))$ (Restricted form of the principle of Specification);
8. $t = t$

9. $Sim\_Meas_{\varphi,s}(t,t_1) \rightarrow (\varphi(t//s) \rightarrow \varphi(t_1//s))$;
   where, in $\varphi(t//s)$ and $\varphi(t_1//s)$, t and t_1 are substituted in the same occurrences of s.
10. $\forall x E!x$;
11. (a) $\overline{r} \rightarrow \overline{k} \equiv \overline{r \Rightarrow k}$
    (b) $\neg\overline{r} \equiv \overline{1-r}$, where r, k are rationals from $[0,1]$.

In axiom 7, $\varphi(t/x)$ stands for replacing each free occurrence of the variable x in $\varphi(x)$ with t. This principle states that, if a term t refers to any existing object in the domain of discourse, then only it can be used to replace the occurrences of the variable x in the formula φ. Axiom 7 is the restricted form of principle of Specification, which is the key behind the success of free logic for dealing with empty names.

In axiom 9, $\varphi(t//s)$ stands for replacing zero or more (but not necessarily all) occurrences of s in φ by t. In the axiom, $Sim\_Meas_{\varphi,s}(t,t_1)$ replaces the identity $(t = t_1)$ occurring in the relevant axiom in classical free logic. This depicts that the t can substitute t_1, not only when they are *identical*; but also when they both possess certain set of properties *relevant to* the wff φ. For instance, if the consequent is concerned about a property such as *beautiful*, then the substitutivity of t by t_1 would depend upon their similarity based on their beauty only. This relaxes the substitutivity condition.

The axiom 10, stipulates that the quantifiers range over all objects that satisfy $E!$.

Axiom 11 corresponds to the "bookkeeping axioms" for truth constants.

2.6 Deduction Rules

The deduction rule of *FFDS* is modus ponens (*MP*) (from φ, $\varphi \rightarrow \psi$ infer ψ). There is another derived deduction rule in *RPL*∀ [4] as stated below;

$$\frac{(\varphi, r) \qquad (\varphi \rightarrow \psi, k)}{(\psi, r \odot k)}$$

Here, $r, k, r \odot k$ are rational elements from $[0,1]$. The pair of the form (φ, r) is a graded formula such that φ is a formula and is a shorthand representation of $\overline{r} \longrightarrow \varphi$ and '$\odot$' is Łukasiewicz t-norm.

3 Soundness of the System

Now we prove the soundness of the system *FFDS*. The truth degree and provability degree are defined following *RPL*∀.

Definition 2: *For a theory T (a closed set of formulas) in FFDS and a formula* φ*;*

1. *The truth degree of* φ *over T is* $\|\varphi\|_T = inf\{\|\varphi\|_M | M \text{ is a model structure of } T\}$*, where,* $\|\varphi\|_M = inf\{\|\varphi\|_{M,v} | v \text{ is M-evaluation}\}$
2. *The provability degree of* φ *over T is* $|\varphi|_T = sup\{r | T \vdash (\varphi, r)\}$.

Theorem 1: For each theory, T in the system and each formula, φ, we have,

$$|\varphi|_T \leq \|\varphi\|_T$$

Proof: It immediately follows from if $T \vdash \psi$ then $\|\psi\|_{M,v} = 1$, for each model of φ [4]. The soundness of deduction rules are proved in [4]. Thus, for proving soundness of the system we will have to prove that for any axiom α, $\|\alpha\|_{M,v} = 1$. Axioms 1 to 4 and axiom 8 and 11 are taken directly from *RPL*∀; hence for them the proof is already well-established [[4], Ch 5]. For the rest of the axioms, proofs are presented here.

<u>**Axiom 5:**</u> $(\varphi \to \forall x\varphi)$ [x is not free in φ]

$$\begin{aligned}
\|\varphi \to \forall x\varphi\|_{M,v} &= \|\varphi\|_{M,v} \Rightarrow \|\forall x\varphi\|_{M,v};\\
&= min(1 - \|\varphi\|_{M,v} + \|\forall x\varphi\|_{M,v}, 1);\\
&= min(1 - \|\varphi\|_{M,v} + inf\{\|\varphi\|_{M,v'} \,|\, v \equiv_x v' \text{ and } v'(x) \in D_i\}, 1);\\
&= min(1 - \|\varphi\|_{M,v} + \|\varphi\|_{M,v}, 1); \quad [\text{since, } x \text{ is not free in } \varphi]\\
&= 1.
\end{aligned}$$

<u>**Axiom 6**</u>: $(\forall x(\varphi \to \psi)) \to (\forall x\varphi \to \forall x\psi)$

To prove, $\|(\forall x(\varphi \to \psi)) \to (\forall x\varphi \to \forall x\psi)\|_{M,v} = 1$
we need to show;

$$\|\forall x\varphi \to \forall x\psi\|_{M,v} \geq \|\forall x(\varphi \to \psi)\|_{M,v}$$

$$\text{or, } (\|\forall x\varphi\|_{M,v} \Rightarrow \|\forall x\psi\|_{M,v}) \geq \|\forall x(\varphi \to \psi)\|_{M,v};$$

$$\begin{aligned}
\text{or, } & min(1 - \|\forall x\varphi\|_{M,v} + \|\forall x\psi\|_{M,v}, 1)\\
&\qquad \geq inf\{\|\varphi \to \psi\|_{M,v'} \,|\, v \equiv_x v' \text{ and } v'(x) \in D_i\};\\
\text{or, } & min(1 - inf\{\|\varphi\|_{M,v'} \,|\, v \equiv_x v' \text{ and } v'(x) \in D_i\}\\
&\qquad + inf\{\|\psi\|_{M,v'} \,|\, v \equiv_x v' \text{ and } v'(x) \in D_i\}, 1)\\
&\qquad \geq inf\{\|\varphi \to \psi\|_{M,v'} \,|\, v \equiv_x v' \text{ and } v'(x) \in D_i\}
\end{aligned}$$

Now suppose, for some constants c_1, c_2, with $d_{c_1}, d_{c_2} \in D_i$;

$$inf\{\|\varphi\|_{M,v'} \,|\, v \equiv_x v' \text{ and } v'(x) \in D_i\} = \|\varphi\|_{M,v(x|c_1)}$$
$$inf\{\|\psi\|_{M,v'} \,|\, v \equiv_x v' \text{ and } v'(x) \in D_i\} = \|\psi\|_{M,v(x|c_2)}$$

where, $v(x|c)$ is same as v, except at the variable x, which is assigned the value d_c, i.e.,

$$v(x|c)(x_1) = \begin{cases} v(x_1), & x_1 \neq x \\ d_c, & x_1 = x \end{cases}$$

So, $$min(1 - inf\{\|\varphi\|_{M,v'} \,|\, v \equiv_x v' \text{ and } v'(x) \in D_i\} + inf\{\|\psi\|_{M,v'} \,|\, v \equiv_x v' \text{ and } v'(x) \in D_i\}, 1)$$
$$= min(1 - \|\varphi\|_{M,v(x|c_1)} + \|\psi\|_{M,v(x|c_2)}, 1)$$
$$\geq min(1 - \|\varphi\|_{M,v(x|c_2)} + \|\psi\|_{M,v(x|c_2)}, 1)$$
$$[Since, \|\varphi\|_{M,v(x|c_2)} \geq \|\varphi\|_{M,v(x|c_1)}]$$
$$= \|\varphi \rightarrow \psi\|_{M,v(x|c_2)}$$
$$\geq inf\{\|\varphi \rightarrow \psi\|_{M,v'} \,|\, v \equiv_x v' \text{ and } v'(x) \in D_i\}$$

Hence, $\|(\forall x(\varphi \rightarrow \psi)) \rightarrow (\forall x\varphi \rightarrow \forall x\psi)\|_{M,v} = 1$

Axiom 7:

$\forall x\varphi \rightarrow (E!t \rightarrow \varphi(t/x))$

It is to be proved;

$$\|\forall x\varphi \rightarrow (E!t \rightarrow \varphi(t/x))\|_{M,v} = 1;$$
$$\text{or, } (\|\forall x\varphi\|_{M,v} \Rightarrow \|E!t \rightarrow \varphi(t/x)\|_{M,v}) = 1;$$
$$\text{or, } \|E!t \rightarrow \varphi(t/x)\|_{M,v} \geq \|\forall x\varphi\|_{M,v};$$
$$\text{or, } (\|E!t\|_{M,v} \Rightarrow \|\varphi(t/x)\|_{M,v}) \geq \|\forall x\varphi\|_{M,v};$$
$$\text{or, } min(1,\ 1 - \|E!t\|_{M,v} + \|\varphi(t/x)\|_{M,v}) \geq inf\{\|\varphi\|_{M,v'} \,|\, v \equiv_x v' \text{ and } v'(x) \in D_i\}$$

Now consider the following two cases:

Case-1:

Let, $\|E!t\|_{M,v} = 1$, i.e., $\|t\|_{M,v} \in D_i$

Then,

$$L.H.S = min(1,\ 1 - \|E!t\|_{M,v} + \|\varphi(t/x)\|_{M,v})$$
$$= min(1,\ 1 - 1 + \|\varphi(t/x)\|_{M,v})$$
$$= min(1, \|\varphi(t/x)\|_{M,v})$$
$$= \|\varphi(t/x)\|_{M,v}$$
$$\geq inf\{\|\varphi\|_{M,v'} \,|\, v \equiv_x v' \text{ and } v'(x) \in D_i\} = R.H.S$$

Case-2: Let, $\|E!t\|_{M,v} = 0$, i.e., $\|t\|_{M,v} \in D_o$;

Then,

$$L.H.S = min(1,\ 1 - \|E!t\|_{M,v} + \|\varphi(t/x)\|_{M,v})$$
$$= min(1, 1 - 0 + \|\varphi(t/x)\|_{M,v})$$
$$= min(1, 1 + \|\varphi(t/x)\|_{M,v})$$
$$= 1$$
$$\geq inf\{\|\varphi\|_{M,v'} \,|\, v \equiv_x v' \text{ and } v'(x) \in D_i\}$$
$$= R.H.S$$

Hence, it is proved that,

$$\|E!t \to \varphi(t/x)\|_{M,v} \geq \|\forall x\varphi\|_{M,v}$$
$$\text{and, } \|\forall x\varphi \to (E!t \to \varphi(t/x))\|_{M,v} = 1$$

Axiom 9:
$Sim\_Meas_{\varphi,s}(t,t_1) \to (\varphi(t//s) \to \varphi(t_1//s))$ It is to show that,

$$\left\|Sim\_Meas_{\varphi,s}(t,t_1) \to (\varphi(t//s) \to \varphi(t_1//s))\right\|_{M,v} = 1$$
$$\text{or, } \left\|Sim\_Meas_{\varphi,s}(t,t_1)\right\|_{M,v} \Rightarrow \left(\|\varphi(t//s)\|_{M,v} \Rightarrow \|\varphi(t_1//s)\|_{M,v}\right) = 1$$
$$\text{or, } \left(\|\varphi(t//s)\|_{M,v} \Rightarrow \|\varphi(t_1//s)\|_{M,v}\right) \geq \left\|Sim\_Meas_{\varphi,s}(t,t_1)\right\|_{M,v}$$
$$\text{or, } min\left(1, 1 - \|\varphi(t//s)\|_{M,v} + \|\varphi(t_1//s)\|_{M,v}\right) \geq \left\|Sim\_Meas_{\varphi,s}(t,t_1)\right\|_{M,v}$$

Case-1:
Suppose, $\|\varphi(t//s)\|_{M,v} \leq \|\varphi(t_1//s)\|_{M,v}$. So,

$$\begin{aligned} L.H.S &= min(1 - \|\varphi(t//s)\|_{M,v} + \|\varphi(t_1//s)\|_{M,v}, 1) \\ &= 1 \\ &\geq \left\|Sim\_Meas_{\varphi,s}(t,t_1)\right\|_{M,v} = R.H.S \end{aligned}$$

Case-2:
Suppose, $\|\varphi(t//s)\|_{M,v} > \|\varphi(t_1//s)\|_{M,v}$
Then, we need to show that,

$$\begin{aligned} &min(1, 1 - \|\varphi(t//s)\|_{M,v} + \|\varphi(t_1//s)\|_{M,v}) \\ &= 1 - \|\varphi(t//s)\|_{M,v} + \|\varphi(t_1//s)\|_{M,v} \\ &\geq \left\|Sim\_Meas_{\varphi,s}(t,t_1)\right\|_{M,v} \end{aligned}$$

Now,

$$\begin{aligned} &1 - \|\varphi(t//s)\|_{M,v} + \|\varphi(t_1//s)\|_{M,v} \\ &\geq 1 - \|\varphi(t//s)\|_{M,v} + (\left\|Sim\_Meas_{\varphi,s}(t,t_1)\right\|_{M,v} \odot \|\varphi(t//s)\|_{M,v}) \\ &\qquad [\text{replacing } \|\varphi(t_1//s)\|_{M,v} \text{ using Lemma 1.}] \\ &= 1 - \|\varphi(t//s)\|_{M,v} + max(0, \left\|Sim\_Meas_{\varphi,s}(t,t_1)\right\|_{M,v} + \|\varphi(t//s)\|_{M,v} - 1) \\ &= max(0, \|\varphi(t//s)\|_{M,v} + \left\|Sim\_Meas_{\varphi,s}(t,t_1)\right\|_{M,v} - 1) - (\|\varphi(t//s)\|_{M,v} - 1) \\ &= max(0, \|\varphi(t//s)\|_{M,v} + \left\|Sim\_Meas_{\varphi,s}(t,t_1)\right\|_{M,v} - 1) \\ &\quad - (\|\varphi(t//s)\|_{M,v} + \left\|Sim\_Meas_{\varphi,s}(t,t_1)\right\|_{M,v} - 1) \\ &\qquad + \left\|Sim\_Meas_{\varphi,s}(t,t_1)\right\|_{M,v} \\ &\geq \left\|Sim\_Meas_{\varphi,s}(t,t_1)\right\|_{M,v} \text{ [Since, } max(0,a) - a \geq 0]. \end{aligned}$$

Hence, it is proved that,

$$\|Sim\_Meas_{\varphi,s}(t,t_1) \rightarrow (\varphi(t//s) \rightarrow \varphi(t_1//s))\|_{M,v} = 1.$$

Axiom 10: $\forall x E!x$

$$\begin{aligned}\|\forall x E!x\|_{M,v} &= inf\{\|E!x\|_{m,v'} | v \equiv_x v' \text{ and } v'(x) \in D_i\};\\ &= inf\{\|E!c\| \,|d_c \in D_i\};\\ &= 1.\end{aligned}$$

This completes the proof of soundness of the $FFDS$ system.

4 Conclusion

In this paper a fuzzy version of positive free logic is proposed that is a unification of positive free logic with dual-domain semantics and predicate rational Pavelka logic. Having in it the properties of both the systems, the resulting system, namely $FFDS$, can resolve the problems that arise in classical logic and fuzzy logic due to inclusion of referentless singular terms corresponding to non-existent objects and also supports graded valuation which can handle vague concepts. This makes the system more appealing for reasoning in ordinary discourse of which non-referring empty names and vagueness are inseparable parts.

Another very crucial improvement, that makes the system more general and flexible is adding the notion of a graded similarity measure along with strict identity. This allows comparing two different objects based on their common properties and assign a degree of similarity, which can't be captured by identity. This notion of similarity is an important characteristic of logic for information systems. This system has accommodated more than one similarity measure so that similarity between any two objects can be judged from different perspectives. The system $FFDS$ is sound. The investigation of completeness is aimed to be our future work. This type of fuzzy free logic based on Gödel logic and product logic, instead of $RPL\forall$, can be further studied.

So the new system $FFDS$ filters out the limitations of positive free logic as well as fuzzy logic. In our daily lives this system can be used for dealing with reasoning in ordinary discourse. This system presents a vast scope of applications.

Acknowledgement. The authors are indebted to Dr. Mihir Kumar Chakraborty and Dr. Purbita Jana for their valuable comments and suggestions.

References

1. Banerjee, M., Chakraborty, M.K.: Foundations of vagueness: a category-theoretic approach. Electron. Notes Theor. Comput. Sci. **82**(4), 10–19 (2003)

2. Běhounek, L., Dvořák, A.: Non-denoting terms in fuzzy logic: an initial exploration. In: Kacprzyk, J., Szmidt, E., Zadrożny, S., Atanassov, K.T., Krawczak, M. (eds.) IWIFSGN/EUSFLAT -2017. AISC, vol. 641, pp. 148–158. Springer, Cham (2018). https://doi.org/10.1007/978-3-319-66830-7_14
3. Bencivenga, E.: Free semantics. In: Dalla Chiara, M.L. (ed.) Italian Studies in the Philosophy of Science. Boston Studies in the Philosophy of Science, vol. 47, pp. 31–48. Springer, Dordrecht (1980). https://doi.org/10.1007/978-94-009-8937-5_3
4. Hájek, P.: Metamathematics of Fuzzy Logic, vol. 4. Springer, Heidelberg (2013)
5. Hájek, P., Paris, J., Shepherdson, J.: Rational Pavelka predicate logic is a conservative extension of łukasiewicz predicate logic. J. Symb. Log. **65**, 669–682 (2000)
6. Khan, M.A., Banerjee, M., Rieke, R.: An update logic for information systems. Int. J. Approximate Reason. **55**(1), 436–456 (2014)
7. Lambert, K.: Free Logic: Selected Essays. Cambridge University Press, Cambridge (2002)
8. Lambert, K.: The philosophical foundations of free logic. In: Lambert, K. (ed.) Free Logic: Selected Essays, pp. 122–175. Cambridge University Press, Cambridge (2004)
9. Lambert, K., et al.: Existential import revisited. Notre Dame J. Formal Log. **4**(4), 288–292 (1963)
10. Lambert, K., et al.: Free logic and the concept of existence. Notre Dame J. Formal Log. **8**(1–2), 133–144 (1967)
11. Lehmann, S.: Strict Fregean free logic. J. Philos. Log. 307–336 (1994)
12. Morscher, E., Simons, P.: Free logic: a fifty-year past and an open future. In: Morscher, E., Hieke, A. (eds.) New Essays in Free Logic. Applied Logic Series, vol. 23, pp. 1–34. Springer, Dordrecht (2001). https://doi.org/10.1007/978-94-015-9761-6_1d
13. Nolt, J.: Free logics. In: Jacquette, D. (ed.) Philosophy of Logic, pp. 1023–1060. Elsevier, Amsterdam (2007)
14. Sessa, M.I.: Approximate reasoning by similarity-based SLD resolution. Theoret. Comput. Sci. **275**(1–2), 389–426 (2002)

A New Dimension of Imperative Logic

Manidipa Sanyal(✉) and Prabal Kumar Sen(✉)

Department of Philosophy, University of Calcutta, Kolkata, India
smanidipa@yahoo.co.in, pksen.cu@gmail.com

Abstract. A full-fledged theory of imperative logic is found in the writings of Peter Vranas. An unconditional prescription is an ordered pair with satisfaction as the first member, and violation as the second member. A conditional prescription is a set of mutually exclusive and jointly exhaustive three values – satisfaction, violation and avoidance. An argument is valid, only if, necessarily, if its premises merit endorsement, then its conclusion merits endorsement. The phrase "meriting endorsement" is interpreted as 'supported by a proposition/prescription'. Among different schools of Indian philosophy, the Mīmāṁsā system offers an analysis of imperative sentences, where actions, guided by instructions, play an important role. *Vidhi* or normal injunctive statements is studied intensely and recently arguments involving '*vidhi'* has been used in special education and in the domain of Robotics. Imperative, discussed in this sense, is however not the only type of imperatives, it is only one variety of different kinds of imperative. Such varieties are very well recognized by Indian grammarians and philosophers as well as by western thinkers. These imperatives also deserve the status of the premise or the conclusion of an inference. The present paper focuses upon unveiling such varieties of imperative sentences from both perspectives—Indian and Western.

Keywords: Imperative · Advice · Prayer · Acceptance-table · Validity

Conversation – verbal or written – is a main source of communication. A necessary condition of proper communication is the use of reason or argument. So,it is language, which plays a vital role even in the field of logic. As argument is primarily inferential, a study of the nature of sentences constituting such inferences is required. Towards the end of the 20th century, attention has been given to inferences constituted of sentences, which are not declarative in nature. Instead of reductionism – imperative sentences reduced to declarative sentences – the standpoint of non-reductionism has been successfully developed in the West, though it was already present in some schools of Indian tradition.

1 Logic of Imperatives - Western

A full-fledged theory of imperative logic is found in the writings of Peter B. M. Vranas [1–4]. An imperative sentence, occurring either as premise or as conclusion of an inference expresses a prescription, which is neither true nor false. Vranas introduced three values to study prescriptions. An unconditional prescription, however, is an ordered pair with satisfaction as the first member, and violation as the second member –

$$I = \langle s, v \rangle$$

M. Banerjee and A. V. Sreejith (Eds.): ICLA 2023, LNCS 13963, pp. 143–158, 2023.
https://doi.org/10.1007/978-3-031-26689-8_11

A conditional prescription is a set of three values – viz., satisfaction, violation and avoidance, which are mutually exclusive and jointly exhaustive. The condition is treated as context, which is the union of the set of satisfaction and that of violation. A conditional imperative "If you trust him, help him" is

i) satisfied if you trust him and help him,
ii) violated if you trust him but don't help him,
iii) avoided if you don't trust him, no matter whether you help him or not.

In the vocabulary of the system formulated by Vranas,
Conditional prescription = <s, v, a> or <<s, v>, a>

context = (s υ v)
avoidance = ~ (s υ v)

We can represent the unconditional prescription using the identical symbolic form, instead of limiting it to an ordered pair of s and v i.e., <s, v>.

Let me prove the case with illustrations of all connectives:

Negation

Unconditional prescription– "help her".
negation– "Don't help her"

you don't help her (satisfied),
you help her (violated),
you remain indifferent (avoided).

It is to be noted that this state of indifference is not the same as being unmindfully indifferent to a passer-by, who may need some help. I may be indifferent to her, because I am mentally otherwise engaged at that moment. But the present case of indifference is a state of conscious indifference, even after hearing somebody giving me the instruction "help her".

Conjunction

Unconditional prescription: "Trust me and touch me".
You trust me and you touch me (satisfied),
you do not trust me or you do not touch me or both (violated)
[i.e., you neither trust me, nor touch me],
you are simply present as a stranger, who denies all
acquaintance (avoided).

In the case of avoidance, the presence of the person for whom the imperative is uttered is important. This presence is accompanied by an awareness of the conjunctive imperative without having a deliberation to violate it. So it is not to be understood as violation, though it appears to be so. In fact, in understanding an imperative statement, it is not enough to depend only on physical observation of the worldly affairs. Unlike descriptive or declarative proposition, it connects us with the total attitude of the agent – utterer or hearer – of the imperative statement.

Disjunction

Unconditional prescription - "Write to me or talk to me".

You write to me or you talk to me (satisfied),

you do not write to me and you do not talk to me (violated),

you are simply present as a stranger, who denies all acquaintance (avoided).

Here, the case is the same as found in the case of conjunction. The illustrations offered, if found cogent, show nevertheless the distinction between imperative logic and standard two-valued logic in a sharper way. This is the status of unconditional prescriptions.

The definition of validity is technically stated in the following way:

D(2) An argument is valid, only if, necessarily, if its premises merit endorsement, then its conclusion merits endorsement. The phrase "meriting endorsement" is interpreted as 'supported by a proposition/prescription'. This interpretation can be made clear if we consider the original definition of validity mentioned by Vranas in a comparatively naive way:

An argument is valid exactly if, necessarily, every fact that sustains every premise of the argument also sustains the conclusion of the argument. Since a conditional imperative premise normally has a proposition as antecedent and prescription as a consequence, "meriting endorsement" in the sense of "being sustained by a fact" is understood in the following way:

a) guaranteed by some fact (in case of a proposition),
b) supported by some reason (in case of a prescription).

Now the term 'reason' covers different cases of application of reason, i.e. reasons for acting, feeling, believing etc. It implies that an imperative does not pertain to direct action only, it also involves feeling, believing and other attitudes which precede an action.

2 Logic of Imperatives - Indian

Among different schools of Indian philosophy, the Mīmāṁsā system, which provides the rules for interpreting Vedic sentences, offers an analysis of imperative sentences; where actions, guided by instructions, play an important role [5]. There are five types of Vedic sentences, of which only the first is in the imperative form:

(i) *Vidhi* or normal injunctive statements (dictating one to perform actions)
(ii) *Mantra* or hymns (recited during sacrifice)
(iii) *Nāmadheya* or titles of the sacrifice (account of names of sacrifices)
(iv) *Niṣedha* or prohibitions (prohibiting the performance of an action)
(v) *Arthavāda* or corroborative statements (encouraging performance of actions that are enjoined by *vidhi,* and discouraging performance of actions that are prohibited by *niṣedha*).

The distinction between narratives and injunctions is distinctly made [6]. *Vidhi* is classified into five types:

1. Principal injunction (*Utpattividhi*): Injunction enjoining an act that is either principal, or auxiliary, or a procedure.
2. Injunction enjoining auxiliaries (*Guṇavidhi).*
3. Restrictive injunction (*Niyamavidhi*): Injunctions making one method mandatory, out of two or more methods which are available for reaching a goal.
4. Exclusive injunction (*Parisaṅkhyāvidhi)*: Injunctions excluding one item from a number of items which are simultaneously present.
5. Injunction setting forth result (*Phalavidhi)*: Injunctions that indicate results. For example, "One who desires heaven should perform fire-sacrifice".

The Mīmāṁsakas are more concerned about the explanation of *Vidhivākya*-s (imperatives/prescriptions) in the context of ritualistic sacrifice [7]. So the imperative here is authoritative (*prāmāṇyavākya)* in nature. Unlike the western thought, the Indian thinkers opine that an imperative points both to the person to whom the command is given, and to the action that is supposed to be produced by that command. The Bhāṭṭas consider *bhāvanā* (not to be confused with motivation) as the meaning of the statement. It is something that is conducive to the execution of the expected result. The causative verbal noun *bhāvanā* ("causing to be") was introduced into Mīmāṁsā hermeneutics by Śabara [8]. The term is a causative verbal noun which denotes the undertaking of an activity by a person. According to the Prābhākaras, what is to be done (ought) is prescribed by the Vedic injunctions. This 'ought' is something, such that it cannot be known (Apūrva) by any other means of knowledge [9].

The inspiration derived from the Vedic sacrifices (MIRA formalism) has been aptly used by in special education [10], and in the domain of Robotics [11], as shown by Bama Srinivasan and Ranjini Parthasarathi. In this interpretation, imperatives are treated either as conditional or as unconditional. From another perspective, imperatives may be affirmative or negative. Conditional imperatives often speak of goal, reason, or sequence of actions. Imperatives are expressed sometimes in terms of binary connectives, viz., conjunction, mutually exclusive disjunction, implication, etc.

Let i and m be two imperatives

(a) Conjunction: i Λ m [Do i and do m]
(b) Disjunction: i ∨ m [Do i or do m]
(c) Sequence of action: i = > im [Do i, then do m]
(d) Ground for performing an action: Ʈ → r φ [If Ʈ then φ]
    (where Ʈ is a ground for an action to be performed indicated by the imperative φ)
(e) Imperative regarding actions to be performed for achieving a goal: φ → p ɵ [Do φ in order to do ɵ]
    (if φ is an imperative indicating an action, such that when performed, it leads to the goal ɵ)

Three values of imperatives have been suggested, viz., "S" (satisfaction), "V" (violation), and "Gn" (absence of goal). Let us take an example to illustrate the ascription of values:

Take a pen to write.

S is the evaluation if the intention to reach the goal of writing is present, and the action is performed.

V is the value ascribed to the imperative A, if the said intention is present, and the action is not performed.

Gn is the ascribed value, if the intention is not present, irrespective of the performance of the action.

This system has introduced the third value "absence of goal" (which is the same as absence of intention to reach the goal) in place of "avoidance" introduced by Vranas, and, unlike Vranas, it enjoys the facility of applying three values both to the unconditional and conditional imperative.

The syntax consists of a language of imperatives, which includes a set of imperatives I such that $\{i_1, i_2, \ldots. i_n\}$, a set of reasons R $\{r_1, r_2, \ldots. r_n\}$, and a set of purpose in terms of goals P $\{p_1, p_2, \ldots. p_n\}$. There are formation rules and several deduction rules including introduction and elimination rules in respect of the connectives. The semantics has been developed in respect of imperatives enjoining goals ($\varphi \rightarrow p\theta$), reason ($\tau \rightarrow r\varphi$), and temporal actions ($i_1 \rightarrow i\, i_2$), respectively. By repeated application of deduction rules, a conclusion ψ can be deduced from a set of premises $\varphi_1, \varphi_2, \ldots. \varphi_n$. It is shown in the following way:

$$\varphi_1, \varphi_2, \ldots \varphi_n \vdash \psi$$

Soundness and completeness of this system have been proved to show that any imperative provable by MIRA formalism (2014) is also satisfied during the performance of action. In proving soundness, it attempts to show that the deduction of a conclusion from a set of premises is valid in terms of the values held by the premises and conclusion.

Soundness Theorem 1.

Let $\varphi_1, \varphi_2, \ldots \varphi_n$ and ψ be imperative or propositional formulas. If $\varphi_1, \varphi_2, \ldots \varphi_n \vdash \psi$, then $\varphi_1, \varphi_2, \ldots \varphi_n | = \psi$ Holds.

The proof for soundness includes one inductive step and proofs for each of the deduction rules.

Completeness Theorem 2.

Let $\varphi_1, \varphi_2, \ldots \varphi_n$ and ψ be imperative or propositional formulas. If $\varphi_1, \varphi_2, \ldots \varphi_n | = \psi$, then the property of a plan $\varphi_1, \varphi_2, \ldots \varphi_n \vdash \psi$ holds.

The proof for completeness is constructed on the basis of induction and being supported by action performance tables and deduction rules.

Imperative, as discussed both in Indian and Western context, is however not the only type of imperatives, it is only one variety of different kinds of imperatives. Such varieties are very well recognized by Indian grammarians and philosophers as well as by western thinkers. They also deserve the status of premise or conclusion of inference. The present paper focuses upon unveiling such varieties of imperative sentences from both perspectives.

3 Interpretation of Imperatives - Indian

In the texts of Sanskrit grammar and of different schools of Indian Philosophy, an in-depth analysis of (i) imperatives (sentences employed for strongly encouraging someone for doing something) and of (ii) prohibitions (for preventing someone from doing something) is found. Such sentences occur profusely in Sanskrit grammar (*Aṣṭādhyāyī* of Pāṇini), and (i) the Vedic texts (*Brāhmaṇa* and *Upaniṣad*), (ii) Smṛti texts (*Manusaṃhitā, Yājňavalkya Saṃhitā* etc.,) (iii) Epics (*Rāmāyana, Mahābhārata, Bhagavadgītā*), *Purāṇa*-s (*Viṣṇupurāṇa, Skandapurāṇa, Bhagavatapurāṇa* etc.) and didactic literature (*Hitopadeśa, Pańcatantra, Cāṇakyaśloka* etc.)

Sanskrit grammar provides us with some rules governing the formation of injunctions and prohibitions. Like German, Sanskrit is an inflected language, where word-order is free, barring a few exceptions. Sentences are collections of words that are characterized by (i) mutual expectancy (*ākānkṣā*) (ii) contiguity (*āsatti*). (iii) compatibility (*yogyatā*) and (iv) import (*tātparya*). Words again are primarily of two types – nouns and verbs. Without entering into much details of grammar, we can focus upon what is relevant for the present paper. Imperative sentences are usually formed in three ways:

(i) by employing the verb in the imperative mood, e.g., '*satyaṃ vada*'(i.e., 'speak the truth' where the termination '*loṭ*' has been used)
(ii) by employing the verb in the potential mood, e.g., '*svargakāmo yajeta*' (one who desires to attain heaven should perform sacrifice, where the termination *lin* has been used),
(iii) by using, instead of such words, nouns that have been formed by adding to the verbs concerned, any one of the verbal suffixes known as '*kṛtyapralyaya*', that are used for forming potential/future participles; e.g., '*satataṃ kāryaṃ karma samācara*' [i.e., 'always perform the obligatory duty', where the work ' *kārya*' has been formed by the addition of the suffix '*ṇyat'*, which is a '*kṛtyapratyaya*'].

Moreover, from rule no. 3/4/7 (liṅarthe leṭ), and the comment on it by Bhaṭṭojī Dīkṣita in his '*Siddhāntakaumudī*' [12], it can be known that in Vedic texts, instead of 'liṅ' or 'lot' another verbal ending called 'leṭ', which expresses subjunctive mood, may be used for forming imperative sentences. An example of this is '*agnihotraṃ juhoti*' (i.e., 'one should perform the agnihotra sacrifice') [13], where the verbal ending 'leṭ' has been used.

We now proceed to discuss the semantic aspects of them, as found by grammar, as well as rules of interpretation. Some consideration of pragmatics will also be undertaken, by considering

(i) the specific context in which a certain imperative or prohibition is being employed;
(ii) the manner in which an imperative can urge the listener/reader to perform the recommended action, and
(iii) the manner in which a prohibition makes the listener/reader desist from performing the prohibited action.

The rule no 3/3/161 of *Aṣṭādhyāyī* [14] (*vidhi-nimantraṇāmaṇtranādhīṣṭa-saṃpraśna-prārthaneṣu liṅ*) means that the verbal termination called 'liṅ' can be employed for forming sentences that can express

(i) *vidhi* or *ājñā*, i.e., command,(e.g. a master asking his servant to close the door),
(ii) *nimantraṇā*, i.e., an invitation, (such that it is obligatory for the invited person to abide by it),
(iii) *āmaṇtrana*, i.e., an invitation, (such that the invited person can either accept or decline it),
(iv) *adhīṣṭa*, i.e., an entreaty or supplication, where someone is respectfully requested to perform a duty or honour (e.g., investing a boy with the sacred thread)
(v) *saṃpraśna*, i.e., a polite question about what is to be done in the near future (e.g., a student asking the teacher – should I now read grammar?)
(vi) *prārthanā*, i.e., prayer where some request is made with the expectation of receiving some favour (a student saying " this is my prayer that I be permitted to study grammar")

Besides, the rule no. 3/3/162 (*loṭ ca*) means that the termination *loṭ*, which usually expresses permission (*anujñā/anumiti*), can also be used for expressing *vidhi* etc., that are expressed by *liṅ*. All these cases are exhortations (*preraṇā*-s), the aim of which is to produce in the listener/reader some activity that was not so far present in him. Pāṇini was interested in pointing out the varieties of exhortations, which is very relevant for the present paper.

It is not, however, difficult to distinguish between these forms of exhortation. In all such cases, X tells Y to perform the action A; but the status of X and Y is not the same on all these cases. In the case of command, the speaker is superior as compared to the listener. The situation is not the same in the case of invitation – X, who is inviting Y, may or may not be superior to Y. In the case of *adhīṣṭa*, or respectful entreaty, X and Y may be of the same stature, or Y may be superior to X. In the case of questioning and prayer, Y is definitely superior as compared to X. In the case of command, invitation, entreaty and prayer, prior to the utterances of the concerned sentences by the speaker (i.e., X), there is no desire in the listener (i.e., Y) for performing the act A that Y is asked to perform. The very purpose of uttering such imperatives is to produce in Y such a desire; which, in its turn, would lead to the performance of A by Y. In the case of permission [e.g., '*yathecchasi tathā kuru*', i.e., 'do as you like', where the termination '*loṭ*' has been used], even prior to the utterance of the sentence concerned by X, the desire for performing the act A is already present in Y; even though the latter cannot perform A, unless the required permission is given by the former. Thus the utterance of permission, so to say, merely removes the preventive factor (*pratibandhaka*), due to which the performance of the action A had not taken place previously – it is unlike order etc. that positively produces some activity in the person to whom they are addressed.

Here we may note another difference of opinion regarding the nature of injunctions and prohibitions that are found in scriptures like *Veda*-s and *Smṛti*-s. According to Mammaṭa [15], the author of *Kāvyaprakāśa*, scriptures are '*prabhusammita*', i.e., entities that act as taskmasters, since scriptural injunctions and prohibitions are inviolable commands that are carried out by us out of our reverence for the scriptures. But this view does not seem to be admitted by Jaimini [16], the author of *Mīmāṃsāsūtra*-s, and Gautama [17], the author of *Nyāyasūtra*-s, both of whom have employed the word '*upadeśa*' (i.e. advice) while defining verbal testimony (*śabdapramāṇa*). Since the scriptures are prime examples of verbal testimony for both these authors, according to both of them, the scriptural injunctions and prohibitions must also be regarded as advices, and not as commands.

A question may arise: what is the basic difference between a command and an advice? The answer has been given in clear terms by Maṇḍana Miśra [18], a follower of the Bhāṭṭa school of Mīmāṃsā. In his *Vidhiviveka*, he has explained the nature of advice, and in the sequel, also distinguished it from command, prayer and permission. In the cases of all these sentences, some person (say X) utters a sentence S, that prompts another person (say Y) to perform some action (say A). According to Maṇḍana Miśra, when S is either a command or a prayer, the performance of A directly leads to some purpose of X being served; but the interest of Y is not taken into account by X. But in the case of advice, the situation is just the opposite; since in this case, performance of A directly serves some purpose of Y, and not of X. In some instances, this may be true of permission as well, but in such cases, Y is already motivated to perform A, even before Y has been granted the required permission. Advice, however, prompts a person to do something; and before listening to this advice, that person was not already so motivated.

What is implied by this discussion is that imperative logic, so far developed, cannot cover all the kinds of imperatives.

4 Interpretation of imperatives - Western

The study of imperatives, which has been conducted since several decades is regarded as interesting because of two reasons [19]:

i) New theoretical tools are needed to understand the semantically encoded linguistic meaning of an imperative.
ii) In a natural language, the necessity of retaining truth-condition may be reviewed in respect of imperative sentences.

The primary point to note in this study is that imperatives don't determine a function from world to truth-values. In Castaňeda [20], imperatives are studied as part of practical thinking, as distinct from theoretical thinking. Practical thinking deals with duties and the conflict between duties. It tends to guide other people regarding their conduct and decision to act. 'Practitions' are the basic units of practical thinking, which is of two kinds—prescriptions and intentions. According to Castaňeda, though the prescription i.e., the thought-content of order, command, request, suggestion, or advice is the common structure of a relation between an agent and his action, the mandates are different in each case. An intention is the first person correspondent to a prescription.

An imperative is a sentence of the form '!p'. In case of a conditional sentence, the antecedent and the consequent cannot both be prescriptions. It is customary to treat the whole compound sentence as an imperative sentence. In order to understand the significance of different varieties of imperatives separately, it is necessary to refer to the speech-act theory of Austin and Searle. A speech-act is a combination of three acts- an utterance-act a locutionary-act and an illocutionary-act. In Austin, the locutionary act is the act of expressing a certain content. This content has two elements [21]:

i) It is the act of using words with a determinate sense and a determinate reference. In Austin's opinion, context and speaker's intentions play a crucial role in making them determinate.
ii) A broad type of illocutionary force is encoded by the sentence mood.

It shows that only at the illocutionary level, the force is made contextually determinate. Searle however deviates from it, but that discussion is avoided here because of fear of digression. Force-content distinction is defended by Frege and Geach [22, 23] It is never the case that all the occurrences of a sentence expressing the same content ought to have the same force, if force is a part of the content of a sentence.

On the other hand, force-neutral content is considered as a myth according to Hanks [24]. He is of the opinion that the 'unity of the proposition' requires something to tie together the ingredients of content. It is the 'intentional action of the speaker' acts as the glue to provide the unity of the proposition. It depends on a condition that the act is neutral with regard to the issue of illocutionary force. Accordingly to Soames [25], the glue is the act of predication which is performed irrespective of whether the proposition is asserted or not.

An imperative and a declarative have two different illocutionary forces, though they may have different types of the same content. Imperatives express a wide range of speech acts, which are beyond commands. Likewise, different types of imperatives may have the same content, though the type of illocutionary force is different in them. The content is force-neutral. Charlow has referred to several kinds of such expressions [26]:

a. Go ahead, take the day off (permission)
b. Talk to your advisor more often (suggestion, advice)
c. Have a piece of fruit (invitation)
d. Get well soon (well-wish)
e. To get the Union Square, take Broadway (instruction)
f. Go on, throw it, Just you dare, (threat)
g. Complete these by tomorrow (command)
h. Enjoy it!
i. Choose your friends wisely (advice)
j. Shall we sleep? (interrogative permission)
k. Consider the red dress (suggestion)

Charlow also referred to some border line cases [27]:

l. Complete your syllabus by the next month, although you may complete it by this month.
m. Take rest for a day, although of course you may prepare for the next travel unierruptedly.
n. Although she must be at her friend's place tonight, is she helping her mother in preparing dinner?
o. I know you are able to, but can you open the window?

5 Attempt to Accommodate All Imperatives in Logic

Let us now see whether arguments containing imperatives, other than commands, in order to act as premise or conclusion can receive the same treatment in the sphere of existing system of imperative logic as is received by the arguments containing commands. We may consider one argument where an imperative(in the sense of command) is used either as a premise or as a conclusion:

A. Either feel a concern for the needy or remain non-commital.
   Do not remain non-commital.
   Therefore, feel a concern for the needy.

We may consider another example which has an advice as a part of the premise of an argument:

B. Choose your friends wisely or you will invite trouble.
   You will not invite trouble.
   Therefore, choose your friends wisely.

In example A, both disjuncts 'Feel a concern for the needy' and 'remain non-committal' are imperative separately. So also the whole disjunctive sentence that occurs as premise. The conclusion also is fully imperative. But in B, only the disjunct 'Choose your friends wisely' is imperative in the first premise since the other disjunct 'you will invite trouble' is a descriptive sentence. So imperative in example B occurs as a part of a premise.The first case is intuitively valid, and it is justified by the definition of validity provided by Vranas. The second, however is neither intuitively valid, nor is it justified by the definition of validity. So, it is not the structure, but the meaning which is important for deciding the status of the argument.

Another point to note is that in many cases of advice, there may be a temporal element and it deserves a different rule of validity, in case it appears as a premise. It may be made clear by citing two examples. The first example contains a premise, which is a command:

C. Wait for me and don't go alone.

   i) Therefore, wait for me.
   ii) Therefore, don't go alone.

In this case, both the conclusions are derivable by simplification from the premise, because, in both the cases, the reason that sustains the premise, also sustains the conclusion. Both are valid arguments. Consider another argument containing advice as a premise:

D. You should wash your hand and eat.

   i) Therefore, You should wash your hand.

ii) Therefore, you should eat.

Here, the portion 'you should wash your hand' may also be replaced by ' wash your hand', the difference between a command and an advice being discernible from the tone in which the sentence is uttered. In the case of first conclusion, the reason sustaining the premise, sustains the conclusion, and it is intuitively valid. But it is not the case with the second conclusion. As per the definition of validity introduced by Vranas, the argument containing (ii) as the conclusion is valid, though it is not intuitively valid. This is so, because if the addressee begins to eat without washing his/her hand, he/she cannot be said to abide by the advice given to him/her. There is an inbuilt temporal element, which does not allow (ii) to be derived from the premise.

The case is similar with making a wish or request, which is another variety of imperative:

E. Enjoy the art-exhibition, and write a comment in the record-book!

i) Therefore, enjoy the art-exhibition.
ii) Therefore, write a comment in the record-book.

Here, too, we cannot say that the request made by the speaker has been abided (or honoured) by the addressee, if the latter writes a comment in the record book without even visiting the art-exhibition.

In fact, the three criteria attached to a command-imperative is not always applicable in case of other imperatives, i.e., suggestion, invitation, request or advice. The reason is this. In all cases of imperatives, the dictates are connected with actions. But the demand for execution of the acts is different in different cases of imperatives. The same spirit is found in Indian thought also. As in suggestion, invitation, request and advice, so in case of *āmantraṇa*, *adhīṣṭa* and *prārthanā*, there is no inbuilt compulsion to execute the act. So, in case of the action being executed or obeyed, the criterion of satisfaction is fulfilled, but nobody can meaningfully employ the term 'violated' if advice or *prārthanā* is not followed or granted respectively. The deeper reason lies in the fact that some purpose of the addressee is fulfilled by uttering advice or prayer, while no such purpose of the addressee is fulfilled in command-imperative. Secondly, there is a subtle difference between the motivation with which the imperative is fulfilled. Keeping in mind these two factors, an attempt may be made to bring imperatives of all types under a single interpretation.

It is better to suggest here four values of imperatives which are hierarchically arranged in the following way:

Four values of imperatives I (Command, request, prayer, advice etc.).
RA (rational or strong acceptance)
CA (Courtesy or weak acceptance) I = <RA, CA, AV, V>
AV (Avoidance or weak denial)
D (Denial or strong denial)

Negation

RA ------ D
CA ------ AV
AV ------ CA
D ------ RA

Conjunction Enjoy the show and be happy

	RA	CA	AV	D
RA	RA	CA	AV	D
CA	CA	CA	AV	D
AV	AV	AV	AV	D
D	D	D	D	D

Disjunction (inclusive) Be attentive to the lecture or take notes of the lecture

	RA	CA	AV	D
RA	RA	RA	RA	RA
CA	RA	CA	CA	CA
AV	RA	CA	AV	AV
D	RA	CA	AV	D

Disjunction (exclusive) Choose your friends wisely or you will suffer

	RA	CA	AV	D
RA	D	AV	CA	RA
CA	AV	AV	CA	CA
AV	CA	CA	AV	AV
D	RA	CA	AV	D

Disjunction (Inclusive)

~ p	v	p
D	RA	RA
AV	CA	CA
CA	CA	AV
RA	RA	D

Disjunction (exclusive)

~ p	v	p
D	RA	RA
AV	CA	CA
CA	CA	AV
RA	RA	D

The acceptance table in the case of both inclusive and exclusive 'Or' is the same. The final column of ' ~p v p' in both the cases shows the value 'acceptance'.

From the acceptance tables mentioned before it is obvious that

a) X or Y = max of X and Y,
   X & Y = min of X and Y.
b) The rule of double negation is not accepted as a rule for this system.

In respect of (a) it is clearly mentioned that this is applicable for X and Y in some cases of imperatives. Often we get such cases where both disjuncts are imperative, and not a combination of declarative and imperative. For example,

Be attentive to the lecture or take notes of the lecture.

Here both or any one of the two can be satisfied. But it can not happen in case of the following case:

Choose your friends wisely or you will suffer.

Here the connective 'Or' can only be meaningfully used exclusively.It is to be further noted that though there are two uses of 'Or' in respect of imperative sentences, but a conditional sentence need not be interpreted in terms of acceptance table of any of the two uses of 'Or', for a conditional statement is a combination of declarative and imperative sentences. For such a statement we need separate table which will be given later.

As we have referred earlier, an imperative statement may have a declarative constituent part, e.g., the first premise in argument B. So we have to know also the conjunction and disjunction table for declarative and imperative statement together.

Let us take a statement C & T (C = declarative, T = Imperative)

Conjunction

	RA	CA	AV	D
T	RA	CA	AV	D
F	D	D	D	D

Disjunction

	RA	CA	AV	D
T	RA	CA	CA	AV
F	RA	CA	D	D

Implication

	RA	CA	AV	D
T	RA	CA	D	D
F	RA	CA	CA	AV

Now let us take the previous argument B. The symbolic form of the argument is as follows:

$$C \vee T$$
$$\sim T/\text{Therefore, } C.$$

So far of the issues of deduction and validity are concerned, i.e., '$x \rightarrow y$' represents premise-conclusion relation, it is to be noted that deduction in imperative logic can not be interpreted in the same way as that in case of ordinary two valued logic which is concerned with descriptive statements. In case of a valid inference '$x \rightarrow y$' in two valued logic, it can be said that y is <u>deduced</u> from x. But here deduction is understood in terms of truth. In case of a valid inference '$x \rightarrow y$', if x is true, y can not be false. In case of imperative inference, validity is defined in terms of satisfaction. If an inference '$x \rightarrow y$' is valid, then satisfaction of x is definitely followed by satisfaction of y.

Now we can test the validity of the argument by constituting a hypothetical statement containing conjunction of the premises as antecedent and the conclusion as the consequent. We may, however, retain the rules of inference and the definition of validity as proposed by Vranas. But it is important to note that it is not at all an extension of Vranas's theory. As per our present criterion of four-value measurement, the requirement of a valid argument is that the value 'D' is not present in the final column of the measurement-table of a valid argument. Let us consider the following table:

[(C	v	T)	&	~T]	⟶	C
RA	RA	T	D	F	RA	RA
RA	RA	F	RA	T	RA	RA
RA	CA	T	D	F	RA	CA
RA	CA	F	CA	T	CA	CA
AV	CA	T	D	F	RA	AV
AV	D	F	D	T	RA	AV
D	AV	T	D	F	RA	D
D	D	F	D	T	RA	D

In the same manner we can test all arguments involving imperative of any type by applying the values mentioned before. It is possible to justify the acceptance tables by applying them to other standard tautologies i.e., $p \rightarrow p$, or $[p \rightarrow (q \rightarrow r)] \rightarrow [(p \rightarrow q) \rightarrow (p \rightarrow r)]$ etc. The task remains to show the soundness and completeness of the system, which will be undertaken in future.

References

1. Vranas, P.B.M.: New foundations for imperative logic i : logical connectives, consistency, and quantifiers. Nous **42**, 529–572 (2008)
2. Vranas, P.B.M.: In defense of imperative inference. J. Philos. Log. **39**, 59–71 (2010)
3. Vranas, P.B.M.: New foundations for imperative logic: pure imperative inference. Mind **120**, 369–446 (2011)
4. Vranas, P.B.M.: New foundations for imperative logic III: A general definition of argument validity. Synthese **193**(6), 1703–1753 (2016)
5. Matilal, B.K.: Perception: An Essay on Classical Indian Theories of Knowledge. Clarendon Press, Oxford (1986)
6. Pandurangi, K.T.: Purvamīmāṁsā from an interdisciplinary point of view. History of Science, Philosophy and Culture in Indian Civilization, II, 6, Motilal Banarsidass, Jawahar Nagar, Delhi (2006)
7. Matilal, B.K., Sen, P.K.: The context principle and some Indian controversies over meaning. Mind **97**(385), 73–97 (1988)
8. Śabarasvāmin, Śābarabhāṣya: Jaimini Mīmāṁsāsūtra. Pune: Anandashram Sanskaran (1976)
9. Rāmānujācārya: Tantrarahasya: A Primer of Prābhākara Mīmāṁsā. In: Rāmaswami Śāstri Śiromaṇi, K.S. (ed.) Oriental Institute, Baroda, IV 11.1 (1956)
10. Srinivasan, B., Ranjani, P.: An intelligent task analysis approach for special education based on MIRA. J. Appl. Log. **11**(1), 137–145 (2013)
11. Srinivasan, B., Ranjani, P.: Scan enabled applications for persons with mobility impairment (SABARI). In: IEEE Region 10 humanitarian technology conference (R10HTC), pp. 105–110 (2014)
12. Dīkṣita, Bhaṭṭojī, Siddhāntakaumudi, Srisa Chandra Vasu, The Panini Office, Allahabad (1904)
13. Taittirīya Saṃhitā Alladi Mahadeva Sastry, Bharatiya Kala Prakashan, 1/5/9/1 (2004)
14. Pāṇīni,The Aṣṭādhyāyī of Pāṇīni, Sumitra M. Katre, Motilal Banarsidass, New Delhi (1989)
15. Mammata Kāvyaprakāśa, Ganganath Jha, Chaukhamba Prakashan, Varanasi (2022)
16. Jaimini, Mīmāṃsāsūtra, Mohan Lal Sandal, Motilal Banarsidass (1980)
17. Gautama, Nyāyasūtra, Mahamahopadhyaya Satis Chandra Vidyabhusan, The Panini Office, Bhuvaneshvari Ashram, Bahadurganj, Allahabad (1913)
18. Miśra Maṇḍana, Vidhiviveka, Kanchana Natarajan, Sri Satguru Publications (1995)

19. Charlow, N.: The meaning of imperatives. Philos. Compass **9**(8), 540–555 (2014)
20. Castañeda, H.N.: Thinking and Doing. D. Reidel Publishing Co., Dordrecht & Boston (1975)
21. Recanati, F.: Content Mood and Force. Philos. Compass **8**(7), 622–632 (2013). p. 623
22. Geach P.: Ascriptivism. Reprinted in Logic Matters, pp. 250–254. Blackwell, Oxford, (1960). 1972
23. Geach P.: Assertion. Reprinted in Logic Matters, pp. 254–269. Blackwell, Oxford, (1965). 1972
24. Hanks, P.: The content-force distinction. Philos. Stud. **134**, 141–164 (2007)
25. Soames, S.: Philosophy of Language. Princeton University Press, Princeton (2010)
26. Charlow, N.: p. 542 (2014)
27. Charlow, N.: pp. 543–544 (2014)

Quasi-Boolean Based Models in Rough Set Theory: A Case of Covering

Masiur Rahaman Sardar(✉)

Department of Mathematics, City College, Kolkata 700009, West Bengal, India
masiur_sardar@citycollegekolkata.org

Abstract. Rough set theory has already been algebraically investigated for decades and quasi-Boolean algebra has formed a basis for several structures related to rough sets. An initiative has been taken in the paper [17] to obtain rough set models for some of these structures. These models have been constructed by defining a g-approximation space $\langle U, R^g\rangle$ out of a generalised approximation space $\langle U, R\rangle$ and an involution g. In this paper, as a continuation of [17], we have thrown light on covering cases and constructed a set model for the algebra IqBa2 [17].

Keywords: Rough set theory · Pre-rough algebra · Quasi-Boolean algebra · Modal logic

1 Introduction

Rough set theory has already been algebraically investigated for decades and quasi-Boolean algebra (qBa) has formed a base for a number of abstract algebras emerging out of rough sets [11]. Pre-rough algebra, amongst them, is one and it was defined by Banerjee and Chakraborty in [3]. The base of pre-rough algebra is a quasi-Boolean algebra which is a more general structure than a Boolean algebra as the law of excluded middle ($x \vee \sim x = 1$) and law of contradiction ($x \wedge \sim x = 0$) generally do not hold in a qBa. Topological quasi-Boolean algebra (tqBa) and topological quasi-Boolean algebra with modal axiom S_5 (tqBa5) come naturally as predecessors of pre-rough algebra.

Later, from different motivations, many abstract algebras stronger than qba but weaker than pre-rough algebra were developed. As for example, systemI algebra, systemII algebra [14] etc. have been introduced in order to access the rough implication $\rightarrow$ which was defined as $x \rightarrow y \equiv (\neg Ix \vee Iy) \wedge (\neg Cx \vee Cy)$ in pre-rough and rough algebras [2,3], where $C \equiv \neg I \neg$. On the other hand, three intermediate algebras IA1 (intermediate algebra of type 1), IA2 (intermediate algebra of type 2) and IA3 (intermediate algebra of type 3) [15,19] are defined based on three intermediate properties viz. $\neg Ix \vee Ix = 1$, for all x (IP1), $I(x \vee y) = Ix \vee Iy$, for all x, y (IP2) and $Ix \leq Iy$ and $Cx \leq Cy$ imply $x \leq y$, for all x, y (IP3) which play a crucial role to define rough implication.

Besides this, 3-valued Łukasiewicz (Monteiro) algebra [4], 3-valued Łukasiewicz (Moisil) algebra [5], Tetravalent Modal Algebra (TMA) [7] are some

M. Banerjee and A. V. Sreejith (Eds.): ICLA 2023, LNCS 13963, pp. 159–171, 2023.
https://doi.org/10.1007/978-3-031-26689-8_12

of the well-established algebraic structures based on quasi-Boolean algebra. It has been established in [1,14] that 3-valued Łukasiewicz (Monteiro) algebra and 3-valued Łukasiewicz (Moisil) algebra are equivalent to pre-rough algebra. Whereas in [14], it was observed that TMA is stronger than tqBa5 but weaker than a pre-rough algebra. In the same paper [14], it has been mentioned that the algebra MDS5 [6] is equivalent to IA2 if lattice distributivity is added to MDS5. A relationship diagram amongst the aforesaid algebras is shown in Fig. 1. For details of these algebras and their logics we refer to [2,3,14,18].

In the paper [3], a rough set model has been constructed for the abstract pre-rough algebra. It was developed in the context of rough set theory specially based on the notions of rough equality and rough inclusion. It has been described in [3] as follows. Let $\langle U, R\rangle$ be an approximation space. Two subsets P and Q of U are said to be roughly equal if $\underline{P}_R = \underline{Q}_R$ and $\overline{P}^R = \overline{Q}^R$ where $\underline{P}_R$ and $\overline{P}^R$ are Pawlakian lower and upper approximations of P respectively. An equivalence relation $\preccurlyeq$ is defined in 2^U, the power set of U, as $P \preccurlyeq Q$ if and only if P and Q are roughly equal. Each equivalence class $[P]_{\preccurlyeq}$ of $2^U/_{\preccurlyeq}$ is called a rough set. Using these rough sets and suitable operations $\sqcap, \sqcup, \neg$ and I, $\langle 2^U/_{\preccurlyeq}, \sqcap, \sqcup, \neg, I, [\emptyset]_{\preccurlyeq}, [U]_{\preccurlyeq}\rangle$ is a model of an abstract pre-rough algebra. The operations $\sqcap, \sqcup, \neg$ and I are defined as

$$[P]_{\preccurlyeq} \sqcap [Q]_{\preccurlyeq} = [P \sqcap Q]_{\preccurlyeq},$$
$$[P]_{\preccurlyeq} \sqcup [Q]_{\preccurlyeq} = [P \sqcup Q]_{\preccurlyeq},$$
$$\neg[P]_{\preccurlyeq} = [\neg P]_{\preccurlyeq},$$
$$I[P]_{\preccurlyeq} = [IP]_{\preccurlyeq},$$

where

$$P \sqcap Q = (P \cap Q) \cup (P \cap \overline{Q}^R \cap (\overline{P \cap Q}^R)^c),$$
$$P \sqcup Q = (P \cup Q) \cap (P \cup \underline{Q}_R \cup (\underline{P \cup Q}_R)^c),$$
$$\neg P = P^c,$$
$$IP = \underline{P}_R,$$

$\cap, \cup$ and c being the set theoretic intersection, union and complementation. The lattice order $\sqsubseteq$ in the above pre-rough algebra is $[P]_{\preccurlyeq} \sqsubseteq [Q]_{\preccurlyeq}$ if and only if P is roughly included in Q, i.e., $\underline{P}_R \subseteq \underline{Q}_R$ and $\overline{P}^R \subseteq \overline{Q}^R$.

But, there are no proper set theoretic rough set models of the abstract algebras shown in Fig. 1 which are really weaker than pre-rough algebras. The phrase 'proper set theoretic rough set model' means that it should be a set model and should not reduce to a pre-rough algebra. A step has been taken in this regard in the paper [18]. In this paper, set models of System0, stqBa, stqBa-D, stqBa-T, stqBa-B, tqBa, tqBa5 and IA1 have been developed using the relation-based rough set theory.

Another direction of work was initiated in the papers [15,17]. In these papers, the authors have considered those algebras where an implication ($\rightarrow$) satisfying the property ($P_{\rightarrow}$): $x \leq y$ if and only if $x \rightarrow y = 1$, for all x, y, can not be defined or not available till now. It is to be noted that an implication $\rightarrow$ satisfying the property ($P_{\rightarrow}$) is required in an algebra to develop the Hilbert-type logic system corresponding to the algebra. For construction of the said logic system, following Rasiowa, algebras are defined by imposing an implication $\rightarrow$ obeying the property ($P_{\rightarrow}$). These algebras are shown in Fig. 2 and for further information about the algebras and their logics one may see the papers [15,17]. Rough set models of some of the algebras IqBaO, IqBaT, IqBa4, IqBa5, IqBa1, IqBa1,T, IqBa1,4 and IqBa1,5 have been presented in [17].

This current paper deals with a parallel type of research that has been initiated in our earlier papers [17,18]. In fact, in this paper, covering cases are considered and one set model has been developed using "deleted neighborhood", in other words, anti-reflexive neighborhood that has importance in a number of areas of computer applications, e.g., the field of computer security [9].

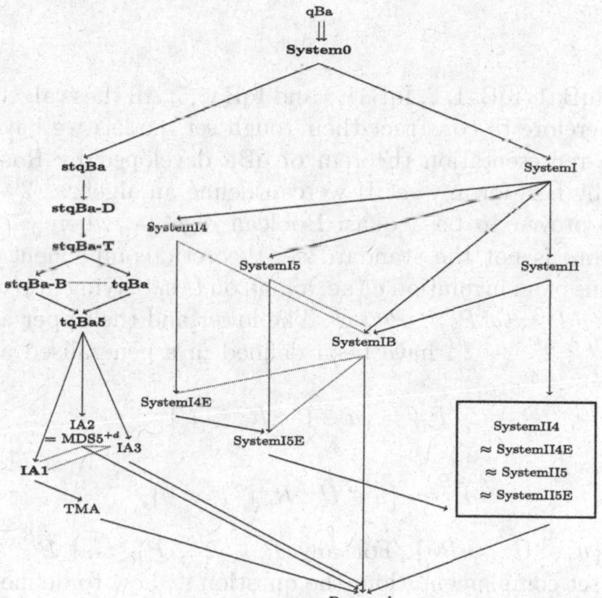

Fig. 1. Structures based on qBa: $P \rightrightarrows Q$ stands for the algebra Q has one more operator and some axioms for the new operator than the algebra P. $P \longrightarrow Q$ stands for both the algebras P and Q have the same operations and the algebra Q is always the algebra P. $P \cdots Q$ stands for the algebras P and Q are independent.

2 Rough Set Models - Relational Approach

In the papers [17,18], rough set models have been presented for the algebras System0, stqBa, stqBa-D, stqBa-T, stqBa-B, tqBa, tqBa5, IA1, IqBaO, IqBaT,

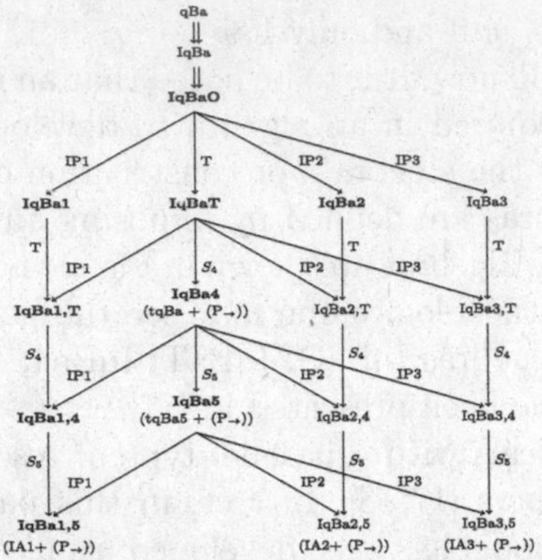

Fig. 2. Algebras with imposed implications: $P \rightrightarrows Q$ stands for the algebra Q has one more operator and some axioms for the new operator than the algebra P. $P \longrightarrow Q$ stands for both the algebras P and Q have the same operations and the algebra Q is always the algebra P.

IqBa4, IqBa5, IqBa1, IqBa1,T, IqBa1,4 and IqBa1,5. All these algebras are based on qBa and therefore to construct their rough set models we have focused our attention on a representation theorem of qBa developed by Rasiowa [13]. As demonstrated by her, for any set U we can define an algebra $\langle 2^U, \cap, \cup, \neg, \emptyset, U\rangle$ which may be proved to be a quasi Boolean algebra, where $\neg$, called quasi-complementation, is not the standard set-theoretic complementation ^{c} but is defined by means of an involution g (i.e. a map on U satisfying $g(g(u)) = u$, for all $u \in U$) namely $\neg P = (g(P))^c$, $P \subseteq U$. The lower and the upper approximation operators $\_{R}, {}^{-R} : 2^U \rightarrow 2^U$ have been defined in a generalised approximation space $\langle U, R\rangle$ by

$$\underline{P}_R = \{u \in U : R_u \subseteq P\}$$

and

$$\overline{P}^R = \{u \in U : R_u \cap P \neq \emptyset\},$$

where $R_u = \{v \in U : uRv\}$. For any $P \in 2^U$, $\underline{P}_R$ and $\overline{P}^R$ are dual with respect to the set complementation; the question is, how to define the algebraic counterparts of these operators in the aforementioned quasi-Boolean algebra, so as to make them dual with respect to the quasi-complementation $\neg$. The issue has been resolved by defining a g-approximation space $\langle U, R^g\rangle$ out of a generalised approximation space $\langle U, R\rangle$ and an involution g on U.

Let $\langle U, R\rangle$ be a generalised approximation space and $g : U \rightarrow U$ be an involution. A binary relation R^g on U has been defined as follows:

$$\text{for any two elements } u \text{ and } v \in U, uR^g v \text{ if and only if } g(u)Rg(v). \quad (1)$$

That is, two elements $u, v \in U$ are related with respect to a new relation R^g if and only if their g-images are related in the relation R. We call $\langle U, R^g\rangle$ a g-generalised approximation space or simply, a g-approximation space.

As g is an involution on U, R can be obtained from R^g as follows:

$$\text{for any two elements } u \text{ and } v \in U, uRv \text{ if and only if } g(u)R^g g(v). \quad (2)$$

Similarly, it says that two elements $u, v \in U$ will be related in the relation R if and only if their g-images are so in the relation R^g.

In this g-approximation space $\langle U, R^g \rangle$, we have defined g-lower approximation and g- upper approximation $\underline{\ }_g, \overline{\ }^g : 2^U \to 2^U$ as follows:
for any $P \in 2^U$,

$$\underline{P}_g = \{u \in U : R^g_u \subseteq P\}$$

and

$$\overline{P}^g = \{u \in U : R^g_{g(u)} \cap g(P) \neq \emptyset\}$$

where $R^g_u = \{v \in U : uR^g v\}$. Using these lower-upper approximations and imposing conditions like reflexivity, symmetric, transitivity etc. on R^g proper rough set models of System0, stqBa, stqBa-D, stqBa-T, stqBa-B, tqBa, tqBa5, IA1 have been constructed in [18].

To construct rough set models for the algebras IqBaO, IqBaT, IqBa4, IqBa5 shown in Fig. 2, a suitable operation that corresponds to $\to$ (available in the above algebras) is needed. Boolean implication $P \to Q (\equiv P^c \cup Q)$, in one way, serves the purpose smoothly. On the other hand, g image of Boolean implication $g(P \to Q)(\equiv P \to_1 Q)$ also fulfils the property $(\mathrm{P}_\to)$. With their help, rough set models of IqBaO, IqBaT, IqBa4, IqBa5 have been presented in [17].

A pair of new approximation operators $\underline{\ }_{g,1}, \overline{\ }^{g,1} : 2^U \to 2^U$ has been defined [17] in order to obtain set models for the algebras IqBa1, IqBa1,T, IqBa1,4 and IqBa1,5 as follows:

$$\underline{P}_{g,1} = \{u \in U : R^g_u \subseteq P\} \cap \{u \in U : R^g_{g(u)} \subseteq P\}$$

and

$$\overline{P}^{g,1} = \{u \in U : R^g_{g(u)} \cap g(P) \neq \emptyset\} \cup \{u \in U : R^g_u \cap g(P) \neq \emptyset\}.$$

For details, one may see the paper [17].

3 Rough Set Model - Covering Based Approach

In this section we shall discuss the covering based rough sets and incorporate the involution g to construct lower-upper approximations so that they will be dual approximations with respect to the quasi-complementation. As we have constructed two types of lower-upper approximations based on a binary relation, some natural questions may arise on covering cases in the following form:

- How can a parallel study be introduced on covering based rough set theory and what would be outcomes in that case?
- Is it possible to develop rough set models of some of the remaining algebras through this study?

In response to the first question, we have defined a g-covering approximation space out of a covering approximation space and an involution g. Thereafter, the basic notions like *Friends* of u, *Neighborhood* of u etc. are introduced in a g-covering approximation space in the same way as they have been defined in a covering approximation space. Relationships between these two spaces and the above-mentioned notions are studied.

For the last question, a new type of collection at each point of a g-covering approximation space has been developed. We call it a "deleted neighborhood". For the importance of this neighborhood, we have taken the following words as it is from the paper [9]: "a neighborhood N(p) of p may be punctured or empty; by that we mean the neighborhood does not contain its center p or is an empty set. Such a neighborhood is called an anti-reflexive neighborhood, including the case of empty neighborhood. It is useful in many applications, e.g., in computer security. We may consider a set of "my" enemies as a neighborhood. Surely, "myself" is not included in that set".

With the help of this deleted neighborhood or anti-reflexive neighborhood, lower-upper approximations have been defined. A rough set model of IqBa2 has been presented using these lower-upper approximations.

3.1 Basics in a Covering Approximation Space

Definition 1 [16] *(Covering of a set): Let* U *be a non empty set and* $\mathcal{C} = \{U_i(\neq \emptyset) \subseteq U : i \in I\}$*, where* I *is an index set, is said to be a covering of* U *if*

$$\bigcup_{i\in I} U_i = U.$$

Definition 2 [16] *(Covering approximation space): Let* U *be a non empty set and* $\mathcal{C}$ *be a covering of* U*. Then, the ordered pair* $\langle U, \mathcal{C}\rangle$ *is called a covering approximation space.*

Definition 3. *Let* $\langle U, \mathcal{C}\rangle$ *be a covering approximation space. For each* $u \in U$*,*

1. *(Friends of* u*):* [16] *Friends of* u *is defined by*

$$F^{\mathcal{C}}(u) = \bigcup_{u\in U_i} U_i.$$

   It is also called the indiscernible neighborhood of u [10].

2. *(Neighborhood of* u*):* [16] *Neighborhood of* u *is defined by*

$$N^{\mathcal{C}}(u) = \bigcap_{u\in U_i} U_i.$$

3. *(Friends' enemy of* u*):* [10,16] *Friends' enemy of* u *is defined by*

$$FE^{\mathcal{C}}(u) = U - F^{\mathcal{C}}(u).$$

4. *(Kernel of u):* [16] *Kernel of u is defined by*

$$K^{\mathcal{C}}(u) = \{y \in U : \forall U_i (u \in U_i \Leftrightarrow y \in U_i)\}.$$

Let $\mathcal{P}^{\mathcal{C}} = \{K^{\mathcal{C}}(u) : u \in U\}$. *Then,* $\mathcal{P}^{\mathcal{C}}$ *is a partition of* U *and called a partition generated by the covering* $\mathcal{C}$.

5. *(Minimal description and Maximal description of u):* [12,20] *Minimal description and Maximal description of u are defined respectively as*

$$md^{\mathcal{C}}(u) = \{U_i \in \mathcal{C} : u \in U_i \text{ and } \forall U \in \mathcal{C}(u \in U \subseteq U_i \text{ implies } U = U_i)\}$$

and

$$Md^{\mathcal{C}}(u) = \{U_i \in \mathcal{C} : u \in U_i \text{ and } \forall U \in \mathcal{C}(U \supseteq U_i \text{ implies } U = U_i)\}.$$

We are now going to define a g-covering approximation space in the following way.

3.2 g-covering Approximation Space

Proposition 1. *Let* $\langle U, \mathcal{C}\rangle$ *be a covering approximation space and* $g: U \to U$ *be an involution, i.e.,* $g(g(u)) = u$, *for all* $u \in U$. *Then* $g(\mathcal{C}) = \{g(U_i) : U_i \in \mathcal{C}\}$ *is a covering of* U.

Proof. Let $u \in U$. Then, $g(u) \in U_i$, for some $i \in I$ (As, $\mathcal{C} = \{U_i(\neq \emptyset) \subseteq U : i \in I\}$ is a covering of U). Then, by the definition of g, $u \in g(U_i)$ and hence $g(\mathcal{C}) = \{g(U_i) : U_i \in \mathcal{C}\}$ is a covering of U.

From the above proposition, we now define a g-covering approximation space below.

Definition 4. *Let* $\langle U, \mathcal{C}\rangle$ *be a covering approximation space and* g *be an involution on* U. *Then,* $\langle U, g(\mathcal{C})\rangle$ *will be called a* g*-covering approximation space.*

In general, $\mathcal{C} \neq g(\mathcal{C})$. The following example supports the statement.

Example 1. Let $U = \{a, b, c, d, e\}$, $\mathcal{C} = \{U_1 = \{a, b\}, U_2 = \{d, e\}, U_3 = \{c, e\}\}$ be a covering of U and $g : U \to U$ be an involution defined by $g(a) = c, g(b) = d, g(c) = a, g(d) = b, g(e) = e$. Now, g($\mathcal{C}$) $= \{g(U_1) = \{c, d\}, g(U_2) = \{b, e\}, g(U_3) = \{a, e\}\}$ and hence $\mathcal{C} \neq g(\mathcal{C})$.

The following is a necessary and sufficient condition that reveals when $\mathcal{C}$ and $g(\mathcal{C})$ coincide.

Proposition 2. *Let* $\langle U, g(\mathcal{C})\rangle$ *be a* g*-covering approximation space. Then* $\mathcal{C} = g(\mathcal{C})$ *if and only if for each* $i \in I$, $g(U_i) = U_j$, *for some* $j \in I$.

Proof. Let $\mathcal{C} = g(\mathcal{C})$ and $U_i \in \mathcal{C}$, for any $i \in I$. Then $U_i \in g(\mathcal{C})$ [as $\mathcal{C} = g(\mathcal{C})$]. This gives, $U_i = g(U_j)$, for some $j \in I$. Conversely, let for each $i \in I$ there exist $j \in I$ such that $g(U_i) = U_j$. We have to show that $\mathcal{C} = g(\mathcal{C})$. Let $U_i \in \mathcal{C}$. Then by the hypothesis $g(U_i) = U_j$, for some $j \in I$. Then by the definition of g, $U_i = g(U_j)$. As $U_j \in \mathcal{C}$, $g(U_j) \in g(\mathcal{C})$, i.e., $U_i \in g(\mathcal{C})$. Thus, $\mathcal{C} \subseteq g(\mathcal{C})$. Let $Y \in g(\mathcal{C})$. Then $Y = g(U_j)$, for some $U_j \in \mathcal{C}$. Then by the hypothesis $g(U_j) = U_k$, for some $k \in I$. Thus, $Y = U_k \in \mathcal{C}$ and hence $g(\mathcal{C}) \subseteq \mathcal{C}$.

Note 1. If $g(U_i) = U_i$, for all $i \in I$ then $\mathcal{C} = g(\mathcal{C})$. But the converse, i.e., $\mathcal{C} = g(\mathcal{C})$ implies $g(U_i) = U_i$, for all $i \in I$, is not true as shown by an example given below.

Example 2. Let U and g be the same as mentioned in Example 1. Let $\mathcal{C} = \{U_1 = \{a,e\}, U_2 = \{c,e\}, U_3 = \{b\}, U_4 = \{d\}\}$ be a covering of U. Then, $g(\mathcal{C}) = \{g(U_1) = \{c,e\}, g(U_2) = \{a,e\}, g(U_3) = \{d\}, g(U_4) = \{b\}\}$ and hence $\mathcal{C} = g(\mathcal{C})$ but for none of $i = 1,2,3,4$, $g(U_i) = U_i$.

Now, we define the notions of *Friends* of u, *Neighborhood* of u etc. in a g-covering approximation space $\langle U, g(\mathcal{C})\rangle$.

Definition 5. *Let $\langle U, g(\mathcal{C})\rangle$ be a g-covering approximation space. Then for each $u \in U$,*

1. *Friends of u is defined by*

$$F^{g(\mathcal{C})}(u) = \bigcup_{u \in g(U_i)} g(U_i),$$

2. *Neighborhood of u is defined by*

$$N^{g(\mathcal{C})}(u) = \bigcap_{u \in g(U_i)} g(U_i),$$

3. *Friends' enemy of u is defined by*

$$FE^{g(\mathcal{C})}(u) = U - F^{g(\mathcal{C})}(u),$$

4. *Kernel of u is defined by*

$$K^{g(\mathcal{C})}(u) = \{y \in U : \forall g(U_i)(u \in g(U_i) \Leftrightarrow y \in g(U_i))\}.$$

   Let $\mathcal{P}^{g(\mathcal{C})} = \{K^{g(\mathcal{C})}(u) : u \in U\}$. Then, $\mathcal{P}^{g(\mathcal{C})}$ is a partition of U and hence it will be called partition generated by the covering $g(\mathcal{C})$.
5. *Minimal description of u is defined by* $md^{g(\mathcal{C})}(u) = \{g(U_i) \in g(\mathcal{C}) : u \in g(U_i)$ *and* $\forall X \in g(\mathcal{C})(u \in X \subseteq g(U_i)$ *implies* $X = g(U_i))\}$.
6. *Maximal description of u is defined by* $Md^{g(\mathcal{C})}(u) = \{g(U_i) \in g(\mathcal{C}) : u \in g(U_i)$ *and* $\forall X \in g(\mathcal{C})(X \supseteq g(U_i)$ *implies* $X = g(U_i))\}$.

The following example is considered to show that *Friends* of u, *Neighborhood* of u etc. in a covering approximation space are generally not the same with *Friends* of u, *Neighborhood* of u etc. in the g-covering approximation space.

Example 3. Let $U, \mathcal{C}$ and g be the same as defined in Example 1. Considering $u = c$ we get

1. $F^{g(\mathcal{C})}(c) = g(U_1) = \{c,d\} \neq F^{\mathcal{C}}(c) = \{c,e\}$,
2. $N^{g(\mathcal{C})}(c) = g(U_1) = \{c,d\} \neq N^{\mathcal{C}}(c) = \{c,e\}$,
3. $FE^{g(\mathcal{C})}(c) = U - F^{g(\mathcal{C})}(c) = \{a,b,e\} \neq FE^{\mathcal{C}}(c) = \{a,b,d\}$,

4. $K^{g(\mathcal{C})}(c) = \{y \in U : \forall g(U_i)(c \in g(U_i) \Leftrightarrow y \in g(U_i))\} = \{c,d\} \neq K^{\mathcal{C}}(c) = \{c\}$,
5. $md^{g(\mathcal{C})}(c) = \{g(U_i) \in g(\mathcal{C}) : c \in g(U_i)$ and $\forall X \in g(\mathcal{C})(c \in X \subseteq g(U_i)$ implies $X = g(U_i))\} = \{g(U_1) = \{c,d\}\} \neq md^{\mathcal{C}}(c) = \{U_3 = \{c,e\}\}$,
6. $Md^{g(\mathcal{C})}(c) = \{g(U_i) \in g(\mathcal{C}) : c \in g(U_i)$ and $\forall X \in g(\mathcal{C})(U \supseteq g(U_i)$ implies $X = g(U_i))\} = \{g(U_1) = \{c,d\}\} \neq Md^{\mathcal{C}}(c) = \{U_3 = \{c,e\}\}$.
7. $\mathcal{P}^{g(\mathcal{C})} = \{K^{g(\mathcal{C})}(u) : u \in U\} = \{K^{g(\mathcal{C})}(a) = \{a\}, K^{g(\mathcal{C})}(b) = \{b\}, K^{g(\mathcal{C})}(c) = \{c,d\} = K^{g(\mathcal{C})}(d), K^{g(\mathcal{C})}(e) = \{e\}\} \neq \mathcal{P}^{\mathcal{C}} = \{K^{\mathcal{C}}(a) = \{a,b\} = K^{\mathcal{C}}(b), K^{\mathcal{C}}(c) = \{c\}, K^{\mathcal{C}}(d) = \{d\}, K^{g(\mathcal{C})}(e) = \{e\}\}$

Proposition 3. *Let $\langle U, g(\mathcal{C})\rangle$ be a g-covering approximation space. Then,*

1. $F^{\mathcal{C}}(u) = g(F^{g(\mathcal{C})}(g(u)))$ *and* $F^{g(\mathcal{C})}(u) = g(F^{\mathcal{C}}(g(u)))$, *for all* $u \in U$,
2. $N^{\mathcal{C}}(u) = g(N^{g(\mathcal{C})}(g(u)))$ *and* $N^{g(\mathcal{C})}(u) = g(N^{\mathcal{C}}(g(u)))$, *for all* $u \in U$,
3. $FE^{\mathcal{C}}(u) = g(FE^{g(\mathcal{C})}(g(u)))$ *and* $FE^{g(\mathcal{C})}(u) = g(FE^{\mathcal{C}}(g(u)))$, *for all* $u \in U$,
4. $K^{\mathcal{C}}(u) = g(K^{g(\mathcal{C})}(g(u)))$ *and* $K^{g(\mathcal{C})}(u) = g(K^{\mathcal{C}}(g(u)))$, *for all* $u \in U$,
5. $md^{\mathcal{C}}(u) = g(md^{g(\mathcal{C})}(g(u)))$ *and* $md^{g(\mathcal{C})}(u) = g(md^{\mathcal{C}}(g(u)))$, *for all* $u \in U$,
6. $Md^{\mathcal{C}}(u) = g(Md^{g(\mathcal{C})}(g(u)))$ *and* $Md^{g(\mathcal{C})}(u) = g(Md^{\mathcal{C}}(g(u)))$, *for all* $u \in U$,
7. $\mathcal{P}^{\mathcal{C}} = g(\mathcal{P}^{g(\mathcal{C})})$ *and* $\mathcal{P}^{g(\mathcal{C})} = g(\mathcal{P}^{\mathcal{C}})$, *where* $g(\mathcal{P}^{g(\mathcal{C})}) = \{g(Y) : Y \in \mathcal{P}^{g(\mathcal{C})}\}$ *and similarly for* $g(\mathcal{P}^{\mathcal{C}})$.

Proof. 1. Let $y \in F^{\mathcal{C}}(u)$. Then, $y, u \in U_j$, for some $j \in I$ and hence $g(y), g(u) \in g(U_j)$. This gives, $g(y) \in F^{g(\mathcal{C})}(g(u))$ and hence $g(g(y)) \in g(F^{g(\mathcal{C})}(g(u)))$, i.e., $y \in g(F^{g(\mathcal{C})}(g(u)))$. Thus $F^{\mathcal{C}}(u) \subseteq g(F^{g(\mathcal{C})}(g(u)))$. Let $y \in g(F^{g(\mathcal{C})}(g(u)))$. Then $y = g(z)$, where $z \in F^{g(\mathcal{C})}(g(u))$. This implies, $z, g(u) \in g(U_k)$, for some $k \in I$ and therefore $g(z)$, $g(g(u)) \in g(g(U_k))$, i.e., $y = g(z)$, $u \in U_k$. This gives, $y \in F^{\mathcal{C}}(u)$ and therefore $g(F^{g(\mathcal{C})}(g(u))) \subseteq F^{\mathcal{C}}(u)$. Thus, $F^{\mathcal{C}}(u) = g(F^{g(\mathcal{C})}(g(u)))$. Proofs of 2, 3, 4, 5, 6 and 7 can be done similarly.

It is time now to define deleted neighborhood or anti-reflexive neighborhood of an element u in U in order to develop a rough set model for the algebra IqBa2.

Definition 6. *Let $\langle U, g(\mathcal{C})\rangle$ be a g-covering approximation space. For each $u \in U$, deleted Neighbourhood of u in the covering approximation space $\langle U, \mathcal{C}\rangle$ and in the g-covering approximation space $\langle U, g(\mathcal{C})\rangle$, denoted by $N_d^{\mathcal{C}}(u)$ and $N_d^{g(\mathcal{C})}(u)$ respectively, are defined by $N_d^{\mathcal{C}}(u) = N^{\mathcal{C}}(u) - \{u\}$ and $N_d^{g(\mathcal{C})}(u) = N^{g(\mathcal{C})}(u) - \{u\}$.*

Note 2. For each $u \in U$, u does not belong to $N_d^{\mathcal{C}}(u)$ and $N_d^{g(\mathcal{C})}(u)$. Moreover, $N_d^{\mathcal{C}}(u)$ or $N_d^{g(\mathcal{C})}(u)$ may be empty for some $u \in U$.

Proposition 4. *Let $\langle U, g(\mathcal{C})\rangle$ be a g-covering approximation space. Then for each $u \in U$, $N_d^{\mathcal{C}}(u) = g(N_d^{g(\mathcal{C})}(g(u)))$ and $N_d^{g(\mathcal{C})}(u) = g(N_d^{\mathcal{C}}(g(u)))$.*

Proof.

$$N_d^{g(\mathcal{C})}(g(u)) = N^{g(\mathcal{C})}(g(u)) - \{g(u)\} \text{ [from Definition 6]}$$

$$\begin{aligned}\text{Then, } g(N_d^{g(\mathcal{C})}(g(u))) &= g(N^{g(\mathcal{C})}(g(u)) - \{g(u)\}) \\ &= g(N^{g(\mathcal{C})}(g(u))) - g(\{g(u)\}) \text{ [as } g(A-B) = g(A) - g(B)] \\ &= N^{\mathcal{C}}(u) - \{u\} \text{ [by 2 of Proposition 3]} \\ &= N_d^{\mathcal{C}}(u) \text{ [from Definition 6]}\end{aligned}$$

Similarly, the other part can be proved.

3.3 Rough Set Model for IqBa2

For a set-theoretic rough set model of the algebra IqBa2, we have to develop a pair of lower-upper approximations which must be dual with respect to the quasi-complementation and satisfies the property IP2: $I(a \vee b) = Ia \vee Ib$. Due to this reason, we define a new pair of lower-upper approximations as follows.

Definition 7. *Let $\langle U, g(\mathcal{C})\rangle$ be a g-covering approximation space. Then for any subset A of U, $\underline{A}_{g(\mathcal{C}),2}$, the $g,2$ lower approximation of A and $\overline{A}^{g(\mathcal{C}),2}$, the $g,2$ upper approximation of A are defined by*

$$\underline{A}_{g(\mathcal{C}),2} = \{u \in U : N_d^{g(\mathcal{C})}(u) \subseteq A\} \tag{3}$$

and

$$\overline{A}^{g(\mathcal{C}),2} = \{u \in U : N_d^{\mathcal{C}}(u) \cap A \neq \emptyset\}. \tag{4}$$

Proposition 5. *In a g-covering approximation space $\langle U, g(\mathcal{C})\rangle$, $\underline{A}_{g(\mathcal{C}),2}$ and $\overline{A}^{g(\mathcal{C}),2}$ are dual approximations with respect to the quasi-complementation $\neg$ defined through g.*

Proof.

$$\begin{aligned}
\neg\left(\underline{\neg A}_{g(\mathcal{C}),2}\right) &= \neg\left(\underline{g(A)^c}_{g(\mathcal{C}),2}\right) \text{ [as } \neg A = (g(A))^c] \\
&= \neg\{u \in U : N_d^{g(\mathcal{C})}(u) \subseteq g(A)^c\} \text{ [by Definition 7]} \\
&= U - \{g(u) : N_d^{g(\mathcal{C})}(u) \subseteq g(A)^c\} \text{ [as } \neg A = U - g(A)] \\
&= U - \{u \in U : N_d^{g(\mathcal{C})}(g(u)) \subseteq g(A)^c\} \text{ [taking } g(u) \text{ as } u] \\
&= \{u \in U : N_d^{g(\mathcal{C})}(g(u)) \cap g(A) \neq \emptyset\} \\
&= \{u \in U : g(N_d^{\mathcal{C}}(u)) \cap g(A) \neq \emptyset\} \text{ [by Proposition 4]} \\
&= \{u \in U : g(N_d^{\mathcal{C}}(u) \cap A) \neq \emptyset\} \text{ [as } g(A \cap B) = g(A) \cap g(B)] \\
&= \{u \in U : N_d^{\mathcal{C}}(u) \cap A \neq \emptyset\} \text{ [as } g \text{ is an involution]} \\
&= \overline{A}^{g(\mathcal{C}),2}.
\end{aligned}$$

As $\neg\neg A = A$, hence $\underline{A}_{g(\mathcal{C}),2}$ and $\overline{A}^{g(\mathcal{C}),2}$ are dual approximations with respect to the quasi-complementation $\neg$ defined through g.

Proposition 6. *In a g-covering approximation space $\langle U, g(\mathcal{C})\rangle$, the following results hold.*

1. $\underline{X}_{g(\mathcal{C}),2} = U$ *and* $\overline{\emptyset}^{g(\mathcal{C}),2} = \emptyset$.
2. *If* $A \subseteq B \subseteq U$ *then* $\underline{A}_{g(\mathcal{C}),2} \subseteq \underline{B}_{g(\mathcal{C}),2}$ *and* $\overline{A}^{g(\mathcal{C}),2} \subseteq \overline{B}^{g(\mathcal{C}),2}$.
3. $\underline{A \cap B}_{g(\mathcal{C}),2} = \underline{A}_{g(\mathcal{C}),2} \cap \underline{B}_{g(\mathcal{C}),2}$ *and* $\overline{A \cup B}^{g(\mathcal{C}),2} = \overline{A}^{g(\mathcal{C}),2} \cup \overline{B}^{g(\mathcal{C}),2}$, *for all* $A, B \subseteq U$.

Proof. Proof of 1 is straightforward.
For proof of 2, let $x \in \underline{A}_{g(\mathcal{C}),2}$. Then by Definition 7, $N_d^{g(\mathcal{C})}(x) \subseteq A$ and hence $N_d^{g(\mathcal{C})}(x) \subseteq B$ (as $A \subseteq B$). This gives, $x \in \underline{B}_{g(\mathcal{C}),2}$ and hence $\underline{A}_{g(\mathcal{C}),2} \subseteq \underline{B}_{g(\mathcal{C}),2}$. Similarly, $\overline{A}^{g(\mathcal{C}),2} \subseteq \overline{B}^{g(\mathcal{C}),2}$ holds.
Proof of 3: To show $\underline{A \cap B}_{g(\mathcal{C}),2} = \underline{A}_{g(\mathcal{C}),2} \cap \underline{B}_{g(\mathcal{C}),2}$, we have to prove $\underline{A}_{g(\mathcal{C}),2} \cap \underline{B}_{g(\mathcal{C}),2} \subseteq \underline{A \cap B}_{g(\mathcal{C}),2}$. Let $x \in \underline{A}_{g(\mathcal{C}),2} \cap \underline{B}_{g(\mathcal{C}),2}$. Then, $N_d^{g(\mathcal{C})}(x) \subseteq A$ and B. Therefore, $N_d^{g(\mathcal{C})}(x) \subseteq A \cap B$ and hence $x \in \underline{A \cap B}_{g(\mathcal{C}),2}$. Thus, $\underline{A}_{g(\mathcal{C}),2} \cap \underline{B}_{g(\mathcal{C}),2} \subseteq \underline{A \cap B}_{g(\mathcal{C}),2}$ and hence the result is proved. Similarly, the other part of 3 can be proved.

Theorem 1. *In a g-covering approximation space* $\langle U, g(\mathcal{C})\rangle$, $\underline{A \cup B}_{g(\mathcal{C}),2} = \underline{A}_{g(\mathcal{C}),2} \cup \underline{B}_{g(\mathcal{C}),2}$ *holds for all* $A, B \subseteq U$ *if and only if for each* $u \in U$, $N_d^{g(\mathcal{C})}(u)$ *contains at most one element of* U.

Proof. Let $\underline{A \cup B}_{g(\mathcal{C}),2} = \underline{A}_{g(\mathcal{C}),2} \cup \underline{B}_{g(\mathcal{C}),2}$, for all $A, B \subseteq U$. It is to be proved that $N_d^{g(\mathcal{C})}(u)$ contains at most one element of U. If possible, let $N_d^{g(\mathcal{C})}(u)$ contain more than one element of U. Then, there are at least two distinct elements $y, z \in N_d^{g(\mathcal{C})}(u)$ where $y \neq u$ and $z \neq u$ [as $u \notin N_d^{g(\mathcal{C})}(u)$]. Let $A = \{y\}$ and $B = N_d^{g(\mathcal{C})}(u) - \{y\}$. Then $z \in B \neq \emptyset$. Then by hypothesis, $\underline{A \cup B}_{g(\mathcal{C}),2} = \underline{A}_{g(\mathcal{C}),2} \cup \underline{B}_{g(\mathcal{C}),2}$ holds, where $A = \{y\}$ and $B = N_d^{g(\mathcal{C})}(u) - \{y\}$. This gives, $\underline{N_d^{g(\mathcal{C})}(u)}_{g(\mathcal{C}),2} = \underline{A}_{g(\mathcal{C}),2} \cup \underline{B}_{g(\mathcal{C}),2}$. As $N_d^{g(\mathcal{C})}(u)$ is a subset of itself so $u \in \underline{N_d^{g(\mathcal{C})}(u)}_{g(\mathcal{C}),2} = \underline{A}_{g(\mathcal{C}),2} \cup \underline{B}_{g(\mathcal{C}),2}$. This implies, either $u \in \underline{A}_{g(\mathcal{C}),2}$ or $u \in \underline{B}_{g(\mathcal{C}),2}$, i.e., either $N_d^{g(\mathcal{C})}(u) \subseteq \{y\}$ or $N_d^{g(\mathcal{C})}(u) \subseteq N_d^{g(\mathcal{C})}(u) - \{y\}$. But we have $z \in N_d^{g(\mathcal{C})}(u) \nsubseteq \{y\}$ and $y \in N_d^{g(\mathcal{C})}(u) \nsubseteq N_d^{g(\mathcal{C})}(u) - \{y\}$. Thus, $N_d^{g(\mathcal{C})}(u)$ contains at most one element of U, for all $u \in U$.
Conversely, let us assume that each $N_d^{g(\mathcal{C})}(u)$ contains at most one element of U. We have to prove that $\underline{A \cup B}_{g(\mathcal{C}),2} = \underline{A}_{g(\mathcal{C}),2} \cup \underline{B}_{g(\mathcal{C}),2}$ holds for all $A, B \subseteq U$. By 2 of Proposition 6, it is sufficient to show that $\underline{A \cup B}_{g(\mathcal{C}),2} \subseteq \underline{A}_{g(\mathcal{C}),2} \cup \underline{B}_{g(\mathcal{C}),2}$. Let $u \in \underline{A \cup B}_{g(\mathcal{C}),2}$. Then, $N_d^{g(\mathcal{C})}(u) \subseteq A \cup B$. As $N_d^{g(\mathcal{C})}(u)$ contains at most one element, so, it follows that either $N_d^{g(\mathcal{C})}(u) \subseteq A$ or $N_d^{g(\mathcal{C})}(u) \subseteq B$ and hence $u \in \underline{A}_{g(\mathcal{C}),2} \cup \underline{B}_{g(\mathcal{C}),2}$. Thus, $\underline{A \cup B}_{g(\mathcal{C}),2} = \underline{A}_{g(\mathcal{C}),2} \cup \underline{B}_{g(\mathcal{C}),2}$, for all $A, B \subseteq U$.

Remark 1. As $\underline{A}_{g(\mathcal{C}),2}$ and $\overline{A}^{g(\mathcal{C}),2}$ are dual approximations with respect to the quasi-complementation $\neg$ and $A \cap B = \neg(\neg A \cup \neg B)$ so, $\overline{A \cap B}^{g(\mathcal{C}),2} = \overline{A}^{g(\mathcal{C}),2} \cap \overline{B}^{g(\mathcal{C}),2}$ holds for all $A, B \subseteq U$ if and only if for each $u \in U$, $N_d^{g(\mathcal{C})}(u)$ contains at most one element of U.

The following example is considered to show that $\underline{A}_{g(\mathcal{C}),2}$ may not be a subset of A, for some $A \subseteq U$.

Example 4. Let U, g and $\mathcal{C}$ be the same as defined in Example 1. Then, $N_d^{g(\mathcal{C})}(a) = \{e\}, N_d^{g(\mathcal{C})}(b) = \{e\}, N_d^{g(\mathcal{C})}(c) = \{d\}, N_d^{g(\mathcal{C})}(d) = \{c\}, N_d^{g(\mathcal{C})}(e) = \emptyset$. Let $A = \{e\}$. Then, $\underline{A}_{g(\mathcal{C}),2} = \{a, b, e\} \nsubseteq A = \{e\}$.

Rough Set model for IqBa2: Let $\langle U, g(\mathcal{C})\rangle$ be a g-covering approximation space with for each $u \in U$, $N_d^{g(\mathcal{C})}(u)$ contains at most one element of U. Now, $\langle 2^U, \cap, \cup, \neg, \emptyset, U\rangle$ is a qBa, where $\neg A = (g(A))^c$, for all $A \in 2^U$. We define $\rightarrow$ in 2^U as follows
$A \rightarrow B \doteq A^c \cup B$.
Then, it is obvious that $A \rightarrow B = U$ if and only if $A \subseteq B$ and consequently $\langle 2^U, \cap, \cup, \rightarrow, \neg, \emptyset, U\rangle$ becomes an IqBa. We now define IA, for all $A \subseteq U$ as $IA = \underline{A}_{g(\mathcal{C}),2}$. Then by Proposition 6 and Theorem 1, $\langle 2^U, \cap, \cup, \rightarrow, \neg, I, \emptyset, U\rangle$ is an IqBa2.

Remark 2.

1. If we define implication as $A \rightarrow_1 B = g(A \rightarrow B) = \neg A \cup g(B)$, for all $A, B \in 2^U$ then $\langle 2^U, \cap, \cup, \rightarrow_1, \neg, I/I_1, \emptyset, U\rangle$ becomes a different model for IqBa2 with respect to the implication $\rightarrow_1$.
2. By Example 4, modal axiom T: $Ia \leq a$ [8] does not hold and hence $\langle 2^U, \cap, \cup, \rightarrow, \neg, I, \emptyset, U\rangle$ is not a model for IqBa2,T.

4 Conclusion and Future Work

We may summarise the contents of this paper and indicate some future directions of work as follows.

- A g-covering approximation space has been developed out of a covering approximation space and an involution g. A necessary and sufficient condition is obtained so that these two spaces coincide.
- Familiar notions that are available in a covering approximation space have been introduced in a g-covering approximation space and relationships between them are studied.
- Deleted neighborhood or anti-reflexive neighborhood has been incorporated in this theory. Basically, they are not granules but their importance has been mentioned [9] in the field of computer security.
- A pair of lower-upper approximations has been introduced which are dual with respect to the quasi-complementation in a g-covering approximation space. Using them, a rough set model of IqBa2 has been presented.
- In covering based rough set theory, there are many lower-upper approximations of a set in various literature. Some of them are dual with respect to the set-theoretic complementation whereas other pairs are not so. A study may be continued on them so that the notion of quasi-complementation can be incorporated and rough set models of remaining algebras may be constructed.

Acknowledgement. The author would like to thank Professor Mihir Kumar Chakraborty for checking the article and providing valuable suggestions that helped to improve the article substantially.

References

1. Banerjee, M.: Rough sets and 3-valued Łukasiewicz logic. Fundam. Informaticae **31**, 213–220 (1997)
2. Banerjee, M., Chakraborty, M.K.: Rough algebra. Bull. Pol. Acad. Sci. (Math) **41**(4), 293–297 (1993)
3. Banerjee, M., Chakraborty, M.K.: Rough sets through algebraic logic. Fundam. Informaticae **28**(3–4), 211–221 (1996)
4. Bechhio, D.: Sur les definitions des algebres trivalentes de Łukasiewicz donnees par A. Monteiro. Logique et Anal. **16**, 339–344 (1973)
5. Boicescu, V., Filipoiu, A., Georgescu, G., Rudeano, S.: Łukasiewicz-Moisil Algebras. North Holland, Amsterdam (1991)
6. Cattaneo, G., Ciucci, D., Dubois, D.: Algebraic models of deviant modal operators based on De Morgan and Kleene lattices. Inf. Sci. **181**, 4075–4100 (2011)
7. Font, J., Rius, M.: An abstract algebraic logic approach to tetravalent modal logics. J. Symbolic Logic **65**(2), 481–518 (2000)
8. Hughes, G.E., Cresswell, M.J.: A New Introduction to Modal Logic. Routledge, London (1996)
9. Lin, T.Y., Liu, G., Chakraborty, M.K., Ślęzak, D.: From topology to anti-reflexive topology. In: IEEE International Conference on Fuzzy Systems, pp. 1–7 (2013)
10. Liu, J., Liao, Z.: The sixth type of covering-based rough sets. In: Granular Computing - GrC, IEEE International Conference on Granular Computing 2008, Hangzhou 26–28 August 2008, pp. 438–441 (2008)
11. Pawlak, Z.: Rough sets. Int. J. Comput. Inf. Sci. **11**(5), 341–356 (1982)
12. Qin, K., Gao, Y., Pei, Z.: On covering rough sets. In: Yao, J.T., Lingras, P., Wu, W.-Z., Szczuka, M., Cercone, N.J., Ślęzak, D. (eds.) RSKT 2007. LNCS (LNAI), vol. 4481, pp. 34–41. Springer, Heidelberg (2007). https://doi.org/10.1007/978-3-540-72458-2_4
13. Rasiowa, H.: An Algebraic Approach to Non-classical Logics. North-Holland Publishing Company, Amsterdam (1974)
14. Saha, A., Sen, J., Chakraborty, M.K.: Algebraic structures in the vicinity of pre-rough algebra and their logics. Inf. Sci. **282**, 296–320 (2014)
15. Saha, A., Sen, J., Chakraborty, M.K.: Algebraic structures in the vicinity of pre-rough algebra and their logics II. Inf. Sci. **333**, 44–60 (2016)
16. Samanta, P., Chakraborty, M.K.: Covering based approaches to rough sets and implication lattices. In: Sakai, H., Chakraborty, M.K., Hassanien, A.E., Ślęzak, D., Zhu, W. (eds.) RSFDGrC 2009. LNCS (LNAI), vol. 5908, pp. 127–134. Springer, Heidelberg (2009). https://doi.org/10.1007/978-3-642-10646-0_15
17. Sardar, M.R., Chakraborty, M.K.: Some implicative topological quasi-Boolean algebras and rough set models. Int. J. Approximate Reasoning **148**, 1–22 (2022). https://doi.org/10.1016/j.ijar.2022.05.008. https://www.sciencedirect.com/science/article/pii/S0888613X22000779
18. Sardar, M.R., Chakraborty, M.K.: Rough set models of some abstract algebras close to pre-rough algebra. Inf. Sci. **621**, 104–118 (2023). https://doi.org/10.1016/j.ins.2022.11.095. https://www.sciencedirect.com/science/article/pii/S002002552201386X
19. Sen, J.: Some Embeddings In Linear Logic And Related Issues. Ph.D. thesis, University of Calcutta, India (2001)
20. Zhu, W., Wang, F.Y.: Reduction and axiomization of covering generalized rough sets. Inf. Sci. **152**, 217–230 (2003)

Labelled Calculi for the Logics of Rough Concepts

Ineke van der Berg[1,3], Andrea De Domenico[1], Giuseppe Greco[1], Krishna B. Manoorkar[1](✉), Alessandra Palmigiano[1,2], and Mattia Panettiere[1]

[1] School of Business and Economics, Vrije Universiteit Amsterdam, Amsterdam, The Netherlands
{i.van.der.berg,a.de.domenico,g.greco,k.b.manoorkar,alessandra.palmigiano, m.panettiere}@vu.nl

[2] Department of Mathematics and Applied Mathematics, University of Johannesburg, Johannesburg, South Africa

[3] Department of Mathematical Sciences, Stellenbosch University, Stellenbosch, South Africa

Abstract. We introduce sound and complete labelled sequent calculi for the basic normal non-distributive modal logic **L** and some of its axiomatic extensions, where the labels are atomic formulas of the first order language of *enriched formal contexts*, i.e., relational structures based on formal contexts which provide complete semantics for these logics. We also extend these calculi to provide a proof system for the logic of *rough formal contexts*.

Keywords: Rough formal contexts · Non-distributive modal logic · Labelled calculi · Proof calculi

1 Introduction

In structural proof theory, powerful solutions to the problem of introducing analytic calculi for large classes of normal modal logics hinge on incorporating information about the relational semantics of the given logics into the calculi. This strategy is prominently used in the design of *labelled calculi* [8, 13, 14], a proof-theoretic format using which, analytic calculi have been introduced for the axiomatic extensions of the basic normal modal logic defined by modal axioms corresponding to geometric implications in the first order language of Kripke frames.

Labelled calculi for classical modal logics manipulate sequents $\Gamma \vdash \Delta$ such that Γ and Δ are multisets of atomic formulas xRy in the first order language of Kripke frames and labelled formulas $x : A$ interpreted on Kripke frames as $x \Vdash A$, i.e. as the condition that the modal formula A be satisfied (or forced) at the state x of a given Kripke frame. The labelled calculus **G3K** for the basic normal modal logic **K** is obtained by expanding the propositional fragment of the Gentzen calculus **G3c** with introduction rules for the modal operators obtained by reading off the interpretation clauses of $\Box$- and $\Diamond$-formulas on Kripke frames. Labelled calculi for axiomatic extensions of **K** defined by Sahlqvist axioms (including the modal logics T, K4, KB, S4, B, S5) are obtained in [13] by

M. Banerjee and A. V. Sreejith (Eds.): ICLA 2023, LNCS 13963, pp. 172–188, 2023.
https://doi.org/10.1007/978-3-031-26689-8_13

augmenting **G3K** with the rules generated by reading off the first order conditions on Kripke frames corresponding to the given axioms.

In the present paper, we extend the design principles for the generation of labelled calculi to *normal non-distributive modal logics*, a class of normal LE-logics (cf. [5]) the propositional fragment of which coincides with the logic of lattices in which the distributive laws are not necessarily valid. In [3,4], non distributive modal logics are used as the underlying environment for an epistemic logic of categories and formal concepts, and in [2] as the logical environment of a theory unifying Formal Concept Analysis [9] and Rough Set Theory [15].

Specifically, making use of the fact that the basic normal non-distributive modal logic is sound and complete w.r.t. *enriched formal contexts* (i.e., relational structures based on formal contexts from FCA) [3,4], and that modal axioms of a certain syntactic shape [5] define elementary (i.e. first order definable) subclasses of enriched formal contexts, we introduce relational labelled calculi for the basic non-distributive modal logic and some of its axiomatic extensions.

Moreover, we adapt and specialize these calculi for capturing the logic of relational structures of a related type, referred to as *rough formal contexts*, which were introduced by Kent in [11] as a formal environment for unifying Formal Concept Analysis and Rough Set Theory. In [10], a sound and complete axiomatization for the non-distributive modal logic of rough formal contexts was introduced by circumventing a technical difficulty which in the present paper is shown to be an impossibility, since two of the three first order conditions characterizing rough formal contexts turn out to be *not modally definable* in the modal signature which the general theory would associate with them (cf. Lemma 4). However, in the richer language of labelled calculi, these first order conditions can still be used to define structural rules which capture the axiomatization introduced in [10] for the logic of rough formal contexts.

Structure of the Paper. Section 2 recalls preliminaries on the logic of enriched and rough formal contexts, Sect. 3 presents a labelled calculus for the logic of enriched formal contexts and its extensions. Section 4 proves soundness and completeness results for the calculus for the logic of rough formal contexts. We conclude in Sect. 5.

2 Preliminaries

In the present section, we recall the definition and relational semantics of the basic normal non-distributive modal logic in the modal signature $\{\Box, \Diamond, \rhd\}$ and some of its axiomatic extensions. This logic and similar others have been studied in the context of a research program aimed at introducing the logical foundations of categorization theory [2–4]. In this context, $\Box c$ and $\Diamond c$ and $\rhd c$ can be given e.g. the epistemic interpretation of the categories of the objects which are *certainly*, *possibly*, and *certainly not* members of category c, respectively. Motivated by these ideas, in [6], possible interpretations of (modal) non-distributive logics are systematically discussed also in their connections with their classical interpretation.

2.1 Basic Normal Non-distributive Modal Logic and Some of Its Axiomatic Extensions

Let Prop be a (countable or finite) set of atomic propositions. The language $\mathcal{L}$ is defined as follows:

$$\varphi ::= \bot \mid \top \mid p \mid \varphi \wedge \varphi \mid \varphi \vee \varphi \mid \Box\varphi \mid \Diamond\varphi \mid \rhd\varphi,$$

where $p \in$ Prop. The *basic*, or *minimal normal $\mathcal{L}$-logic* is a set $\mathbf{L}$ of sequents $\varphi \vdash \psi$, with $\varphi, \psi \in \mathcal{L}$, containing the following axioms:

$$\begin{array}{cccccc} p \vdash p & \bot \vdash p & p \vdash p \vee q & p \wedge q \vdash p & \top \vdash \Box\top & \Box p \wedge \Box q \vdash \Box(p \wedge q) \\ & p \vdash \top & q \vdash p \vee q & p \wedge q \vdash q & \Diamond\bot \vdash \bot & \Diamond(p \vee q) \vdash \Diamond p \vee \Diamond q \\ & & & & \top \vdash \rhd\bot & \rhd p \wedge \rhd q \vdash \rhd(p \vee q) \end{array}$$

and closed under the following inference rules:

$$\frac{\varphi \vdash \chi \quad \chi \vdash \psi}{\varphi \vdash \psi} \quad \frac{\varphi \vdash \psi}{\varphi(\chi/p) \vdash \psi(\chi/p)} \quad \frac{\chi \vdash \varphi \quad \chi \vdash \psi}{\chi \vdash \varphi \wedge \psi} \quad \frac{\varphi \vdash \chi \quad \psi \vdash \chi}{\varphi \vee \psi \vdash \chi} \quad \frac{\varphi \vdash \psi}{\Box\varphi \vdash \Box\psi} \quad \frac{\varphi \vdash \psi}{\Diamond\varphi \vdash \Diamond\psi} \quad \frac{\varphi \vdash \psi}{\rhd\psi \vdash \rhd\varphi}$$

An *$\mathcal{L}$-logic* is any extension of $\mathbf{L}$ with $\mathcal{L}$-axioms $\varphi \vdash \psi$. In what follows, for any set Σ of $\mathcal{L}$-axioms, we let $\mathbf{L}.\Sigma$ denote the axiomatic extension of $\mathbf{L}$ generated by Σ. Throughout the paper, we will consider all subsets Σ of the set of axioms listed in the table below. Some of these axioms are well known from classical modal logic, and have also cropped up in [2] in the context of the definition of relational structures simultaneously generalizing Formal Concept Analysis and Rough Set Theory. In Proposition 1, we list their first-order correspondents w.r.t. the relational semantics discussed in the next section.

$\Diamond\Diamond A \vdash \Diamond A$	$\Box A \vdash \Box\Box A$	$A \vdash \Box\Diamond A$	$\Diamond\Box A \vdash A$
$\Box A \vdash A$	$A \vdash \Diamond A$	$A \vdash \rhd\rhd A$	

2.2 Relational Semantics of $\mathcal{L}$-logics

The present subsection collects notation, notions and facts from [2,6]. For any binary relation $T \subseteq U \times V$, and any $U' \subseteq U$ and $V' \subseteq V$, we let T^c denote the set-theoretic complement of T in $U \times V$, and

$$T^{(1)}[U'] := \{v \mid \forall u(u \in U' \Rightarrow uTv)\} \qquad T^{(0)}[V'] := \{u \mid \forall v(v \in V' \Rightarrow uTv)\}. \qquad (1)$$

Well known properties of this construction (cf. [7, Sections 7.22–7.29]) are stated in the following lemma.

Lemma 1. *For any sets U, V, U' and V', and for any families of sets $\mathcal{V}$ and $\mathcal{U}$,*

1. *$X_1 \subseteq X_2 \subseteq U$ implies $T^{(1)}[X_2] \subseteq T^{(1)}[X_1]$, and $Y_1 \subseteq Y_2 \subseteq V$ implies $T^{(0)}[Y_2] \subseteq T^{(0)}[Y_1]$.*
2. *$U' \subseteq T^{(0)}[V']$ iff $V' \subseteq T^{(1)}[U']$.*
3. *$U' \subseteq T^{(0)}[T^{(1)}[U']]$ and $V' \subseteq T^{(1)}[T^{(0)}[V']]$.*
4. *$T^{(1)}[U'] = T^{(1)}[T^{(0)}[T^{(1)}[U']]]$ and $T^{(0)}[V'] = T^{(0)}[T^{(1)}[T^{(0)}[V']]]$.*
5. *$T^{(0)}[\bigcup \mathcal{V}] = \bigcap_{V' \in \mathcal{V}} T^{(0)}[V']$ and $T^{(1)}[\bigcup \mathcal{U}] = \bigcap_{U' \in \mathcal{U}} T^{(1)}[U']$.*

If $R \subseteq U \times V$, and $S \subseteq V \times W$, then the composition $R; S \subseteq U \times W$ is defined as follows:

$$u(R; S)w \quad \text{iff} \quad u \in R^{(0)}[S^{(0)}[w]] \quad \text{iff} \quad \forall v(vSw \Rightarrow uRv).$$

In what follows, we fix two sets A and X, and use a, b (resp. x, y) for elements of A (resp. X), and B, C, A_j (resp. Y, W, X_j) for subsets of A (resp. of X).

A *polarity* or *formal context* (cf. [9]) is a tuple $\mathbb{P} = (A, X, I)$, where A and X are sets, and $I \subseteq A \times X$ is a binary relation. In what follows, for any such polarity, we will let $J \subseteq X \times A$ be defined by the equivalence xJa iff aIx. Intuitively, formal contexts can be understood as abstract representations of databases [9], so that A represents a collection of *objects*, X a collection of *features*, and for any object a and feature x, the tuple (a, x) belongs to I exactly when object a has feature x.

As is well known, for every formal context $\mathbb{P} = (A, X, I)$, the pair of maps

$$(\cdot)^{\uparrow} : \mathcal{P}(A) \to \mathcal{P}(X) \quad \text{and} \quad (\cdot)^{\downarrow} : \mathcal{P}(X) \to \mathcal{P}(A),$$

respectively defined by the assignments $B^{\uparrow} := I^{(1)}[B]$ and $Y^{\downarrow} := I^{(0)}[Y]$, form a Galois connection (cf. Lemma 1.2), and hence induce the closure operators $(\cdot)^{\uparrow\downarrow}$ and $(\cdot)^{\downarrow\uparrow}$ on $\mathcal{P}(A)$ and on $\mathcal{P}(X)$ respectively.[1] The fixed points of these closure operators are referred to as *Galois-stable* sets. For a formal context $\mathbb{P} = (A, I, X)$, a *formal concept* of $\mathbb{P}$ is a tuple $c = (B, Y)$ such that $B \subseteq A$ and $Y \subseteq X$, and $B = Y^{\downarrow}$ and $Y = B^{\uparrow}$. The subset B (resp. Y) is referred to as the *extension* (resp. the *intension*) of c and is denoted by $[\![c]\!]$ (resp. $(\![c]\!)$). By Lemma 1.3, the sets B and Y are Galois-stable. It is well known (cf. [9]) that the set of formal concepts of a formal context $\mathbb{P}$, with the order defined by

$$c_1 \leq c_2 \quad \text{iff} \quad [\![c_1]\!] \subseteq [\![c_2]\!] \quad \text{iff} \quad (\![c_2]\!) \subseteq (\![c_1]\!),$$

forms a complete lattice, namely the *concept lattice* of $\mathbb{P}$, which we denote by $\mathbb{P}^+$.

For the language $\mathcal{L}$ defined in the previous section, an *enriched formal $\mathcal{L}$-context* is a tuple $\mathbb{F} = (\mathbb{P}, R_\Box, R_\Diamond, R_\rhd)$, where $R_\Box \subseteq A \times X$ and $R_\Diamond \subseteq X \times A$ and $R_\rhd \subseteq A \times A$ are *I-compatible* relations, that is, for all $a, b \in A$, and all $x \in X$, the sets $R_\Box^{(0)}[x]$, $R_\Box^{(1)}[a]$, $R_\Diamond^{(0)}[a]$, $R_\Diamond^{(1)}[x]$, $R_\rhd^{(0)}[b]$, $R_\rhd^{(1)}[a]$ are Galois-stable in $\mathbb{P}$. As usual in modal logic, these relations can be interpreted in different ways, for instance as the epistemic attributions of features to objects by agents.

A *valuation* on such an $\mathbb{F}$ is a map $V\colon \mathsf{Prop} \to \mathbb{P}^+$. For every $p \in \mathsf{Prop}$, we let $[\![p]\!] := [\![V(p)]\!]$ (resp. $(\![p]\!) := (\![V(p)]\!)$) denote the extension (resp. the intension) of the interpretation of p under V. A *model* is a tuple $\mathbb{M} = (\mathbb{F}, V)$ where $\mathbb{F} = (\mathbb{P}, R_\Box, R_\Diamond, R_\rhd)$ is an enriched formal context and V is a valuation on $\mathbb{F}$. For every $\varphi \in \mathcal{L}$, the following 'forcing' relations can be recursively defined as follows:

$\mathbb{M}, a \Vdash p$	iff $a \in [\![p]\!]_\mathbb{M}$	$\mathbb{M}, x \succ p$	iff $x \in (\![p]\!)_\mathbb{M}$
$\mathbb{M}, a \Vdash \top$	always	$\mathbb{M}, x \succ \top$	iff aIx for all $a \in A$
$\mathbb{M}, x \succ \bot$	always	$\mathbb{M}, a \Vdash \bot$	iff aIx for all $x \in X$
$\mathbb{M}, a \Vdash \varphi \wedge \psi$	iff $\mathbb{M}, a \Vdash \varphi and \mathbb{M}, a \Vdash \psi$	$\mathbb{M}, x \succ \varphi \wedge \psi$	iff $(\forall a \in A)(\mathbb{M}, a \Vdash \varphi \wedge \psi \Rightarrow aIx)$
$\mathbb{M}, x \succ \varphi \vee \psi$	iff $\mathbb{M}, x \succ \varphi and \mathbb{M}, x \succ \psi$	$\mathbb{M}, a \Vdash \varphi \vee \psi$	iff $(\forall x \in X)(\mathbb{M}, x \succ \varphi \vee \psi \Rightarrow aIx)$.

[1] When $B = \{a\}$ (resp. $Y = \{x\}$) we write $a^{\uparrow\downarrow}$ for $\{a\}^{\uparrow\downarrow}$ (resp. $x^{\downarrow\uparrow}$ for $\{x\}^{\downarrow\uparrow}$).

As to the interpretation of modal formulas:

$$
\begin{array}{ll}
\mathbb{M}, a \Vdash \Box\varphi \text{ iff}(\forall x \in X)(\mathbb{M}, x \succ \varphi \Rightarrow aR_{\Box}x) & \mathbb{M}, x \succ \Box\varphi \text{ iff}(\forall a \in A)(\mathbb{M}, a \Vdash \Box\varphi \Rightarrow aIx) \\
\mathbb{M}, x \succ \Diamond\varphi \text{ iff for all } a \in A, \textit{if} \mathbb{M}, a \Vdash \varphi \text{ then } xR_{\Diamond}a & \mathbb{M}, a \Vdash \Diamond\varphi \text{ iff}(\forall x \in X)(\mathbb{M}, x \succ \Diamond\varphi \Rightarrow aIx) \\
\mathbb{M}, a \Vdash \rhd\varphi \text{ iff}(\forall b \in A)(\mathbb{M}, b \Vdash \varphi \Rightarrow aR_{\rhd}b) & \mathbb{M}, x \succ \rhd\varphi \text{ iff}(\forall a \in A)(\mathbb{M}, a \Vdash \rhd\varphi \Rightarrow aIx).
\end{array}
$$

The definition above ensures that, for any $\mathcal{L}$-formula φ,

$$\mathbb{M}, a \Vdash \varphi \text{ iff } a \in [\![\varphi]\!]_{\mathbb{M}}, \quad \text{and} \quad \mathbb{M}, x \succ \varphi \text{ iff } x \in (\![\varphi]\!)_{\mathbb{M}}.$$

Finally, as to the interpretation of sequents:

$$\mathbb{M} \models \varphi \vdash \psi \quad \text{iff} \quad [\![\varphi]\!]_{\mathbb{M}} \subseteq [\![\psi]\!]_{\mathbb{M}} \quad \text{iff} \quad (\![\psi]\!)_{\mathbb{M}} \subseteq (\![\varphi]\!)_{\mathbb{M}}.$$

A sequent $\varphi \vdash \psi$ is *valid* on an enriched formal context $\mathbb{F}$ (in symbols: $\mathbb{F} \models \varphi \vdash \psi$) if $\mathbb{M} \models \varphi \vdash \psi$ for every model $\mathbb{M}$ based on $\mathbb{F}$. The basic non-distributive logic **L** is sound and complete w.r.t. the class of enriched formal contexts (cf. [2]).

Then, via a general canonicity result (cf. [5]), the following proposition (cf. [2, Proposition 4.3]) implies that, for any subset Σ of the set of axioms at the end of Sect. 2.1, the logic $\mathbf{L}.\Sigma$ is complete w.r.t. the class of enriched formal contexts defined by those first-order sentences in the statement of the proposition below corresponding to the axioms in Σ.

These first order sentences are compactly represented as inclusions of relations defined as follows. For any enriched formal context $\mathbb{F} = (\mathbb{P}, R_{\Box}, R_{\Diamond}, R_{\rhd})$, the relations $R_{\blacklozenge} \subseteq X \times A$, $R_{\blacksquare} \subseteq A \times X$ and $R_{\blacktriangleright} \subseteq A \times A$ are defined by $xR_{\blacklozenge}a$ iff $aR_{\Box}x$, and $aR_{\blacksquare}x$ iff $xR_{\Diamond}a$, and $aR_{\blacktriangleright}b$ iff $bR_{\rhd}a$. Moreover, for all relations $R, S \subseteq A \times X$ we let $R; S \subseteq A \times X$ be defined[2] by $a(R; S)x$ iff $a \in R^{(0)}[I^{(1)}[S^{(0)}[x]]]$, and for all relations $R, S \subseteq X \times A$ we let $R; S \subseteq X \times A$ be defined by $x(R; S)a$ iff $x \in R^{(0)}[I^{(0)}[S^{(0)}[a]]]$.

Proposition 1. *For any enriched formal context* $\mathbb{F} = (\mathbb{P}, R_{\Box}, R_{\Diamond}, R_{\rhd})$*:*

1. $\mathbb{F} \models \Box\varphi \vdash \varphi$ *iff* $R_{\Box} \subseteq I$.
2. $\mathbb{F} \models \varphi \vdash \Diamond\varphi$ *iff* $R_{\Diamond} \subseteq J$.
3. $\mathbb{F} \models \Box\varphi \vdash \Box\Box\varphi$ *iff* $R_{\Box} \subseteq R_{\Box} ; R_{\Box}$.
4. $\mathbb{F} \models \varphi \vdash \rhd\rhd\varphi$ *iff* $R_{\rhd} = R_{\blacktriangleright}$.
5. $\mathbb{F} \models \Diamond\Diamond\varphi \vdash \Diamond\varphi$ *iff* $R_{\Diamond} \subseteq R_{\Diamond} ; R_{\Diamond}$.
6. $\mathbb{F} \models \varphi \vdash \Box\Diamond\varphi$ *iff* $R_{\Diamond} \subseteq R_{\blacklozenge}$.
7. $\mathbb{F} \models \Diamond\Box\varphi \vdash \varphi$ *iff* $R_{\blacklozenge} \subseteq R_{\Diamond}$.

The proposition above motivated the introduction of the notion of conceptual approximation space in [2], as a subclass of the enriched formal contexts modelling the $\rhd$-free fragment of the language $\mathcal{L}$. A *conceptual approximation space* is an enriched formal context $\mathbb{F} = (\mathbb{P}, R_{\Box}, R_{\Diamond})$ verifying the first order sentence $R_{\Box}; R_{\blacksquare} \subseteq I$. Such an $\mathbb{F}$ is *reflexive* if $R_{\Box} \subseteq I$ and $R_{\Diamond} \subseteq J$, is *symmetric* if $R_{\Diamond} = R_{\blacklozenge}$ or equivalently if $R_{\blacksquare} = R_{\Box}$, and is *transitive* if $R_{\Box} \subseteq R_{\Box} ; R_{\Box}$ and $R_{\Diamond} \subseteq R_{\Diamond} ; R_{\Diamond}$ (cf. [1,2] for a discussion on terminology).

---

[2] These compositions and those defined in Sect. 2.2 are pairwise different, since each of them involves different types of relations. However, the types of the relations involved in each definition provides a unique reading of such compositions, which justifies our abuse of notation.

2.3 The Logic of Rough Formal Contexts

Examples of conceptual approximation spaces have cropped up in the context of Kent's proposal for a simultaneous generalization of approximation spaces from RST and formal contexts from FCA [12]. Specifically, Kent introduced *rough formal contexts* as tuples $\mathbb{G} = (\mathbb{P}, E)$ such that $\mathbb{P} = (A, X, I)$ is a polarity, and $E \subseteq A \times A$ is an equivalence relation. The relation E induces two relations $R_\Box, S_\Box \subseteq A \times X$ defined as follows: for every $a \in A$ and $x \in X$,

$$aR_\Box x \text{ iff } \exists b(aEb \,\&\, bIx) \qquad aS_\Box x \text{ iff } \forall b(aEb \Rightarrow bIx) \tag{2}$$

The reflexivity of E implies that $S_\Box \subseteq I \subseteq R_\Box$; hence, $R_\Box$ and $S_\Box$ can respectively be regarded as the *lax*, or *upper*, and as the *strict*, or *lower*, approximation of I relative to E. For any rough formal context $\mathbb{G} = (\mathbb{P}, E)$, let $S_\blacklozenge \subseteq X \times A$ be defined by the equivalence $xS_\blacklozenge a$ iff $aS_\Box x$,

Lemma 2. *If $\mathbb{G} = (\mathbb{P}, E)$ is a rough formal context, then $S_\blacklozenge = J; E$.*

Proof. For any $a \in A$ and $x \in X$,

$xS_\blacklozenge a$	iff $aS_\Box x$	Definition of $S_\blacklozenge$
	iff $\forall b(bEa \Rightarrow bIx)$	Definition of $S_\Box$
	iff $\forall b(bEa \Rightarrow xJb)$	Definition of J
	iff $E^{(0)}[a] \subseteq J^{(1)}[x]$	notation $T^{(0)}[-]$ and $T^{(1)}[-]$
	iff $x \in J^{(0)}[E^{(0)}[a]]$	Lemma 1.2
	iff $x(J; E)a.$	Definition of $J; E$

In [2, Section 5] and [10, Section 3], the logic of rough formal contexts was introduced, based on the theory of enriched formal contexts as models of non-distributive modal logics, the characterization results reported on in Proposition 1, and the following:

Lemma 3. *([2, Lemma 5.3]) For any polarity $\mathbb{P} = (A, X, I)$, and any I-compatible relation $E \subseteq A \times A$ such that its associated $S_\Box \subseteq A \times X$ (defined as in (2)) is I-compatible,*[3]

E is reflexive iff $S_\Box \subseteq I$; and E is transitive iff $S_\Box \subseteq S_\Box ; S_\Box$.

These results imply that the characterizing properties of rough formal contexts can be taken as completely axiomatised in the modal language $\mathcal{L}$ via the following axioms:

$$\Box\varphi \vdash \varphi \qquad \Box\varphi \vdash \Box\Box\varphi \qquad \varphi \vdash \rhd\rhd\varphi.$$

Clearly, any rough formal context $\mathbb{G} = (\mathbb{P}, E)$ such that E is I-compatible is an enriched formal $\mathcal{L}_\rhd$-context, where $\mathcal{L}_\rhd$ is the $\{\Box, \Diamond\}$-free fragment of $\mathcal{L}$. However, interestingly, it is impossible to capture the reflexivity and transitivity of E by means of $\mathcal{L}_\rhd$-axioms, as the next lemma shows:

---

[3] Notice that E being I-compatible does not imply that $S_\Box$ is. Let $\mathbb{G} = (\mathbb{P}, E)$ s.t. $A := \{a, b\}$, $X := \{x, y\}$, $I := \{(a, x), (a, y), (b, y)\}$, and $E := A \times A$. Then E is I-compatible. However, $S_\Box = \{(a, y), (b, y)\}$ is not, as $S_\Box^{(0)}[x] = \varnothing$ is not Galois stable, since $\varnothing^{\uparrow\downarrow} = X^\downarrow = \{a\}$. In [10], it was remarked that $S_\Box$ being I-compatible does not imply that E is.

Lemma 4. *The class of enriched formal $\mathcal{L}_{\triangleright}$-contexts $\mathbb{F} = (\mathbb{P}, R_{\triangleright})$ such that $R_{\triangleright} \subseteq A \times A$ is reflexive (resp. transitive) is not modally definable in its associated language $\mathcal{L}_{\triangleright}$.*

Proof. Assume for contradiction that $\mathcal{L}_{\triangleright}$-axioms $\varphi \vdash \psi$ and $\chi \vdash \xi$ exist such that $\mathbb{F} \models \varphi \vdash \psi$ iff $R_{\triangleright}$ is reflexive, and $\mathbb{F} \models \chi \vdash \xi$ iff $R_{\triangleright}$ is transitive for any enriched formal $\mathcal{L}_{\triangleright}$-context $\mathbb{F} = (\mathbb{P}, R_{\triangleright})$. Then, these equivalences would hold in particular for those special formal $\mathcal{L}_{\triangleright}$-contexts $\mathbb{F} = (\mathbb{P}_W, R_{\triangleright})$ such that $\mathbb{P}_W = (W_A, W_X, I_{\Delta^c})$ such that $W_A = W_X = W$ for some set W, and $aI_{\Delta^c}x$ iff $a \neq x$, and $R_{\triangleright} := H_{R^c}$ is defined as $aH_{R^c}b$ iff $(a, b) \notin R$ for some binary relation $R \subseteq W \times W$. By construction, letting $\mathbb{X} = (W, R)$, the following chain of equivalences holds: $\mathbb{F} \models \varphi \vdash \psi$ iff $[\![\varphi]\!]_V \subseteq [\![\psi]\!]_V$ for every valuation $V : \mathsf{Prop} \to \mathbb{P}^+$. However, by construction, $\mathbb{P}^+ \cong \mathcal{P}(W)$ (cf. [2, Proposition 3.4]). Moreover, the definition of the forcing relation $\Vdash$ on $\mathbb{F}$ implies that

$$
\begin{aligned}
[\![\triangleright\varphi]\!] = R_{\triangleright}^{(0)}[[\![\varphi]\!]] = H_{R^c}^{(0)}[[\![\varphi]\!]] &= \{b \in W_A \mid \forall a(a \Vdash \varphi \Rightarrow aR^c b)\} \\
&= \{b \in W_A \mid \forall a(aRb \Rightarrow a \nVdash \varphi)\}
\end{aligned}
$$

That is, restricted to the class of $\mathcal{L}_{\triangleright}$-contexts which arise from classical Kripke frames $\mathbb{X} = (W, R)$ in the way indicated above, the interpretation of $\triangleright$-formulas coincides with the interpretation of $\Box\neg$-formulas in the language of classical modal logic, which induces a translation τ, from $\mathcal{L}_{\triangleright}$-formulas to formulas in the language of classical modal logic, which is preserved and reflected from the special formal $\mathcal{L}_{\triangleright}$-contexts $\mathbb{F}$ to the Kripke frames with which they are associated. Therefore, by construction, for any Kripke frame $\mathbb{X} = (X, R)$, R is irreflexive iff H_{R^c} is reflexive iff $\mathbb{F} \models \varphi \vdash \psi$ iff $\mathbb{X} \models \tau(\varphi) \vdash \tau(\psi)$, contradicting the well known fact that the class of Kripke frames $\mathbb{X} = (X, R)$ such that R is irreflexive is not modally definable.

Reasoning similarly, to show the statement concerning transitivity, it is enough to see that the class of Kripke frames $\mathbb{X} = (W, R)$ s.t. R^c is transitive is not modally definable. Consider the Kripke frames $\mathbb{X}_i = (W_i, R_i)$ such that $W_i = \{a_i, b_i\}$, $R_i = \{(a_i, b_i)\}$, for $1 \leq i \leq 2$. Clearly, R_i^c is transitive in $\mathcal{F}_i$, so the two frames satisfy the property. However, their disjoint union $\mathbb{X}_1 \uplus \mathbb{X}_2 = (W, R)$, given by $W = \{a_1, b_1, a_2, b_2\}$ and $R = \{(a_1, b_1), (a_2, b_2)\}$, does not: indeed, $(a_1, a_2), (a_2, b_1) \in R^c$ but $(a_1, b_1) \notin R^c$. Hence, the statement follows from the Goldblatt-Thomason theorem for classical modal logic.

3 Relational Labelled Calculi for $\mathcal{L}$-logics

Below, p, q denote atomic propositions; a, b, c (resp. x, y, z) are labels corresponding to objects (resp. features). Given labels a, x and a modal formula A, well-formed formulas are of the type $a : A$ and $x{::}A$, while φ, ψ are meta-variables for well-formed formulas. Well-formed terms are of any of the following shapes: aIx, $aR_{\Box}x$, $xR_{\Diamond}a$, $aR_{\blacksquare}x$, $xR_{\blacklozenge}a$, and $t_1 \Rightarrow t_2$, where t_1 is of any of the following shapes: $aR_{\Box}x$, $aR_{\blacksquare}x$, $yR_{\Diamond}a$, $yR_{\blacklozenge}a$, $aR_{\triangleright}b$, $aR_{\blacktriangleright}b$, and t_2 is of the form aIy. Relational terms $t_1 \Rightarrow t_2$ are interpreted as $\forall u(t_1 \to t_2)$ where u is the variable shared by t_1 and t_2. A sequent is an expression of the form $\Gamma \vdash \Delta$, where Γ, Δ are meta-variables for multisets of well-formed formulas or terms. For any labels u, v and relations R, S we write $u(R; S)v$ as a shorthand for the term $wSv \Rightarrow uRw$.

3.1 Labelled Calculus R.L for the Basic $\mathcal{L}$-logic

Initial rules and cut rules

$$\mathrm{Id}_{a:p}\ \frac{}{\Gamma, a:p \vdash a:p, \Delta} \qquad \frac{}{\Gamma, x::p \vdash x::p, \Delta}\ \mathrm{Id}_{x::p} \qquad \bot\ \frac{}{\Gamma \vdash x::\bot, \Delta} \qquad \frac{}{\Gamma \vdash a:\top, \Delta}\ \top$$

$$\mathrm{Cut}_{aa}\ \frac{\Gamma \vdash a:A, \Delta \qquad \Gamma', a:A \vdash \Delta'}{\Gamma, \Gamma' \vdash \Delta, \Delta'} \qquad \frac{\Gamma \vdash x::A, \Delta \qquad \Gamma', x::A \vdash \Delta'}{\Gamma, \Gamma' \vdash \Delta, \Delta'}\ \mathrm{Cut}_{xx}$$

Switch rules*

$$\mathrm{S}xa\ \frac{\Gamma, x::B \vdash x::A, \Delta}{\Gamma, a:A \vdash a:B, \Delta} \qquad \frac{\Gamma, a:A \vdash a:B, \Delta}{\Gamma, x::B \vdash x::A, \Delta}\ \mathrm{S}ax$$

$$\mathrm{S}a\Diamond x\ \frac{\Gamma, yR_\Diamond a \Rightarrow bIy \vdash b:A, \Delta}{\Gamma, x::A \vdash xR_\Diamond a, \Delta} \qquad \frac{\Gamma, a:A \vdash aR_\Box x, \Delta}{\Gamma, bR_\Box x \Rightarrow bIy \vdash y::A, \Delta}\ \mathrm{S}a\Box x$$

$$\mathrm{S}x\Box a\ \frac{\Gamma, bR_\Box x \Rightarrow bIy \vdash y::A, \Delta}{\Gamma, a:A \vdash aR_\Box x, \Delta} \qquad \frac{\Gamma, x::A \vdash xR_\Diamond a, \Delta}{\Gamma, yR_\Diamond a \Rightarrow bIy \vdash b:A, \Delta}\ \mathrm{S}x\Diamond a$$

$$\mathrm{S}a\Diamond x\ \frac{\Gamma, b:A \vdash yR_\Diamond a \Rightarrow bIy, \Delta}{\Gamma, xR_\Diamond a \vdash x::A, \Delta} \qquad \frac{\Gamma, a:A \vdash aR_\Box x, \Delta}{\Gamma, y::A \vdash bR_\Box x \Rightarrow bIy, \Delta}\ \mathrm{S}a\Box x$$

$$\mathrm{S}x\Box a\ \frac{\Gamma, y::A \vdash bR_\Box x \Rightarrow bIy, \Delta}{\Gamma, aR_\Box x \vdash a:A, \Delta} \qquad \frac{\Gamma, xR_\Diamond a \vdash x::A, \Delta}{\Gamma, b:A \vdash yR_\Diamond a \Rightarrow bIy, \Delta}\ \mathrm{S}x\Diamond a$$

$$\mathrm{S}x\rhd a\ \frac{\Gamma, bR_\rhd a \Rightarrow bIy \vdash y::A, \Delta}{\Gamma, c:A \vdash cR_\rhd a, \Delta} \qquad \frac{\Gamma, c:A \vdash cR_\rhd a, \Delta}{\Gamma, bR_\rhd a \Rightarrow bIy \vdash y::A, \Delta}\ \mathrm{S}a\rhd x$$

$$\mathrm{S}x\rhd a\ \frac{\Gamma, y::A \vdash bR_\rhd a \Rightarrow bIy, \Delta}{\Gamma, cR_\rhd a \vdash c:A, \Delta} \qquad \frac{\Gamma, cR_\rhd a \vdash c:A, \Delta}{\Gamma, y::A \vdash bR_\rhd a \Rightarrow bIy, \Delta}\ \mathrm{S}a\rhd x$$

*Side condition: the variables x, y, a, b occurring as labels of a formula in the premise of any of these rules must not occur in Γ, Δ.

Switch rules for $R_\blacksquare$, $R_\blacklozenge$, and $R_\blacktriangleright$ are analogous to those for $R_\Box$, $R_\Diamond$, and $R_\rhd$. These rules encode the I-compatibility conditions of $R_\blacksquare$, $R_\blacklozenge$, $R_\blacktriangleright$, $R_\Box$, $R_\Diamond$, and $R_\rhd$ (cf. Remark 2).

Approximation rules*

$$approx_x\ \frac{\Gamma, x::A \vdash aIx, \Delta}{\Gamma \vdash a:A, \Delta} \qquad \frac{\Gamma, a:A \vdash aIx, \Delta}{\Gamma \vdash x::A, \Delta}\ approx_a$$

*Side condition: the variables x, y occurring as labels of a formula in the premise of any of these rules must not occur in Γ, Δ.

For $T, T' \in \{R_\Diamond,\ J,\ J;I,\ J;R_\Box,\ J;R_\rhd,\ R_\blacklozenge,\ J;R_\blacksquare,\ J;R_\blacktriangleright\}$ and $S, S' \in \{R_\Box,\ I,\ I;J,\ I;R_\Diamond,\ I;R_\blacklozenge,\ R_\blacksquare\}$ and for all labels u, v, w of the form a or x, we have the following switch rules:

Pure structure switch rules*

$$\text{S}(I;S)\,\frac{\Gamma, xTu \vdash xT'v, \Delta}{\Gamma, a(I;T')v \vdash a(I;T)u, \Delta} \qquad \frac{\Gamma, aSu \vdash aS'v, \Delta}{\Gamma, x(J;S')v \vdash x(J;S)u, \Delta}\,\text{S}(J;T)$$

$$\text{-S}(I;S)\,\frac{\Gamma, a(I;T')v \vdash a(I;T)u, \Delta}{\Gamma, xTu \vdash xT'v, \Delta} \qquad \frac{\Gamma, x(J;S')v \vdash x(J;S)u, \Delta}{\Gamma, aSu \vdash aS'v, \Delta}\,\text{-S}(J;T)$$

$$\text{Id}(I;J)_R\,\frac{\Gamma \vdash aSu, \Delta}{\Gamma \vdash a(I;(J;S))u, \Delta} \qquad \frac{\Gamma \vdash xTu, \Delta}{\Gamma \vdash x(J;(I;T))u, \Delta}\,\text{Id}(J;I)_R$$

$$\text{Id}(I;J)_L\,\frac{\Gamma, aSu \vdash \Delta}{\Gamma, a(I;(J;S))u \vdash \Delta} \qquad \frac{\Gamma, xTu \vdash \Delta}{\Gamma, x(J;(I;T))u \vdash \Delta}\,\text{Id}(J;I)_L$$

*Side condition: the variable x (resp. a) occurring in the premise of rules S(I; S), -S(I; S) (resp. S(J; T), -S(J; T)) must not occur in Γ, Δ.

The rules above encode the definition of I-composition of relations on formal contexts [2, Definition 3.10].

Adjunction rules

$$\Diamond \dashv \blacksquare\,\frac{\Gamma \vdash xR_{\Diamond}a, \Delta}{\Gamma \vdash aR_{\blacksquare}x, \Delta} \qquad \blacklozenge \dashv \Box\,\frac{\Gamma \vdash aR_{\Box}x, \Delta}{\Gamma \vdash xR_{\blacklozenge}a, \Delta} \qquad \rhd \dashv \blacktriangleright\,\frac{\Gamma \vdash aR_{\rhd}b, \Delta}{\Gamma \vdash bR_{\blacktriangleright}a, \Delta}$$

$$\frac{\Gamma \vdash aR_{\blacksquare}x, \Delta}{\Gamma \vdash xR_{\Diamond}a, \Delta}\,\Diamond \dashv \blacksquare^{-1} \qquad \frac{\Gamma \vdash xR_{\blacklozenge}a, \Delta}{\Gamma \vdash aR_{\Box}x, \Delta}\,\blacklozenge \dashv \Box^{-1} \qquad \frac{\Gamma \vdash aR_{\blacktriangleright}b, \Delta}{\Gamma \vdash bR_{\rhd}a, \Delta}\,\blacktriangleright \dashv \rhd$$

Adjunction rules encode the fact that operators $\Diamond$ and $\blacksquare$, $\blacklozenge$ and $\Box$, and $\rhd$ and $\blacktriangleright$ constitute pairs of adjoint operators.

Invertible logical rules for propositional connectives

$$\wedge_L\,\frac{\Gamma, a : A, a : B \vdash \Delta}{\Gamma, a : A \wedge B \vdash \Delta} \qquad \frac{\Gamma \vdash a : A, \Delta \qquad \Gamma \vdash a : B, \Delta}{\Gamma \vdash a : A \wedge B, \Delta}\,\wedge_R$$

$$\vee_L\,\frac{\Gamma \vdash x :: A, \Delta \qquad \Gamma \vdash x :: B, \Delta}{\Gamma \vdash x :: A \vee B, \Delta} \qquad \frac{\Gamma, x :: A, x :: B \vdash \Delta}{\Gamma, x :: A \vee B \vdash \Delta}\,\vee_R$$

Invertible logical rules for modal connectives*

$$\Box_L\,\frac{\Gamma, a : \Box A \vdash x :: A, aR_{\Box}x, \Delta}{\Gamma, a : \Box A \vdash aR_{\Box}x, \Delta} \qquad \frac{\Gamma, x :: A \vdash aR_{\Box}x, \Delta}{\Gamma \vdash a : \Box A, \Delta}\,\Box_R$$

$$\Diamond_L\,\frac{\Gamma, a : A \vdash xR_{\Diamond}a, \Delta}{\Gamma \vdash x :: \Diamond A, \Delta} \qquad \frac{\Gamma, x :: \Diamond A \vdash a : A, xR_{\Diamond}a, \Delta}{\Gamma, x :: \Diamond A \vdash xR_{\Diamond}a, \Delta}\,\Diamond_R$$

$$\rhd_L\,\frac{\Gamma, a : \rhd A \vdash b : A, aR_{\rhd}b, \Delta}{\Gamma, a : \rhd A \vdash aR_{\rhd}b, \Delta} \qquad \frac{\Gamma, b : A \vdash aR_{\rhd}b, \Delta}{\Gamma \vdash a : \rhd A, \Delta}\,\rhd_R$$

*Side condition: the variable x (resp. a, resp. b) must not occur in the conclusion of $\Box_R$ (resp. $\Diamond_L$, resp. $\rhd_R$).

Logical rules encode the definition of satisfaction and refutation for propositional and modal connectives discussed in Sect. 2.2. The proof of their soundness in Appendix A shows how this encoding works.

3.2 Relational Calculi for the Axiomatic Extensions of the Basic $\mathcal{L}$-logic

The structural rule corresponding to each axiom listed in Table 1 is generated as the read-off of the first-order condition corresponding to the given axiom as listed in Proposition 1. For any nonempty subset Σ of modal axioms as reported in Table 1, we let **R.L**Σ denote the extension of **R.L** with the corresponding rules.

Table 1. Modal axioms and their corresponding rules.

Modal axiom	Relational calculus rule	Modal axiom	Relational calculus rule
$\Box p \vdash p$	$\dfrac{\Gamma \vdash aR_\Box x, \Delta}{\Gamma \vdash aIx, \Delta}$	$p \vdash \Diamond p$	$\dfrac{\Gamma \vdash xR_\Diamond a, \Delta}{\Gamma \vdash aIx, \Delta}$
$p \vdash \Box\Diamond p$	$\dfrac{\Gamma \vdash xR_\Diamond a, \Delta}{\Gamma \vdash xR_\blacklozenge a, \Delta}$	$\Diamond\Box p \vdash p$	$\dfrac{\Gamma \vdash xR_\blacklozenge a, \Delta}{\Gamma \vdash xR_\Diamond a, \Delta}$
$\Box p \vdash \Box\Box p$	$\dfrac{\Gamma \vdash aR_\Box x, \Delta}{\Gamma, bR_\Box x \Rightarrow yJb \vdash aR_\Box y, \Delta}$	$\Diamond p \vdash \Diamond\Diamond p$	$\dfrac{\Gamma \vdash xR_\Diamond a, \Delta}{\Gamma, yR_\Diamond a \Rightarrow bIy \vdash xR_\Diamond b, \Delta}$
$p \vdash \rhd\rhd p$	$\dfrac{\Gamma \vdash aR_\rhd b, \Delta}{\Gamma \vdash bR_\rhd a, \Delta}$		

3.3 The Relational Calculus R.Lρ for the $\mathcal{L}$-logic of Rough Formal Contexts

The calculus **R.L** introduced in Sect. 3.1 can be specialized so as to capture the semantic environment of rough formal contexts by associating the connective $\Box$ (resp. $\blacklozenge$) with relational labels in which $S_\Box$ (resp. $S_\blacklozenge$) occurs, and adding rules encoding the reflexivity and the transitivity of E, rather than the (equivalent, cf. Lemma 3) first-order conditions on $S_\Box$. We need the following set of switching rules encoding the relation between E and I, and the I-compatibility of E and $S_\Box$ (and $S_\blacklozenge$).

Interdefinability rules

$$\text{swSf}^*\ \frac{\Gamma, bS_\Box x \Rightarrow bIy \vdash y :: A, \Delta}{\Gamma, a : A \vdash aS_\Box x, \Delta} \qquad \frac{\Gamma, a : A \vdash aS_\Box x, \Delta}{\Gamma, bS_\Box x \Rightarrow bIy \vdash y :: A, \Delta}\ \text{swSfi}^*$$

$$\text{swSdf}^*\ \frac{\Gamma, x :: A \vdash xS_\blacklozenge a, \Delta}{\Gamma, bEa \vdash b : A, \Delta} \qquad \frac{\Gamma, bEa \vdash b : A, \Delta}{\Gamma, x :: A \vdash xS_\blacklozenge a, \Delta}\ \text{swSdfi}^*$$

$$\text{swES}^*\ \frac{\Gamma, aEc \vdash aS_\Box x, \Delta}{\Gamma, bS_\Box x \Rightarrow bIy \vdash yS_\blacklozenge a, \Delta} \qquad \frac{\Gamma, bS_\Box x \Rightarrow bIy \vdash yS_\blacklozenge a, \Delta}{\Gamma, aEc \vdash aS_\Box x, \Delta}\ \text{swESi}^*$$

$$\text{curryS}^{**}\ \frac{\Gamma \vdash aS_\Box x, \Delta}{\Gamma, bEa \vdash bIx, \Delta} \qquad \frac{\Gamma, bEa \vdash bIx, \Delta}{\Gamma, \vdash aS_\Box x, \Delta}\ \text{uncurryS}^{**}$$

*Side condition: the variables y, a, b occurring as labels to a formula in the premise of any of these rules do not occur in Γ, Δ.

**Side condition: b does not occur Γ, Δ.

Rules for equivalence relations

$$\text{refl}\ \frac{\Gamma, aEa \vdash \Delta}{\Gamma \vdash \Delta} \qquad \frac{\Gamma \vdash aEb, \Delta}{\Gamma \vdash bEa, \Delta}\ \text{sym} \qquad \frac{\Gamma \vdash aEb, bEc\Delta}{\Gamma \vdash aEc, \Delta}\ \text{trans}$$

4 Properties of R.Lρ and R.LΣ

4.1 Soundness

Any sequent $\Gamma \vdash \Delta$ is to be interpreted in any enriched formal $\mathcal{L}$-context $\mathbb{F} = (\mathbb{P}, R_\Box, R_\Diamond, R_\rhd)$ based on $\mathbb{P} = (A, X, I)$ in the following way: for any assignment $V : \mathsf{Prop} \to \mathbb{P}^+$ that can be uniquely extended to an assignment on $\mathcal{L}$-formulas, and for any interpretation of labels $\alpha : \{a, b, c, \ldots\} \to A$ and $\chi : \{x, y, z, \ldots\} \to X$, we let $\iota_{(V,\alpha,\chi)}$ be the interpretation of well-formed formulas and well-formed terms indicated in the following table:

$a : A$	$\alpha(a) \in [\![A]\!]_V$	$x{::}A$	$\chi(x) \in (\!|A|\!)_V$
$aR_\Box x$	$\alpha(a)R_\Box\chi(x)$	$aR_\blacksquare x$	$\alpha(a)R_\blacksquare\chi(x)$
$xR_\Diamond a$	$\chi(x)R_\Diamond\alpha(a)$	$xR_\blacklozenge a$	$\chi(x)R_\blacklozenge\alpha(a)$
$aR_\rhd b$	$\alpha(a)R_\Diamond\alpha(b)$	$aR_\blacktriangleright b$	$\alpha(a)R_\blacklozenge\alpha(b)$
aIx	$\alpha(a)I\chi(x)$	$t_1(u) \Rightarrow t_2(u)$	$\forall u(\iota_{(V,\alpha,\chi)}(t_1(u)) \Rightarrow \iota_{(V,\alpha,\chi)}(t_2(u)))$

Under this interpretation, sequents $\Gamma \vdash \Delta$ are interpreted as follows[4]:

$$\forall V \forall \alpha \forall \chi (\mathop{\&}_{\gamma \in \Gamma} \iota_{(V,\alpha,\chi)}(\gamma) \Longrightarrow \mathop{⅋}_{\delta \in \Delta} \iota_{(V,\alpha,\chi)}(\delta)).$$

In the following, we show the soundness of the interdefinability rules in **R.L**ρ, being the proof of soundness of the (pure structure) switch rules similar. The soundness of the rules for the basic calculus R.L is proved in Appendix A.

Remark 1. Given a polarity $\mathbb{P} = (A, X, I)$, $c \in \mathbb{P}^+$, and $B \subseteq A$, the condition

$$(\forall x \in X)(c \subseteq I^{(0)}[x] \Rightarrow B \subseteq I^{(0)}[x]),$$

can be rewritten using the defining properties of $\bigcap$ as the inclusion

$$B \subseteq \bigcap \left\{ I^{(0)}[x] \mid x \in X, c \subseteq I^{(0)}[x] \right\},$$

which, by Lemma 2, is equivalent to $B \subseteq c$.

Lemma 5. *The rules swSf, swSfi, swSdf, swSdfi, swES, swESi, curryS, uncurryS, refl, sym, and trans are sound with respect to the class of rough formal contexts.*

[4] The symbols & and ⅋ denotes a meta-linguistic conjunction and a disjunction, respectively.

Proof. Under the assumption that E and $S_\Box$ are I-compatible, all the formulae are interpreted as concepts. In what follows, we will refer to the objects (resp. features) occurring in Γ and Δ in the various rules with $\overline{d}$ (resp. $\overline{w}$). For the sake of readability, in what follows we omit an explicit reference to the interpretation maps α and χ.

(swSf and swSfi)

$$\begin{aligned}
&\forall V\forall\overline{d}\forall\overline{w}\forall x\forall y(\& \Gamma \;\&\; \forall b(bS_\Box x \Rightarrow bIy) \Rightarrow y \in (\!|A|\!)_V \parr \bigparr \Delta) && \\
\text{iff } &\forall V\forall\overline{d}\forall\overline{w}\forall x\forall y\big(\& \Gamma \;\&\; S_\Box^{(0)}[x] \subseteq I^{(0)}[y] \Rightarrow y \in (\!|A|\!)_V \parr \bigparr \Delta\big) && \text{Def. of } (\cdot)^{(0)}\\
\text{iff } &\forall V\forall\overline{d}\forall\overline{w}\forall x\forall y\big(\& \Gamma \;\&\; S_\Box^{(0)}[x] \subseteq I^{(0)}[y] \Rightarrow [\![A]\!]_V \subseteq I^{(0)}[y] \parr \bigparr \Delta\big) && V(A) \text{ closed}\\
\text{iff } &\forall V\forall\overline{d}\forall\overline{w}\forall x\big(\& \Gamma \Rightarrow \forall y\big(S_\Box^{(0)}[x] \subseteq I^{(0)}[y] \Rightarrow [\![A]\!]_V \subseteq I^{(0)}[y]\big) \parr \bigparr \Delta\big) && \text{uncurrying + side}\\
\text{iff } &\forall V\forall\overline{d}\forall\overline{w}\forall x\big(\& \Gamma \Rightarrow [\![A]\!]_V \subseteq S_\Box^{(0)}[x] \parr \bigparr \Delta\big) && S\ I\text{-comp, Remark 1}\\
\text{iff } &\forall V\forall\overline{d}\forall\overline{w}\forall x\big(\& \Gamma \Rightarrow \forall a\big(a \in [\![A]\!]_V \Rightarrow a \in S_\Box^{(0)}[x]\big) \parr \bigparr \Delta\big) && \text{Def. of} \subseteq\\
\text{iff } &\forall V\forall\overline{d}\forall\overline{w}\forall x\big(\& \Gamma \;\&\; a \in [\![A]\!]_V \Rightarrow a \in S_\Box^{(0)}[x] \parr \bigparr \Delta\big) && \text{currying}\\
\text{iff } &\forall V\forall\overline{d}\forall\overline{w}\forall x(\& \Gamma \;\&\; a \in [\![A]\!]_V \Rightarrow aS_\Box x \parr \bigparr \Delta) && \text{Def. of } (\cdot)^{(0)}
\end{aligned}$$

(swSdf and swSdfi)

$$\begin{aligned}
&\forall V\forall\overline{d}\forall\overline{w}\forall a\forall x(\& \Gamma \;\&\; x \in (\!|A|\!)_V \Rightarrow xS_\blacklozenge a \parr \bigparr \Delta) && \\
\text{iff } &\forall V\forall\overline{d}\forall\overline{w}\forall a\forall x\big(\& \Gamma \;\&\; [\![A]\!]_V \subseteq I^{(0)}[x] \Rightarrow x \in S_\blacklozenge^{(0)}[a] \parr \bigparr \Delta\big) && V(A) \text{ closed}\\
\text{iff } &\forall V\forall\overline{d}\forall\overline{w}\forall a\forall x\big(\& \Gamma \;\&\; [\![A]\!]_V \subseteq I^{(0)}[x] \Rightarrow I^{(0)}[S_\blacklozenge^{(0)}[a]] \subseteq I^{(0)}[x] \parr \bigparr \Delta\big) && S \text{ is } I\text{-compatible}\\
\text{iff } &\forall V\forall\overline{d}\forall\overline{w}\forall a\big(\& \Gamma \Rightarrow I^{(0)}[S_\blacklozenge^{(0)}[a]] \subseteq [\![A]\!]_V \parr \bigparr \Delta\big) && V(A) \text{ closed, Remark 1}\\
\text{iff } &\forall V\forall\overline{d}\forall\overline{w}\forall a\forall b\big(\& \Gamma \;\&\; b \in I^{(0)}[S_\blacklozenge^{(0)}[a]] \Rightarrow b \in [\![A]\!]_V \parr \bigparr \Delta\big) && \text{Def. of} \subseteq\\
\text{iff } &\forall V\forall\overline{d}\forall\overline{w}\forall a\forall b\big(\& \Gamma \;\&\; b \in I^{(0)}[J^{(0)}[E^{(0)}[a]]] \Rightarrow b \in [\![A]\!]_V \parr \bigparr \Delta\big) && \text{Remark 2}\\
\text{iff } &\forall V\forall\overline{d}\forall\overline{w}\forall a\forall b\big(\& \Gamma \;\&\; b \in I^{(0)}[I^{(1)}[E^{(0)}[a]]] \Rightarrow b \in [\![A]\!]_V \parr \bigparr \Delta\big) && \text{Def. of } J\\
\text{iff } &\forall V\forall\overline{d}\forall\overline{w}\forall a\forall b\big(\& \Gamma \;\&\; b \in E^{(0)}[a] \Rightarrow b \in [\![A]\!]_V \parr \bigparr \Delta\big) && E \text{ is } I\text{-compatible}\\
\text{iff } &\forall V\forall\overline{d}\forall\overline{w}\forall a\forall b(\& \Gamma \;\&\; bEa \Rightarrow b \in [\![A]\!]_V \parr \bigparr \Delta) && \text{Def. of } (\cdot)^{(0)}
\end{aligned}$$

(curryS and uncurryS)

$$\begin{aligned}
&\forall V\forall\overline{d}\forall\overline{w}\forall a\forall x(\& \Gamma \Rightarrow aS_\Box x \parr \bigparr \Delta) && \\
\text{iff } &\forall V\forall\overline{d}\forall\overline{w}\forall a\forall x(\& \Gamma \Rightarrow \forall b(bEa \Rightarrow bIx) \parr \bigparr \Delta) && \text{Def. of } S_\Box\\
\text{iff } &\forall V\forall\overline{d}\forall\overline{w}\forall a\forall x\forall b(\& \Gamma \Rightarrow (bEa \Rightarrow bIx \parr \bigparr \Delta)) && \text{side condition}\\
\text{iff } &\forall V\forall\overline{d}\forall\overline{w}\forall a\forall x\forall b(\& \Gamma \;\&\; bEa \Rightarrow bIx \parr \bigparr \Delta) && \text{currying}
\end{aligned}$$

(swES and swESi) The proof is similar to the previous ones. The soundness of rules refl, sym, and trans follows from the fact that relation E is equivalence relation in a rough formal context.

Remark 2. The soundness of the switch rules is proved exactly as the soundness of the interdefinability rules in Lemma 5 by the I-compatibility of the relations in enriched formal contexts. More in general, these rules encode *exactly* the I-compatibility of such relations. Let us show this for $R_\Box$, as the others are proved similarly. One of the two I-compatibility conditions can be rewritten as

$$\begin{aligned}
&I^{(0)}[I^{(1)}[R_\Box^{(0)}[x]]] \subseteq R_\Box^{(0)}[x] && \\
\text{iff } &\forall y(y \in I^{(1)}[R_\Box^{(0)}[x]] \Rightarrow aIy) \Rightarrow aR_\Box x && \text{Def. of } I^{(0)}[\cdot]\\
\text{iff } &\forall y(\forall b(bR_\Box x \Rightarrow bIy) \Rightarrow aIy) \Rightarrow aR_\Box x && \text{Def. of } I^{(1)}[\cdot]
\end{aligned}$$

In what follows we are not assuming that $R_\Box$ is I-compatible; hence the valuation of an arbitrary formula does not need to be closed, but rather just a pair containing an arbitrary set of objects and its intension, or a an arbitrary set of features and its extension. Ignoring the contexts for readability, the rule $S_{x\Box a}$ is interpreted as

$$\begin{array}{lll}
& \forall V, a, x\, (\forall y\, (\forall b (bR_\Box x \Rightarrow bIy) \Rightarrow y \in \llparenthesis A \rrparenthesis_V) \Longrightarrow (a \in \llbracket A \rrbracket_V \Rightarrow aR_\Box x)) & \\
\text{iff} & \forall V, a, x \left(\forall y\, (\forall b (bR_\Box x \Rightarrow bIy) \Rightarrow y \in \llparenthesis A \rrparenthesis_V) \Longrightarrow (\llbracket A \rrbracket_V \subseteq R_\Box^{(0)}[x])\right) & \text{Def. of } R_\Box^{(0)}[\cdot] \\
\text{iff} & \forall V, a, x \left(\forall y \left(y \in I^{(1)}[R_\Box^{(0)}[x]] \Rightarrow y \in \llparenthesis A \rrparenthesis_V\right) \Longrightarrow (\llbracket A \rrbracket_V \subseteq R_\Box^{(0)}[x])\right) & \text{Def. of } I^{(1)}[\cdot] \\
\text{iff} & \forall V, a, x \left(\forall y \left(y \in I^{(1)}[R_\Box^{(0)}[x]] \Rightarrow y \in \llparenthesis A \rrparenthesis_V\right) \Longrightarrow (\llbracket A \rrbracket_V \subseteq R_\Box^{(0)}[x])\right) & \text{Def. of } I^{(1)}[\cdot] \\
\text{implies} & \forall V, a, x \left(\llbracket A \rrbracket_V \subseteq I^{(0)}[I^{(1)}[R_\Box^{(0)}[x]]] \Longrightarrow (\llbracket A \rrbracket_V \subseteq R_\Box^{(0)}[x])\right) & I^{(0)}[\cdot])\text{antitone}^{8} \\
\text{iff} & \forall V, a, x \left(I^{(0)}[I^{(1)}[R_\Box^{(0)}[x]]] \subseteq R_\Box^{(0)}[x]\right) & I^{(0)}[\cdot])\text{Def. of } \subseteq
\end{array}$$

[5]The second I-compatibility condition for $R_\Box$ is proved similarly using $S_{a\Box x}$.

4.2 Syntactic Completeness of the Basic Calculus and Its Axiomatic Extensions

In the present section, we show that the axioms and rules of **R**.**L**Σ, where Σ is a subset of the set of axioms in Table 1, are derivable in **R**.**L** extended with the corresponding rules. The axioms and rules of the basic logic **L** and some of its axiomatic extensions are discussed in Appendix B. Below, we show how the axioms $\Box p \vdash p$, $\Box p \vdash \Box\Box p$, and $p \vdash \rhd\rhd p$ can be derived using rules refl, sym, and trans respectively.

$$\dfrac{\dfrac{\dfrac{\dfrac{x :: p \vdash x :: p}{b : \Box p, x :: p \vdash bS_\Box x}\,\Box_L}{b : \Box p, x :: p, bEb \vdash bIx}\,\text{curry}}{b : \Box p, x :: p \vdash bIx}\,\text{refl}}{b : \Box p \vdash b : p}\,\text{approx}_x$$

$$\dfrac{\dfrac{\dfrac{\dfrac{\dfrac{\dfrac{\dfrac{x :: p \vdash x :: p}{a : \Box p, x :: p \vdash aS_\Box x}\,\Box_L}{a : \Box p, x :: p, bEa \vdash bIx}\,\text{curry}}{a : \Box p, x :: p, bEc, cEa \vdash bIx}\,\text{trans}}{a : \Box p, x :: p, cEa \vdash cS_\Box x}\,\text{uncurry}}{a : \Box p, cEa \vdash c : \Box p}\,\Box_R}{a : \Box p, y :: \Box p, \vdash yS_\blacklozenge a}\,\text{swSdfi}}{a : \Box p \vdash a : \Box\Box p}\,\Box_R$$

$$\dfrac{\dfrac{\dfrac{b : \rhd p, a : p \vdash a : p, bR_\rhd a}{b : \rhd p, a : p \vdash bR_\rhd a}\,\rhd_L}{b : \rhd p, a : p \vdash aR_\rhd b}\,\text{sym}}{a : p \vdash a : \rhd\rhd p}\,\rhd_R$$

5 Conclusions

In the present paper, we have introduced labelled calculi for a finite set of non-distributive modal logics in a modular way, and we have shown that the calculus associated with each such logic is sound w.r.t. the relational semantics of that logic given by elementary classes of enriched formal contexts, and syntactically complete w.r.t. the given logic. These results showcase that the methodology introduced in [13] for introducing labelled calculi by suitably integrating semantic information in the design of the rules can be extended from classical modal logics to the wider class of non-distributive logics. This methodology has proved successful for designing calculi for classical modal logics enjoying excellent computational properties, such as cut elimination,

[5] And also $\llbracket A \rrbracket_V \subseteq I^0[\llparenthesis A \rrparenthesis_V]$ holds in both the cases: the one where $\llbracket A \rrbracket$ is the extension of an arbitrary set of features, and when $\llparenthesis A \rrparenthesis$ is the intension of $\llbracket A \rrbracket$.

subformula property, being contraction-free, and being suitable for proof-search. Future developments of this work include the proofs of these results for the calculi introduced in the present paper.

A Soundness of the Basic Calculus

Lemma 6. *The basic calculus* **R.L** *is sound for the logic of enriched formal contexts.*

Proof. The soundness of the axioms, cut rules and propositional rules is trivial from the definitions of satisfaction and refutation relation for enriched formal contexts. We now discuss the soundness for the other rules.

Adjunction rules. The soundness of the adjunction rules follows from the fact that $R_{\blacksquare} = R_{\Diamond}^{-1}$, $R_{\blacklozenge} = R_{\Box}^{-1}$ and $R_{\rhd} = R_{\blacktriangleright}^{-1}$.

Approximation rules. We only give proof for $approx_a$. The proof for $approx_x$ is similar. In what follows, we will refer to the objects (resp. features) occurring in Γ and Δ in the various rules with $\overline{d}$ (resp. $\overline{w}$).

$$
\begin{array}{lll}
& \forall V \forall \overline{d} \forall \overline{w} \forall a \forall x (\& \Gamma \,\&\, x \succ A \Rightarrow aIx \parr \parr \Delta) & \\
\text{iff} & \forall V \forall \overline{d} \forall \overline{w} \forall a (\& \Gamma \,\&\, \forall x(x \succ A \Rightarrow aIx) \parr \parr \Delta) & x \text{ does not appear in } \Gamma \text{ or } \Delta \\
\text{iff} & \forall V \forall \overline{d} \forall \overline{w} \forall a \forall x (\& \Gamma \,\&\, x \in (\![V(A)]\!) \Rightarrow aIx \parr \parr \Delta) & \\
\text{iff} & \forall V \forall \overline{d} \forall \overline{w} \forall a \forall x \left(\& \Gamma \,\&\, a \in I^{(0)}(\![V(A)]\!) \parr \parr \Delta\right) & \text{Def. of } (\cdot)^{(0)} \\
\text{iff} & \forall V \forall \overline{d} \forall \overline{w} \forall a \forall x (\& \Gamma \,\&\, a \in [\![V(A)]\!] \parr \parr \Delta) & V(A) \text{ is closed} \\
\text{iff} & \forall V \forall \overline{d} \forall \overline{w} \forall a \forall x (\& \Gamma \,\&\, a \Vdash A \parr \parr \Delta) &
\end{array}
$$

Invertible rules for modal connectives. We only give proofs for $\Box_L$ and $\Box_R$. The proofs for $\Diamond_R$, $\Diamond_L$, $\rhd_R$, and $\rhd_L$ can be given in a similar manner.

$$
\begin{array}{lll}
& \forall V \forall \overline{d} \forall \overline{w} \forall x \forall y (\& \Gamma \,\&\, a \Vdash \Box A \Rightarrow x \succ A \parr aR_{\Box}x \parr \parr \Delta) & \\
\textit{implies} & \forall V \forall \overline{d} \forall \overline{w} \forall x \forall y (\& \Gamma \,\&\, a \Vdash \Box A \Rightarrow \forall b(b \vdash \Box A \Rightarrow bR_{\Box}x) \parr aR_{\Box}x \parr \parr \Delta) & Def.of \Box \\
\textit{implies} & \forall V \forall \overline{d} \forall \overline{w} \forall x \forall y (\& \Gamma \,\&\, a \Vdash \Box A \Rightarrow aR_{\Box}x \parr \parr \Delta) &
\end{array}
$$

The invertibility of the rule $\Box_L$ is obvious from the fact that the premise can be obtained from the conclusion by weakening.

$$
\begin{array}{lll}
& \forall V \forall \overline{d} \forall \overline{w} \forall a \forall x (\& \Gamma \,\&\, x \succ A \Rightarrow aR_{\Box}x \parr \parr \Delta) & \\
\text{iff} & \forall V \forall \overline{d} \forall \overline{w} \forall a (\& \Gamma \,\&\, \forall x(x \succ A \Rightarrow aR_{\Box}x) \parr \parr \Delta) & x \text{ does not appear in } \Gamma \text{ or } \Delta \\
\text{iff} & \forall V \forall \overline{d} \forall \overline{w} \forall a (\& \Gamma \& \Rightarrow a \Vdash \Box A \parr \parr \Delta) & x \text{ Def. of } \Box
\end{array}
$$

Switch rules. Soundness of the rules Sxa and Sax follows from the fact that for any concepts c_1 and c_2 we have

$$[\![c_1]\!] \subseteq [\![c_2]\!] \iff (\![c_2]\!) \subseteq (\![c_1]\!).$$

The soundness of all other switch rules follows from the definition of modal connectives and I-compatibility. As all the proofs are similar we only prove the soundness of $Sa\Diamond x$ as a representative case. Soundness of other rules can be proved in an analogous manner.

$$\begin{array}{lll}
& \forall V\forall\bar{d}\forall\overline{w}\forall a\forall b\,(\&\,\Gamma\;\&\;\forall y(yR_\Diamond a \Rightarrow bIy) \Rightarrow b \Vdash A \parr \bigparr \Delta) & \\
\text{iff} & \forall V\forall\bar{d}\forall\overline{w}\forall a\forall b\left(\&\,\Gamma\;\&\;b \in I^{(0)}[R_\Diamond^{(0)}[a]] \Rightarrow b \Vdash A \parr \bigparr \Delta\right) & \text{Def. of } R_\Diamond^{(0)} \text{ and } I^{(0)} \\
\text{iff} & \forall V\forall\bar{d}\forall\overline{w}\forall a\left(\&\,\Gamma \Rightarrow \forall b(b \in I^{(0)}[R_\Diamond^{(0)}[a]] \Rightarrow b \Vdash A) \parr \bigparr \Delta\right) & b \text{ does not appear in } \Gamma \text{ or } \Delta \\
\text{iff} & \forall V\forall\bar{d}\forall\overline{w}\forall a\left(\&\,\Gamma \Rightarrow I^{(0)}[R_\Diamond^{(0)}[a]] \subseteq [\![V(A)]\!] \parr \bigparr \Delta\right) & \text{b does not appear in } \Gamma \text{ or } \Delta \\
\text{iff} & \forall V\forall\bar{d}\forall\overline{w}\forall a\left(\&\,\Gamma \Rightarrow I^{(1)}[[\![V(A)]\!]] \subseteq I^{(1)}[I^{(0)}[R_\Diamond^{(0)}[a]]] \parr \bigparr \Delta\right) & I^{(1)} \text{ is antitone and } [\![V(A)]\!] \text{ is closed} \\
\text{iff} & \forall V\forall\bar{d}\forall\overline{w}\forall a\left(\&\,\Gamma \Rightarrow I^{(1)}[[\![V(A)]\!]] \subseteq R_\Diamond^{(0)}[a] \parr \bigparr \Delta\right) & R_\Box \text{ is I-compatible} \\
\text{iff} & \forall V\forall\bar{d}\forall\overline{w}\forall a\left(\&\,\Gamma \Rightarrow \forall x(x \in I^{(1)}[[\![V(A)]\!]] \Rightarrow x \in R_\Diamond^{(0)}[a]) \parr \bigparr \Delta\right) & \\
\text{implies} & \forall V\forall\bar{d}\forall\overline{w}\forall a\forall x\left(\&\,\Gamma\;\&\;x \in I^{(1)}[[\![V(A)]\!]] \Rightarrow x \in R_\Diamond^{(0)}[a]) \parr \bigparr \Delta\right) & \\
\text{iff} & \forall V\forall\bar{d}\forall\overline{w}\forall a\forall x\,(\&\,\Gamma\;\&\;x \succ A \Rightarrow xR_\Diamond a \parr \bigparr \Delta) & \text{Def. of } R_\Diamond^{(0)}
\end{array}$$

Soundness of the axiomatic extensions considered in Sect. 3.2 is immediate from the Proposition 1.

B Syntactic completeness

As to the axioms and rules of the basic logic **L**, below, we only derive in **R.L** the axioms and rules encoding the fact that $\Diamond$ is a normal modal operator plus the axiom $p \vdash p \vee q$.

$$\dfrac{\dfrac{\dfrac{\dfrac{\dfrac{}{x :: \Diamond A, x :: \Diamond B, b : A \vdash b : A, xR_\Diamond b}\,\mathrm{Id}_{b:A}}{x :: \Diamond A, x :: \Diamond B, b : A \vdash xR_\Diamond b}\,\Diamond_R}{x :: \Diamond A \vee \Diamond B, b : A \vdash xR_\Diamond b}\,\vee_R}{x :: \Diamond A \vee \Diamond B, b : A \vdash bR_\blacksquare x}\,\Diamond \dashv \blacksquare}{x :: \Diamond A \vee \Diamond B, aR_\blacksquare x \Rightarrow aIy \vdash y : A}\,\mathrm{S}x\blacksquare a^c
\qquad
\dfrac{\dfrac{\dfrac{\dfrac{\dfrac{}{x :: \Diamond A, x :: \Diamond B, b : B \vdash b : B, xR_\Diamond b}\,\mathrm{Id}_{b:B}}{x :: \Diamond A, x :: \Diamond B, b : B \vdash xR_\Diamond b}\,\Diamond_R}{x :: \Diamond A \vee \Diamond B, b : B \vdash xR_\Diamond b}\,\vee_R}{x :: \Diamond A \vee \Diamond B, b : B \vdash bR_\blacksquare x}\,\Diamond \dashv \blacksquare}{x :: \Diamond A \vee \Diamond B, aR_\blacksquare x \Rightarrow aIy \vdash y : B}\,\mathrm{S}x\blacksquare a^c$$

$$\dfrac{\dfrac{\dfrac{\dfrac{x :: \Diamond A \vee \Diamond B, aR_\blacksquare x \Rightarrow aIy \vdash y : A \qquad x :: \Diamond A \vee \Diamond B, aR_\blacksquare x \Rightarrow aIy \vdash y : B}{x :: \Diamond A \vee \Diamond B, aR_\blacksquare x \Rightarrow aIy \vdash y : A \vee B}\,\vee_L}{x :: \Diamond A \vee \Diamond B, a : A \vee B \vdash aR_\blacksquare x}\,\mathrm{S}x\blacksquare a}{x :: \Diamond A \vee \Diamond B, a : A \vee B \vdash xR_\Diamond a}\,\Diamond \dashv \blacksquare^{-1}}{x :: \Diamond A \vee \Diamond B \vdash x :: \Diamond(A \vee B)}\,\Diamond_L$$

$$\dfrac{\dfrac{\dfrac{\dfrac{}{bR_\Box x \Rightarrow bIy, x : \bot \vdash y :: \bot}\,\bot}{a : \bot, x : \bot \vdash aR_\blacksquare x}\,\mathrm{S}x\blacksquare a}{a : \bot, x : \bot \vdash xR_\Diamond a}\,\Diamond \dashv \blacksquare^{-1}}{x : \bot \vdash x : \Diamond\bot}\,\Diamond_L
\qquad
\dfrac{\dfrac{\dfrac{\dfrac{y :: \varphi \vdash y :: \psi}{y :: \varphi, x :: \Diamond\varphi \vdash y :: \psi, xR_\Diamond a}\,\mathrm{W}}{a : \psi, x :: \Diamond\varphi \vdash a : \varphi, xR_\Diamond a}\,\mathrm{S}xa}{a : \psi, x :: \Diamond\varphi \vdash xR_\Diamond a}\,\Diamond_R}{x :: \Diamond\varphi \vdash x :: \Diamond\psi}\,\Diamond_L
\qquad
\dfrac{x : p, x : q \vdash x : p}{x :: p \vee q \vdash x : p}\,\vee_R$$

The syntactic completeness for the other axioms and rules of **L** can be shown in a similar way. In particular, the admissibility of the substitution rule can be proved by induction in a standard manner.

We now consider the reflexivity axiom $p \vdash \Diamond p$ and the transitivity axiom $\Box p \vdash \Box\Box p$. The derivation for dual axioms $\Box p \vdash p$ and $\Diamond\Diamond p \vdash \Diamond p$ can be provided analogously.

$$approx_a\ \dfrac{\mathrm{refl}\ \dfrac{\Diamond_R\ \dfrac{\mathrm{Id}_{a:p}\ \dfrac{}{x :: p, a : p \vdash a : p, aR_{\Diamond}x}}{x :: p, a : p \vdash aR_{\Diamond}x}}{x : \Diamond p, a : p \vdash aIx}}{x : \Diamond p \vdash x : p}$$

$$\Box_R\ \dfrac{\blacklozenge \dashv \Box^{-1}\ \dfrac{Sa\blacklozenge x\ \dfrac{\Box_R\ \dfrac{\dfrac{-\mathrm{S}(J;S)^*\ \dfrac{\mathrm{Id}(J;I)_R\ \dfrac{\Box \dashv \blacklozenge^{-1}\ \dfrac{\dfrac{\mathrm{trans}\ \dfrac{\Box_L\ \dfrac{\mathrm{Id}_{x::p}\ \dfrac{}{a : \Box p, x :: p \vdash x :: p, aR_{\Box}x}}{a : \Box p, x :: p \vdash aR_{\Box}x}}{bR_{\Box}x \Rightarrow zJb, a : \Box p, x :: p \vdash aR_{\Box}z}}{z(J;R_{\Box})x, a : \Box p, x :: p \vdash aR_{\Box}z}}{z(J;R_{\Box})x, a : \Box p, x :: p \vdash zR_{\blacklozenge}a}}{z(J;R_{\Box})x, a : \Box p, x :: p \vdash z(J;(I;R_{\blacklozenge}))a}}{b(I;R_{\blacklozenge})a, a : \Box p, x :: p \vdash bR_{\Box}x}}{yR_{\blacklozenge}a \Rightarrow bIy, a : \Box p, x :: p \vdash bR_{\Box}x}}{yR_{\blacklozenge}a \Rightarrow bIy, a : \Box p \vdash b : \Box p}}{x :: \Box p, a : \Box p \vdash xR_{\blacklozenge}a}}{x :: \Box p, a : \Box p \vdash aR_{\Box}x}}{a : \Box p \vdash a : \Box\Box p}$$

Completeness for the other axiomatic extensions can be shown in a similar way.

References

1. Conradie, W., et al.: Modal reduction principles across relational semantics. arXiv preprint arXiv:2202.00899 (2022)
2. Conradie, W., et al.: Rough concepts. Inf. Sci. **561**, 371–413 (2021)
3. Conradie, W., Frittella, S., Palmigiano, A., Piazzai, M., Tzimoulis, A., Wijnberg, N.M.: Toward an epistemic-logical theory of categorization. In: Electronic Proceedings in Theoretical Computer Science, EPTCS, vol. 251 (2017)
4. Conradie, W., Frittella, S., Palmigiano, A., Piazzai, M., Tzimoulis, A., Wijnberg, N.M.: Categories: how i learned to stop worrying and love two sorts. In: Väänänen, J., Hirvonen, Å., de Queiroz, R. (eds.) WoLLIC 2016. LNCS, vol. 9803, pp. 145–164. Springer, Heidelberg (2016). https://doi.org/10.1007/978-3-662-52921-8_10
5. Conradie, W., Palmigiano, A.: Algorithmic correspondence and canonicity for non-distributive logics. Ann. Pure Appl. Logic **170**(9), 923–974 (2019)
6. Conradie, W., Palmigiano, A., Robinson, C., Wijnberg, N.: Non-distributive logics: from semantics to meaning. In: Rezus, A. (ed.) Contemporary Logic and Computing, volume 1 of Landscapes in Logic, pp. 38–86. College Publications (2020)
7. Davey, B., Priestley, H.: Introduction to Lattices and Order. Cambridge University Press, Cambridge (2002)
8. Gabbay, D.M.: Labelled Deductive Systems: Volume 1. Oxford University Press, Oxford, England (1996)
9. Ganter, B., Wille, R.: Formal Concept Analysis: Mathematical Foundations. Springer Science & Business Media, Berlin (2012)
10. Greco, G., Jipsen, P., Manoorkar, K., Palmigiano, A., Tzimoulis, A.: Logics for rough concept analysis. In: Khan, M.A., Manuel, A. (eds.) ICLA 2019. LNCS, vol. 11600, pp. 144–159. Springer, Heidelberg (2019). https://doi.org/10.1007/978-3-662-58771-3_14
11. Kent, R.E.: Rough concept analysis. In: Ziarko, W.P., et al. (eds.) Rough Sets, Fuzzy Sets and Knowledge Discovery. Workshops in Computing, pp. 248–255. Springer, London (1994)

12. Kent, R.E.: Rough concept analysis: a synthesis of rough sets and formal concept analysis. Fundam. Inf. **27**(2–3), 169–181 (1996)
13. Negri, S.: Proof analysis in modal logic. J. Philos. Log. **34**(5–6), 507–544 (2005)
14. Negri, S., Von Plato, J.: Proof Analysis: A Contribution to Hilbert's Last Problem. Cambridge University Press, Cambridge (2011)
15. Pawlak, Z.: Rough sets. Int. J. Comput. Inf. Sci. **11**(5), 341–356 (1982). https://doi.org/10.1007/BF01001956

An Infinity of Intuitionistic Connexive Logics

Hao Wu and Minghui Ma(✉)

Institute of Logic and Cognition, Sun Yat-sen University, No. 135 Xingangxi Road, Haizhu District, Guangzhou 510275, China
wuhao43@mail2.sysu.edu.cn, mamh6@mail.sysu.edu.cn

Abstract. We develop infinitely many intuitionistic connexive logics $\mathsf{C}_{m,n}$ with $m > 0$ and $n \geq 0$ which are obtained from intuitionistic propositional logic by adding the negation sign $\sim$ which admits principles of connexive implication and $\sim^{2m+n}p \leftrightarrow \sim^{n}p$. We introduce $\langle m, n\rangle$-connexive logics and show that lattices of these connexive logics are isomorphic to lattices of superintuitionistic logics. Furthermore, we give cut-free G3-style sequent calculi for $\langle m, n\rangle$-connexive logics.

Keywords: Connexive logic · Intuitionistic logic · Sequent calculus

1 Introduction

Connexive logic has its roots in ancient time (cf. e.g. [15]). It enters into modern logic mainly by McCall's investigation on the *connexive implication* (cf. e.g. [13,14]). The following Aristotle's theses (A1) and (A2) as well as Boethian theses (B1) and (B2) are taken into account when connexive logics are explored: (A1) $\sim(\sim\varphi \to \varphi)$, (A2) $\sim(\varphi \to \sim\varphi)$, (B1) $(\varphi \to \psi) \to \sim(\varphi \to \sim\psi)$ and (B2) $(\varphi \to \sim\psi) \to \sim(\varphi \to \psi)$. These are not tautologies in classical propositional logic (CPL). Connexive logics containing one of them are called *contra-classical logics* (cf. [5,19]). McCall [14] presented a consistent, independent of classical bivalent logic and Post complete system CC1 to accommodate the connexive implication. After McCall, there are various trends in the study of connexive logic (cf. e.g. [6,7,9,10,12,17–26]). These trends develop connexive logics in both semantical and syntactic aspects. Wansing [8,24] proposed the basic connexive logic $\mathcal{C}$ which enjoys pleasant semantics and proof-theoretic properties. The Hilbert-style system for $\mathcal{C}$ is obtained from positive intuitionistic propositional logic (IPL) by adding (DN) $\sim\sim\varphi \leftrightarrow \varphi$, (M1) $\sim(\varphi \wedge \psi) \leftrightarrow (\sim\varphi \vee \sim\psi)$, (M2) $\sim(\varphi \vee \psi) \leftrightarrow (\sim\varphi \wedge \sim\psi)$ and (BT) $\sim(\varphi \to \psi) \leftrightarrow (\varphi \to \sim\psi)$. Using (DN) and (BT), we can derive Aristotle's and Boethian theses. The double negation laws and Boethius thesis shed lights on the understanding of connexivity. Using them

This work was supported by Chinese National Funding of Social Sciences (Grant no. 18ZDA033).

M. Banerjee and A. V. Sreejith (Eds.): ICLA 2023, LNCS 13963, pp. 189–206, 2023.
https://doi.org/10.1007/978-3-031-26689-8_14

every formula is equivalently transformed into a negation normal form (Definition 3) built from literals p or $\sim p$ where p is a variable. As Wansing [24] observed, the system C is equivalent to the positive fragment of IPL by treating each $\sim p$ as a new variable. And recent studies towards connexive logic in [3, 4] are taken from algebraic approaches.

Inspired by Wansing' system C, the present work develops infinitely many connexive logics which contain at least one of connexive principles. We use the full IPL as the base and hence the desired intuitionistic connexive logics are IPL with a negation operator $\sim$ satisfying additional axioms. For each pair of natural numbers $\langle m, n\rangle$ with $m > 0$ and $n \geq 0$, we generalize the double negation axiom to ($\mathrm{DN}_{m,n}$) $\sim^{2m+n}\varphi \leftrightarrow \sim^{n}\varphi$. This axiom in algebraic form was originally proposed by Berman [1], and it is systematically investigated in the study of weakenings of Belnap-Dunn four-valued logic (cf. [11]). For each pair $\langle m, n\rangle$ with $m > 0$ and $n \geq 0$, we define a connexive logic $\mathsf{C}_{m,n}$ by adding ($\mathrm{DN}_{m,n}$), (M1), (M2) and (BT) to the IPL. Furthermore, if we take an intermediate logic L as basis, we also obtain infinitely many connexive logics.

This paper is organized as follows. Section 2 gives syntax and semantics for intuitionistic connexive logics. Section 3 presents Hilbert-style axiomatic systems and lattices of $\langle m, n\rangle$–connexive logics. Section 4 introduces G3-style Gentzen sequent calculi for connexive logics. Section 5 gives some concluding remarks.

2 Intuitionistic Connexive Logics

Let $\mathbb{Z}$, $\mathbb{Z}^{+}$ and $\mathbb{Z}^{*}$ be sets of all integers, positive integers and non-negative integers respectively. For $k_1, k_2 \in \mathbb{Z}$, let $[k_1, k_2) = \{i \in \mathbb{Z}^{*} : k_1 \leq i < k_2\}$. Let $\mathbb{E}$ and $\mathbb{O}$ be sets of all even and odd numbers in $\mathbb{Z}^{*}$ respectively. The language of intuitionistic connexive logic $\mathscr{L}_C$ consists of a denumerable set of variables $\mathbb{P} = \{p_i : i \in \mathbb{Z}^{*}\}$, intuitionistic connectives $\bot$, $\wedge$, $\vee$, $\rightarrow$ and negation $\sim$. The set of all $\mathscr{L}_C$-formulas $\mathscr{F}$ is defined as follows:

$$\mathscr{F} \ni \varphi ::= p \mid \bot \mid (\varphi_1 \wedge \varphi_2) \mid (\varphi_1 \vee \varphi_2) \mid (\varphi_1 \rightarrow \varphi_2) \mid \sim\varphi$$

where $p \in \mathbb{P}$. If we remove $\sim$ from $\mathscr{L}_C$, we obtain the set $\mathscr{F}_I$ of all formulas for IPL. We use abbreviations $\top := \neg\bot$, $\neg\varphi := \varphi \rightarrow \bot$ and $\varphi \leftrightarrow \psi := (\varphi \rightarrow \psi) \wedge (\psi \rightarrow \varphi)$. For every finite set of formulas Γ, let $\bigwedge\Gamma$ and $\bigvee\Gamma$ be the conjunction and disjunction of all formulas in Γ respectively. Let $\bigwedge\varnothing = \top$ and $\bigvee\varnothing = \bot$. For $k \geq 0$, let $\sim^{k}\varphi$ be defined by $\sim^{0}\varphi = \varphi$ and $\sim^{k+1}\varphi = \sim\sim^{k}\varphi$.

Let $mc(\varphi)$ be the *main connective* of φ. Let $Sub(\varphi)$ be the set of all subformulas of φ. Let $var(\varphi)$ be the set of all propositional variables appearing in φ. The *complexity $c(\varphi)$* of a formula φ is defined inductively as usual. A substitution is a function $s : \mathbb{P} \rightarrow \mathscr{F}$. Let φ^{s} be the formula obtained from φ by substitution s. For all $\varphi, \psi, \chi \in \mathscr{F}$, let $\varphi(\psi_1/\chi_1, \ldots, \psi_n/\chi_n)$ be obtained from φ by substituting ψ_i for one or more occurrences of χ_i in φ.

Definition 1. *A* frame *is a pair $\mathfrak{F} = (W, R)$ where $W \neq \varnothing$ is a set of states and R is a partial order (a reflexive, transitive and anti-symmetric binary relation)* *on W. For every $w \in W$ and $X \subseteq W$, let $R(w) = \{u \in W : wRu\}$ and $R[X] = \bigcup_{w\in X} R(w)$. A subset $X \subseteq W$ is an* upset *in $\mathfrak{F}$ if $R[X] = X$. The set of all upsets in $\mathfrak{F}$ will be denoted by $\mathsf{Up}(\mathfrak{F})$. For each pair $\langle m, n\rangle \in \mathbb{Z}^+ \times \mathbb{Z}^*$, the set of all $\langle m, n\rangle$-*literals *is defined as $\mathbb{X}_{m,n} = \{\sim^k p : k \in [0, 2m + n), p \in \mathbb{P}\} \cup \{\sim^l \bot : 0 < l < 2m + n\}$. An $\langle m, n\rangle$-*valuation *in a frame $\mathfrak{F}$ is a function $V : \mathbb{X}_{m,n} \to \mathsf{Up}(\mathfrak{F})$. An $\langle m, n\rangle$-*model *is a triple $\mathfrak{M} = (\mathfrak{F}, V)$ where $\mathfrak{F}$ is a frame and V is an $\langle m, n\rangle$-valuation in $\mathfrak{F}$.*

Note that $\varnothing, W \in \mathsf{Up}(\mathfrak{F})$ and $\mathsf{Up}(\mathfrak{F})$ is closed under $\cap$ and $\cup$. As in intuitionistic logic, if $X, Y \in \mathsf{Up}(\mathfrak{F})$, then $X \to_R Y = \{w \in W : R(w) \cap X \subseteq Y\}$ also belongs to $\mathsf{Up}(\mathfrak{F})$. For every $\langle m, n\rangle$-model $\mathfrak{M} = (W, R, V)$ and formula φ, the *truth-set* $V(\varphi)$ of φ in $\mathfrak{M}$ is defined inductively as follows:

$$
\begin{aligned}
V(\bot) &= \varnothing & V(\varphi \wedge \psi) &= V(\varphi) \cap V(\psi)\\
V(\varphi \vee \psi) &= V(\varphi) \cup V(\psi) & V(\varphi \to \psi) &= V(\varphi) \to_R V(\psi)\\
V(\sim^{2m+n}\varphi) &= V(\sim^n\varphi) & V(\sim(\varphi \wedge \psi)) &= V(\sim\varphi) \cup V(\sim\psi)\\
V(\sim(\varphi \vee \psi)) &= V(\sim\varphi) \cap V(\sim\psi) & V(\sim(\varphi \to \psi)) &= V(\varphi) \to_R V(\sim\psi)
\end{aligned}
$$

For a set of formulas Γ, let $V(\Gamma) = \bigcap_{\varphi\in\Gamma} V(\varphi)$. A formula φ is true at w in $\mathfrak{M}$ (notation: $\mathfrak{M}, w \models_{m,n} \varphi$) if $w \in V(\varphi)$. We write $\mathfrak{M} \models_{m,n} \varphi$ if $V(\varphi) = W$.

Definition 2. *Let $\mathfrak{F} = (W, R)$ be a frame. A formula φ is $\langle m, n\rangle$-*valid *at w in $\mathfrak{F}$ (notation: $\mathfrak{F}, w \models_{m,n} \varphi$) if $w \in V(\varphi)$ for all $\langle m, n\rangle$-valuations V in $\mathfrak{F}$. A formula φ is $\langle m, n\rangle$-*valid *in $\mathfrak{F}$ (notation: $\mathfrak{F} \models_{m,n} \varphi$) if $\mathfrak{F}, w \models_{m,n} \varphi$ for all $w \in W$. A formula φ is $\langle m, n\rangle$-*valid *(notation: $\models_{m,n} \varphi$) if $\mathfrak{F} \models_{m,n} \varphi$ for all frames $\mathfrak{F}$. The* intuitionistic $\langle m, n\rangle$-connexive logic *is defined as $\mathsf{C}_{m,n} = \{\varphi \in \mathscr{F} : \models_{m,n} \varphi\}$. A formula φ is an $\langle m, n\rangle$-*consequence *of a set of formulas Γ (notation: $\Gamma \models_{m,n} \varphi$) if $V(\Gamma) \subseteq V(\psi)$ for every $\langle m, n\rangle$-model $\mathfrak{M} = (W, R, V)$.*

Lemma 1. *For all $i, j \in [0, 2m + n)$, $\models_{m,n} \sim^i p \leftrightarrow \sim^j p$ iff $i = j$.*

Proof. Clearly $\models_{m,n} \sim^i p \leftrightarrow \sim^j p$ if $i = j$. Assume $\models_{m,n} \sim^i p \leftrightarrow \sim^j p$. Let $\mathfrak{M} = (W, R, V)$ be the $\langle m, n\rangle$-model where $W = \{w\}$, $R = \{\langle w, w\rangle\}$, $V(\sim^i p) = \{w\}$ and $V(\sim^k p) = \varnothing$ for all $k \in [0, 2m + n)$ with $k \neq i$. Then $\mathfrak{M}, w \models_{m,n} \sim^i p$ and so $\mathfrak{M}, w \models_{m,n} \sim^j p$. Then $w \in V(\sim^j p)$. Hence $i = j$. □

Let IPL be the set of all intuitionistic tautologies. By Definition 2, all instances of formulas in IPL are $\langle m, n\rangle$-valid in all frames, and hence belong to $\mathsf{C}_{m,n}$.

Lemma 2. *The following formulas are $\langle m, n\rangle$-valid: $\sim^{2m+n}\varphi \leftrightarrow \sim^n\varphi$; $\sim(\varphi \wedge \psi) \leftrightarrow (\sim\varphi \vee \sim\psi)$; $\sim(\varphi \vee \psi) \leftrightarrow (\sim\varphi \wedge \sim\psi)$; $\sim(\varphi \to \psi) \leftrightarrow (\varphi \to \sim\psi)$; $\sim(\sim\varphi \to \varphi)$.*

Proof. Let $\mathfrak{M}$ be an $\langle m, n\rangle$-model. For the $\langle m, n\rangle$-validity of $\sim(\varphi \to \psi) \leftrightarrow (\varphi \to \sim\psi)$, we have $\mathfrak{M}, w \models_{m,n} \sim(\varphi \to \psi)$ iff $\forall u \in R(w)(\mathfrak{M}, u \models_{m,n} \varphi \Rightarrow \mathfrak{M}, u \models_{m,n} \sim\psi)$, namely, $\mathfrak{M}, w \models_{m,n} \varphi \to \sim\psi$. Other items are shown similarly. □

Corollary 1. *For all $e \in \mathbb{E}$, $o \in \mathbb{O}$ and $k \in \mathbb{Z}^*$, the following are $\langle m, n\rangle$-valid: $\sim^{2km+i}\varphi \leftrightarrow \sim^i\varphi$ for $i \geq n$; $\sim^e(\varphi\wedge\psi) \leftrightarrow (\sim^e\varphi\wedge\sim^e\psi)$; $\sim^o(\varphi\wedge\psi) \leftrightarrow (\sim^o\varphi\vee\sim^o\psi)$; $\sim^e(\varphi\vee\psi) \leftrightarrow (\sim^e\varphi\vee\sim^e\psi)$; $\sim^o(\varphi\vee\psi) \leftrightarrow (\sim^o\varphi\wedge\sim^o\psi)$; $\sim^k(\varphi \to \psi) \leftrightarrow (\varphi \to \sim^k\psi)$.*

Remark 1. By Lemma 2, Aristotle's thesis (A1) and Boethian thesis (B2) are $\langle m,n\rangle$-valid. However, this is not the case in general for (A2) and (B1). Clearly (A2) and (B1) are $\langle 1,0\rangle$-valid. Let $m > 1$ or $n > 0$. Then $2m+n > 2$. We have the following facts: (1) $\not\models_{m,n} \sim(p \to \sim p)$. Clearly $p, \sim\sim p \in \mathbb{X}_{m,n}$. Let $\mathfrak{M} = (W,R,V)$ be the model where $W = \{w\}, R = \{\langle w,w\rangle\}$, $V(p) = \{w\}$ and $V(\sim\sim p) = \varnothing$. Then $\mathfrak{M}, w \models_{m,n} p$ and $\mathfrak{M}, w \not\models_{m,n} \sim\sim p$; (2) $\not\models_{m,n} (p \to q) \to \sim(p \to \sim q)$. Clearly $p, q, \sim\sim q \in \mathbb{X}_{m,n}$. Let $\mathfrak{M} = (W,R,V)$ be the model where $W = \{w\}, R = \{\langle w,w\rangle\}$, $V(p) = \{w\} = V(q)$ and $V(\sim\sim q) = \varnothing$. Then $\mathfrak{M}, w \models_{m,n} p \to q$. By $\mathfrak{M}, w \models_{m,n} p$ and $\mathfrak{M}, w \not\models_{m,n} \sim\sim q$, we have $\mathfrak{M}, w \not\models_{m,n} \sim(p \to \sim\sim q)$. The following table shows the $\langle m,n\rangle$-validity of connexive theses:

$\mathsf{C}_{m,n}$	(A1)	(A2)	(B1)	(B2)
$m = 1$ and $n = 0$	Yes	Yes	Yes	Yes
$m > 1$ or $n > 0$	Yes	No	No	Yes

Note that $\mathsf{C}_{m,n}$ has at least one connexive principle and we call it *connexive*.

Definition 3. *The set NEG-AT of all* negated atoms *is defined as NEG-AT* $= \{\sim^k p : k \in \mathbb{Z}^*, p \in \mathbb{P}\} \cup \{\sim^k \bot : k \in \mathbb{Z}^+\}$. *A formula* φ *is in* negation normal form *(NNF) if it is built from negated atoms using only* $\bot, \wedge, \vee$ *and* $\to$. *Let* $\mathcal{N}$ *be the set all NNFs. Let* $\nu(\varphi)$ *be the set of all negated atoms appearing in a formula* $\varphi \in \mathcal{N}$. *The* degree $d(\varphi)$ *of a formula* $\varphi \in \mathcal{N}$ *is defined inductively by* $d(p) = 0 = d(\bot)$ *and* $d(\varphi \odot \psi) = d(\varphi) + d(\psi) + 1$ *where* $\odot \in \{\wedge, \vee, \to\}$. *Let* $\mathscr{F}_{m,n}$ *be the set of all formulas built from literals in* $\mathbb{X}_{m,n}$ *using only* $\bot, \wedge, \vee$ *and* $\to$. *Note that* $\mathscr{F}_{m,n} \subseteq \mathcal{N}$. *For every* $\varphi \in \mathscr{F}_{m,n}$, *let* $\lambda(\varphi)$ *be the set of all literals in* φ. *We write* $\varphi(\lambda_1, \ldots, \lambda_n)$ *if* $\lambda(\varphi) \subseteq \{\lambda_1, \ldots, \lambda_n\}$.

Lemma 3. *For every formula* $\varphi \in \mathscr{F}$, *there exists a unique formula* $N(\varphi) \in \mathcal{N}$ *such that* $\models_{m,n} \varphi \leftrightarrow N(\varphi)$.

Proof. For all $\psi, \chi, \xi \in \mathscr{F}$, if $\models_{m,n} \psi \leftrightarrow \chi$ and $\models_{m,n} \chi \leftrightarrow \xi$, then $\models_{m,n} \psi \leftrightarrow \chi$. By Lemma 2, the negation $\sim$ goes through $\wedge, \vee$ and $\to$, and we obtain $N(\varphi)$ in NNF such that $\models_{m,n} \varphi \leftrightarrow N(\varphi)$. □

Lemma 4. *For every negated atom* $\alpha \in$ *NEG-AT, there exists a unique* $\lambda \in \mathbb{X}_{m,n}$ *such that* $\models_{m,n} \alpha \leftrightarrow \lambda$.

Proof. Let $\alpha = \sim^k \beta$. If $k \in [0, 2m+n)$, then $\lambda = \sim^k \beta$ is the required literal. Let $k > 2m+n$. Then there exist $j \geq 1$ and $i \in [0, 2m+n)$ with $k = 2jm + i$. By Corollary 1, $\models_{m,n} \sim^k \beta \leftrightarrow \sim^i \beta$ where $\sim^i \beta \in \mathbb{X}_{m,n}$. □

Lemma 5. *For every formula* $\varphi \in \mathcal{N}$, *if* $\models_{m,n} \alpha_1 \leftrightarrow \alpha_2$ *with* $\alpha_1 \in$ *NEG-AT and* $\alpha_2 \in \nu(\varphi)$, *then* $\models_{m,n} \varphi \leftrightarrow \varphi(\alpha_1/\alpha_2)$.

Proof. The proof proceeds by induction on $d(\varphi)$. The case $d(\varphi) = 0$ is trivial. Assume $d(\varphi) > 0$. This is easily shown by induction hypothesis. For example, let $\varphi = \psi \to \chi$. By induction hypothesis, $\models_{m,n} \psi \leftrightarrow \psi(\alpha_1/\alpha_2)$ and $\models_{m,n} \chi \leftrightarrow \chi(\alpha_1/\alpha_2)$. Hence $\models_{m,n} (\psi \to \chi) \leftrightarrow (\psi \to \chi)(\alpha_1/\alpha_2)$. □

Theorem 1. *For every formula $\varphi \in \mathscr{F}$, there exists a unique formula $N_{m,n}(\varphi) \in \mathscr{F}_{m,n}$ such that $\models_{m,n} \varphi \leftrightarrow N_{m,n}(\varphi)$.*

Proof. By Lemma 3, there is a unique $N(\varphi) \in \mathscr{N}$ with $\models_{m,n} \varphi \leftrightarrow N(\varphi)$. By Lemma 4, for every $\alpha \in \nu(N(\varphi))$, there is a unique $\lambda \in \mathbb{X}_{m,n}$ with $\models_{m,n} \alpha \leftrightarrow \lambda$. By Lemma 5, we get a unique formula $N_{m,n}(\varphi) \in \mathscr{F}_{m,n}$ with $\models_{m,n} N(\varphi) \leftrightarrow N_{m,n}(\varphi)$. Then $\models_{m,n} \varphi \leftrightarrow N_{m,n}(\varphi)$. □

Example 1. For every formula $\varphi \in \mathscr{F}$, the unique equivalent formula $N_{m,n}(\varphi)$ can be calculated automatically. Let $m = 2$ and $n = 1$. Consider the formula $\varphi = {\sim}{\sim}(({\sim}^2 p \wedge {\sim}^6 q) \vee {\sim}^4 p)$. We calculate $N_{m,n}(\varphi)$ as follows:

$$\begin{aligned}
\models_{m,n} \varphi &\leftrightarrow {\sim}({\sim}({\sim}^2 p \wedge {\sim}^6 q) \wedge {\sim}{\sim}^4 p) && \text{(Lemma 3)}\\
&\leftrightarrow {\sim}{\sim}({\sim}^2 p \wedge {\sim}^6 q) \vee {\sim}{\sim}{\sim}^4 p && \text{(Lemma 3)}\\
&\leftrightarrow {\sim}({\sim}{\sim}^2 p \vee {\sim}{\sim}^6 q) \vee {\sim}{\sim}{\sim}^4 p && \text{(Lemma 3)}\\
&\leftrightarrow ({\sim}^4 p \wedge {\sim}^8 q) \vee {\sim}^6 p && \text{(Lemma 3)}\\
&\leftrightarrow ({\sim}^4 p \wedge {\sim}^4 q) \vee {\sim}^2 p && \text{(Lemma 5)}
\end{aligned}$$

Note that ${\sim}^4 p, {\sim}^4 q, {\sim}^2 p \in \mathbb{X}_{2,1}$. This completes the calculation.

Proposition 1 (Persistency). *For every $\varphi \in \mathscr{F}_{m,n}$ and $\langle m,n \rangle$-valuation V in a frame $\mathfrak{F}$, $V(\varphi) \in \mathsf{Up}(\mathfrak{F})$. Then for every $\psi \in \mathscr{F}$, $V(\psi) \in \mathsf{Up}(\mathfrak{F})$.*

Proof. The proof proceeds by induction on $d(\varphi)$. The case $d(\varphi) = 0$ is trivial. The case $\varphi = \varphi_1 \odot \varphi_2$ with $\odot \in \{\wedge, \vee, \to\}$ is shown by induction hypothesis. Hence $V(\varphi) \in \mathsf{Up}(\mathfrak{F})$. If $\psi \in \mathscr{F}$, by Theorem 1, $V(\psi) = V(N_{m,n}(\psi))$. □

Proposition 2. *For every formula $\chi \in \mathscr{F}$, if $\{{\sim}^i \varphi \leftrightarrow {\sim}^i \psi : i \in [0, 2m+n)\} \subseteq \mathsf{C}_{m,n}$, then $\chi \leftrightarrow \chi(\varphi/\psi) \in \mathsf{C}_{m,n}$.*

Proof. Assume $\{{\sim}^i \varphi \leftrightarrow {\sim}^i \psi : i \in [0, 2m+n)\} \subseteq \mathsf{C}_{m,n}$. The proof proceeds by induction on $c(\chi)$. The case $\chi \in \mathbb{P} \cup \{\bot\}$ or $\chi = \varphi$ is trivial. The case $\chi = \chi_1 \odot \chi_2$ with $\odot \in \{\wedge, \vee, \to\}$ is shown by induction hypothesis. Suppose $\chi = {\sim}^k \theta$ with $k > 0$ and $mc(\theta) \neq {\sim}$. Suppose $\theta \in \mathbb{P} \cup \{\bot\}$. Then $\psi = {\sim}^j \theta$ for some $j \in [0, k)$. If $k - j < 2m+n$, then ${\sim}^{k-j} \varphi \leftrightarrow {\sim}^{k-j} \psi \in \mathsf{C}_{m,n}$, i.e., ${\sim}^k \theta \leftrightarrow {\sim}^{k-j} \varphi \in \mathsf{C}_{m,n}$ where ${\sim}^{k-j} \varphi = \chi(\varphi/\psi)$. Suppose $k - j \geq 2m+n$. Then ${\sim}^{k-j} \psi \leftrightarrow {\sim}^i \psi \in \mathsf{C}_{m,n}$ for some $i < 2m+n$. Hence $\chi \leftrightarrow {\sim}^i \psi \in \mathsf{C}_{m,n}$. By the assumption, ${\sim}^i \psi \leftrightarrow {\sim}^i \varphi \in \mathsf{C}_{m,n}$. Then $\chi \leftrightarrow {\sim}^i \varphi \in \mathsf{C}_{m,n}$. Clearly ${\sim}^{k-j} \varphi \leftrightarrow {\sim}^i \varphi \in \mathsf{C}_{m,n}$. Hence $\chi \leftrightarrow \chi(\varphi/\psi) \in \mathsf{C}_{m,n}$. Suppose $\theta = \theta_1 \odot \theta_2$ with $\odot \in \{\wedge, \vee, \to\}$. These cases are shown similarly. Here we show only $\odot = \wedge$. The case $\psi = {\sim}^i \theta$ with $i \in [0, k)$ is shown as (1). Suppose $\psi \in Sub(\theta_1)$ or $\psi \in Sub(\theta_2)$. If $k \in \mathbb{E}$, then $\chi \leftrightarrow ({\sim}^k \theta_1 \wedge {\sim}^k \theta_2) \in \mathsf{C}_{m,n}$. If $k \in \mathbb{E}$, then $\chi \leftrightarrow ({\sim}^k \theta_1 \vee {\sim}^k \theta_2) \in \mathsf{C}_{m,n}$. Let $\varphi \in Sub(\theta_1)$. By induction hypothesis, ${\sim}^k \theta_1 \leftrightarrow {\sim}^k \theta_1(\varphi/\psi) \in \mathsf{C}_{m,n}$. Then $\chi \leftrightarrow \chi(\varphi/\psi) \in \mathsf{C}_{m,n}$. □

By Proposition 2, the intuitionistic connexive logic $\mathsf{C}_{m,n}$ is closed under the following rule of replacement of equivalents:

$$\frac{\sim^i\varphi \leftrightarrow \sim^i\psi}{\chi \leftrightarrow \chi(\varphi/\psi))}(\mathrm{RE}_{m,n}),\text{ where } i \in [0, 2m+n).$$

Note that, if $\varphi \leftrightarrow \psi \in \mathsf{C}_{1,0}$ and $\sim\varphi \to \sim\psi \in \mathsf{C}_{1,0}$, then $\chi \leftrightarrow \chi(\varphi/\psi) \in \mathsf{C}_{1,0}$.

3 Lattices of $\langle m, n\rangle$-Connexive Logics

In this section, we first give a Hilbert-style axiomatic system for $\mathsf{C}_{m,n}$ for every $\langle m, n\rangle \in \mathbb{Z}^+ \times \mathbb{Z}^*$. Then we introduce $\langle m, n\rangle$-connexive logics which are extensions of $\mathsf{C}_{m,n}$ and obtain in such a way the lattice of all $\langle m, n\rangle$-connexive logics. Let IPC be the intuitionistic propositional calculus.

Definition 4. *The Hilbert-style axiomatic system* $\mathsf{HC}_{m,n}$ *consists of the following axiom schemata and inference rules:*

(IPC) All axiom schemata of IPC.
(C1) $\sim^{2m+n}\varphi \leftrightarrow \sim^n\varphi$.
(C2) $\sim(\varphi \wedge \psi) \leftrightarrow (\sim\varphi \vee \sim\psi)$.
(C3) $\sim(\varphi \vee \psi) \leftrightarrow (\sim\varphi \wedge \sim\psi)$.
(C4) $\sim(\varphi \to \psi) \leftrightarrow (\varphi \to \sim\psi)$.
(MP) from $\varphi \to \psi$ *and* φ *infer* ψ.

We use $\vdash_{\mathsf{HC}_{m,n}} \varphi$ *if* φ *is a* theorem *of* $\mathsf{HC}_{m,n}$. *Let* $\mathsf{HC}_{m,n}$ *denote the set of all theorems of* $\mathsf{HC}_{m,n}$. *Let* $\Gamma \vdash_{\mathsf{HC}_{m,n}} \psi$ *denote that* ψ *is* deducible *from* Γ *in* $\mathsf{HC}_{m,n}$.

Fact 2. *For all* $e \in \mathbb{E}$, $o \in \mathbb{O}$ *and* $k \in \mathbb{Z}^*$, (1) $\varphi, \Gamma \vdash_{\mathsf{HC}_{m,n}} \psi$ *iff* $\Gamma \vdash_{\mathsf{HC}_{m,n}} \varphi \to \psi$; (2) *the following formulas are theorems of* $\mathsf{HC}_{m,n}$: $\sim^{2km+i}\varphi \leftrightarrow \sim^i\varphi$ *where* $i \geq n$; $\sim^e(\varphi\wedge\psi) \leftrightarrow (\sim^e\varphi\wedge\sim^e\psi)$; $\sim^o(\varphi\wedge\psi) \leftrightarrow (\sim^o\varphi\vee\sim^o\psi)$; $\sim^e(\varphi\vee\psi) \leftrightarrow (\sim^e\varphi\vee\sim^e\psi)$; $\sim^o(\varphi \vee \psi) \leftrightarrow (\sim^o\varphi \wedge \sim^o\psi)$; $\sim^k(\varphi \to \psi) \leftrightarrow (\varphi \to \sim^k\psi)$.

Lemma 6. *For every formula* $\varphi \in \mathscr{F}$, $\vdash_{\mathsf{HC}_{m,n}} \varphi \leftrightarrow N_{m,n}(\varphi)$.

Proof. The proof is similar to the proof of Theorem 1. Using (C1)–(C4), we obtain syntactic versions of Lemma 3–5. Hence $\vdash_{\mathsf{HC}_{m,n}} \varphi \leftrightarrow N_{m,n}(\varphi)$. □

A set of formulas Γ is an $\mathsf{HC}_{m,n}$-*theory* if Γ is closed under $\vdash_{\mathsf{HC}_{m,n}}$, i.e., if $\Gamma \vdash_{\mathsf{HC}_{m,n}} \varphi$, then $\varphi \in \Gamma$. A set of formulas Γ is $\mathsf{HC}_{m,n}$-*consistent* if $\Gamma \nvdash_{\mathsf{HC}_{m,n}} \bot$. An $\mathsf{HC}_{m,n}$-consistent theory Δ is *prime* if $\varphi \vee \psi \in \Delta$ implies $\varphi \in \Delta$ or $\psi \in \Delta$. Let $\mathfrak{T}_{m,n}$ be the set of all prime $\mathsf{HC}_{m,n}$-theories. Let $[\Gamma\rangle = \{\varphi \in \mathscr{F} \mid \Gamma \vdash_{\mathsf{HC}_{m,n}} \varphi\}$ be the $\mathsf{HC}_{m,n}$-theory generated by Γ.

Lemma 7. *If* $\Gamma \nvdash_{\mathsf{HC}_{m,n}} \varphi$, *there exists* $\Delta \in \mathfrak{T}_{m,n}$ *with* $\Gamma \subseteq \Delta$ *and* $\varphi \notin \Delta$.

Proof. Assume $\Gamma \nvdash_{\mathsf{HC}_{m,n}} \varphi$. Let $\Pi = \{\Sigma \mid \Gamma \subseteq \Sigma, \Sigma \text{ is a } \mathsf{HC}_{m,n}\text{-theory}, \varphi \notin \Sigma\}$. Obviously $[\Gamma\rangle \in \Pi$ and every nonempty $\subseteq$-chain in Π has an upper bound. By Zorn's lemma, Π has a $\subseteq$-maximal element Δ. Now we show Δ is prime. Suppose not. There exist formulas ψ_1 and ψ_2 such that $\psi_1 \vee \psi_2 \in \Delta$ but $\psi_1 \notin \Delta$ and $\psi_2 \notin \Delta$. Let $\Delta_1 = [\Delta \cup \{\psi_1\}\rangle$ and $\Delta_2 = [\Delta \cup \{\psi_2\}\rangle$. Obviously $\Gamma \subseteq \Delta_1 \cap \Delta_2$ and Δ_1, Δ_2 are $\mathsf{HC}_{m,n}$-theories. By the $\subseteq$-maximality of Δ, $\Delta_1 \notin \Pi$ and $\Delta_2 \notin \Pi$. Then $\varphi \in \Delta_1$ and $\varphi \in \Delta_2$. Hence $\Delta, \psi_1 \vdash_{\mathsf{HC}_{m,n}} \varphi$ and $\Delta, \psi_2 \vdash_{\mathsf{HC}_{m,n}} \varphi$. Then $\Delta, \psi_1 \vee \psi_2 \vdash_{\mathsf{HC}_{m,n}} \varphi$. Since $\psi_1 \vee \psi_2 \in \Delta$, $\varphi \in \Delta$ which contradicts $\varphi \notin \Delta$. □

Lemma 8. *For every $\Delta \in \mathfrak{T}_{m,n}$ and $\varphi, \psi \in \mathscr{F}$, the following hold:*

(1) $\mathsf{HC}_{m,n} \subseteq \Delta$ *and* $\bot \notin \Delta$.
(2) $\varphi \wedge \psi \in \Delta$ *iff* $\varphi \in \Delta$ *and* $\psi \in \Delta$.
(3) $\varphi \vee \psi \in \Delta$ *iff* $\varphi \in \Delta$ *or* $\psi \in \Delta$.
(4) $\varphi \to \psi \in \Delta$ *iff* $\forall \Sigma \in \mathfrak{T}_{m,n}(\Delta \subseteq \Sigma \;\&\; \varphi \in \Sigma \Rightarrow \psi \in \Sigma)$.
(5) ${\sim}^{2m+n}\varphi \in \Delta$ *iff* ${\sim}^{n}\varphi \in \Delta$.
(6) ${\sim}(\varphi \wedge \psi) \in \Delta$ *iff* ${\sim}\varphi \in \Delta$ *or* ${\sim}\psi \in \Delta$.
(7) ${\sim}(\varphi \vee \psi) \in \Delta$ *iff* ${\sim}\varphi \in \Delta$ *and* ${\sim}\psi \in \Delta$.
(8) ${\sim}(\varphi \to \psi) \in \Delta$ *iff* $\forall \Sigma \in \mathfrak{T}_{m,n}(\Delta \subseteq \Sigma \;\&\; \varphi \in \Sigma \Rightarrow {\sim}\psi \in \Sigma)$.

Proof. (1)–(3) are clear. For (4), assume $\varphi \to \psi \in \Delta$, $\Delta \subseteq \Sigma$ and $\varphi \in \Sigma$. Then $\varphi \to \psi \in \Sigma$. Then $\psi \in \Sigma$. Assume $\varphi \to \psi \notin \Delta$. Then $\varphi, \Delta \nvdash_{\mathsf{HC}_{m,n}} \psi$. By Lemma 7, there exists $\Sigma \in \mathfrak{T}_{m,n}$ with $\Delta \cup \{\varphi\} \subseteq \Sigma$ and $\psi \notin \Sigma$. For (5), assume ${\sim}^{2m+n}\varphi \in \Delta$. By $\vdash_{\mathsf{HC}_{m,n}} {\sim}^{2m+n}\varphi \to {\sim}^{n}\varphi$, we have ${\sim}^{n}\varphi \in \Delta$. The other direction is similar. Note that (6) and (7) are shown as (5). Clearly ${\sim}(\varphi \to \psi) \in \Delta$ iff $\varphi \to {\sim}\psi \in \Delta$. Then (8) follows from (4). □

Definition 5. *The* canonical $\langle m, n\rangle$-model *for* $\mathsf{HC}_{m,n}$ *is defined as the model* $\mathfrak{M}_{m,n} = (W_{m,n}, R_{m,n}, V_{m,n})$ *where (1)* $W_{m,n} = \mathfrak{T}_{m,n}$*; (2)* $\Delta R_{m,n} \Sigma$ *iff* $\Delta \subseteq \Sigma$*; (3)* $V_{m,n}(\lambda) = \{\Delta \in W_{m,n} : \lambda \in \Delta\}$ *for every* $\lambda \in \mathbb{X}_{m,n}$*. The frame* $\mathfrak{F}_{m,n} = (W_{m,n}, R_{m,n})$ *is the* canonical $\langle m, n\rangle$-frame *for* $\mathsf{HC}_{m,n}$.

Lemma 9. *For every $\varphi \in \mathscr{F}$ and $\Delta \in W_{m,n}$, $\mathfrak{M}_{m,n}, \Delta \models_{m,n} \varphi$ iff $\varphi \in \Delta$.*

Proof. By Theorem 1 and Lemma 6, it suffices to show $\mathfrak{M}_{m,n}, \Delta \models_{m,n} N_{m,n}(\varphi)$ iff $N_{m,n}(\varphi) \in \Delta$. We show it by induction on $d(N_{m,n}(\varphi))$. The case $N_{m,n}(\psi) = \lambda \in \mathbb{X}_{m,n}$ is trivial. The case $N_{m,n}(\psi) = \psi_1 \odot \psi_2$ for $\odot \in \{\wedge, \vee, \to\}$ is shown by induction hypothesis and Lemma 8. □

Theorem 3 (Completeness). $\Gamma \vdash_{\mathsf{HC}_{m,n}} \varphi$ *iff* $\Gamma \models_{m,n} \varphi$.

Proof. By Lemma 2, all axioms are $\langle m, n\rangle$-valid. Moreover, (MP) preserves $\langle m, n\rangle$-validity. Then we have the soundness of $\mathsf{HC}_{m,n}$. For the other direction, assume $\Gamma \nvdash_{\mathsf{HC}_{m,n}} \varphi$. By Lemma 7, there exists $\Delta \in \mathfrak{T}_{m,n}$ with $\Gamma \subseteq \Delta$ and $\varphi \notin \Delta$. By Lemma 9, $\mathfrak{M}_{m,n}, \Delta \not\models_{m,n} \varphi$ and $\mathfrak{M}_{m,n}, \Delta \models_{m,n} \psi$ for all $\psi \in \Gamma$. □

Corollary 2. $\mathsf{HC}_{m,n} = \mathsf{C}_{m,n}$.

A *literal substitution* is a function $\iota : \mathbb{X}_{m,n} \to \mathscr{F}$. Every literal substitution ι is homomorphically extended to a function $\iota : \mathscr{F}_{m,n} \to \mathscr{F}$. For every formula $\varphi \in \mathscr{F}_{m,n}$ and literal substitution ι, let φ^ι be the formula obtained from φ by ι. Now we consider the following rules: (US) from φ infer φ^s where s is an arbitrary substitution; and (LS) from φ infer $N_{m,n}(\varphi)^\iota$ where ι is a literal substitution.

Proposition 3. *For every $\varphi \in \mathsf{C}_{m,n}$, (1) $N_{m,n}(\varphi)^\iota \in \mathsf{C}_{m,n}$ for every literal substitution ι; and (2) $\varphi^s \in \mathsf{C}_{m,n}$ for every substitution s.*

Proof. For (1), assume $\varphi \in \mathsf{C}_{m,n}$. By Theorem 1, $N_{m,n}(\varphi) \in \mathsf{C}_{m,n}$. Clearly $N_{m,n}(\varphi) \in \mathscr{F}_{m,n}$. It suffices to show that, for all $\chi \in \mathscr{F}_{m,n}$, if $\chi \in \mathsf{C}_{m,n}$, then $\chi^\iota \in \mathsf{C}_{m,n}$ for every literal substitution ι. Let $\lambda(\chi) = \{\lambda_1, \ldots, \lambda_n\}$ and $\chi^\iota = \chi(\xi_1/\lambda_1, \ldots, \xi_n/\lambda_n)$. Suppose $\not\models_{m,n} \chi^\iota$. There exists a frame $\mathfrak{F} = (W, R)$, an $\langle m, n\rangle$-valuation V in $\mathfrak{F}$ and $w \in W$ such that $w \notin V(\chi^\iota)$. Let V' be an $\langle m, n\rangle$-valuation in $\mathfrak{F}$ such that $V'(\lambda_i) = V(\xi_i)$ for $1 \leq i \leq n$. Clearly $V'(\chi) = V(\chi^\iota)$. Then $w \notin V'(\chi)$ and so $\chi \notin \mathsf{C}_{m,n}$. The proof of (2) is similar to (1). □

An *$\langle m, n\rangle$-connexive logic* is a set L of formulas such that $L \subseteq \mathscr{F}$ and the following conditions hold: (J) $\mathsf{C}_{m,n} \subseteq L$; (MP) if $\varphi, \varphi \to \psi \in L$, then $\psi \in L$; (LS) if $\varphi \in L$, then $\varphi^\iota \in L$ for every literal substitution ι; (US) if $\varphi \in L$, then $\varphi^s \in L$ for every substitution s. A logic L_1 is a *sublogic* of L_2 (or L_2 is an *extension* of L_1) if $L_1 \subseteq L_2$. A logic L_1 is a *proper sublogic* of L_2 (or L_2 is an *proper extension* of L_1) if $L_1 \subsetneq L_2$. Let $\mathsf{Ext}(L)$ be the set of all extensions of L. The smallest $\langle m, n\rangle$-connexive logic generated by a set of formulas Σ over L is defined as $L \oplus \Sigma = \bigcap\{L' \in \mathsf{Ext}(\mathsf{C}_{m,n}) \mid L \cup \Sigma \subseteq L'\}$. Clearly $\mathsf{C}_{m,n}$ is the minimal $\langle m, n\rangle$-connexive logic. For every $\langle m, n\rangle$-connexive logic L, $\mathsf{Ext}(L)$ is closed under $\cap$ and $\oplus$, and hence it forms a lattice with top $\mathscr{F}$ and bottom L.

A formula φ is a *theorem* of L (notation: $\vdash_L \varphi$) if $\varphi \in L$. A formula φ is *deducible* from a set of formulas Γ in L (notation: $\Gamma \vdash_L \varphi$) if there exists a sequence of formulas $\psi_1, \ldots, \psi_n = \varphi$ and each ψ_i is either a member of $L \cup \Gamma$ or derived from previous formula(s) by (MP), (LS) or (US). Obviously the deduction theorem holds for L. An $\langle m, n\rangle$-connexive logic L is *finitely axiomatizable* if there exists a finite $\Sigma \subseteq \mathscr{F}$ with $L = \mathsf{C}_{m,n} \oplus \Sigma$. For all formulas $\varphi, \psi \in \mathscr{F}$ such that $\lambda(N_{m,n}(\varphi)) = \{\lambda_1, \ldots, \lambda_n\}$ and $\lambda(N_{m,n}(\psi)) = \{\lambda_1, \ldots, \lambda_m\}$, the *literal repeatless disjunction* of φ and ψ is defined as $\varphi \veebar \psi = N_{m,n}(\varphi)(\lambda_1, \ldots, \lambda_n) \vee N_{m,n}(\psi)(\lambda_{n+1}, \ldots, \lambda_{n+m})$ where $N_{m,n}(\psi)(\lambda_{n+1}, \ldots, \lambda_{n+m})$ is the formula obtained from $N_{m,n}(\psi)$ by substituting λ_{n+i} for λ_i in $N_{m,n}(\psi)$ with $1 \leq i \leq m$.

Proposition 4. *Let $L_1 = \mathsf{C}_{m,n} \oplus \{\varphi_i : i \in I\}$ and $L_2 = \mathsf{C}_{m,n} \oplus \{\psi_j : j \in J\}$. Then $L_1 \cap L_2 = \mathsf{C}_{m,n} \oplus \{\varphi_i \veebar \psi_j : i \in I, j \in J\}$.*

Proof. By Lemma 6, $L_1 = \mathsf{C}_{m,n} \oplus \{N_{m,n}(\varphi_i) : i \in I\}$ and $L_2 = \mathsf{C}_{m,n} \oplus \{N_{m,n}(\psi_j) : j \in J\}$. Assume $\chi \in L_1 \cap L_2$. Then there are finite subsets $I' \subseteq I$ and $J' \subseteq J$ such that $\bigwedge_{i' \in I'} N_{m,n}(\varphi_i) \to \chi \in \mathsf{C}_{m,n}$ and $\bigwedge_{j' \in J'} N_{m,n}(\psi_j) \to \chi \in \mathsf{C}_{m,n}$. Then $\bigwedge_{i \in I', j \in J'}(N_{m,n}(\varphi_i) \vee N_{m,n}(\psi_j)) \to \chi \in \mathsf{C}_{m,n}$. Clearly

$N_{m,n}(\varphi_i) \vee N_{m,n}(\psi_j)$ is a literal substitution instance of $\varphi_i \veebar \psi_j$. Hence $\chi \in \mathsf{C}_{m,n} \oplus \{\varphi_i \veebar \psi_j : i \in I, j \in J\}$. The other direction is shown easily. □

Recall the lattice $\mathsf{Ext}(\mathsf{IPL})$ of all superintuitionistic logics (cf. [2]). We will show that it is isomorphic to $\mathsf{Ext}(\mathsf{C}_{m,n})$ for every $\langle m, n\rangle \in \mathbb{Z}^+ \times \mathbb{Z}^*$. For every superintuitionistic logic $S \in \mathsf{Ext}(\mathsf{IPL})$, we define the $\langle m, n\rangle$-connexive logic S^* as the set of theorems in the system obtained from $\mathsf{HC}_{m,n}$ by replacing IPC with S. Let $L \in \mathsf{Ext}(\mathsf{C}_{m,n})$. The superintuitionistic logic L_* is defined as follows: Let $N_{m,n}(L) = \{N_{m,n}(\varphi) : \varphi \in L\}$. For each $\lambda \in \mathbb{X}_{m,n}$, let p_λ be the variable associated with λ. For every $\varphi \in L$, let $\lambda(N_{m,n}(\varphi)) = \{\lambda_1, \ldots, \lambda_n\}$ and $\varphi_* = N_{m,n}(\varphi)(p_{\lambda_1}/\lambda_1, \ldots, p_{\lambda_n}/\lambda_n)$. We define $L_* = \{\varphi_*^s : \varphi \in L \text{ and } s : \mathbb{P} \to \mathscr{F}_I\}$.

Lemma 10. *Let* $S \in \mathsf{Ext}(\mathsf{IPL})$ *and* $L \in \mathsf{Ext}(\mathsf{C}_{m,n})$. *Then* (1) $S \subseteq S^*$; (2) $L_* \subseteq L$; *and* (3) L_* *is a superintuitionistic logic.*

Proof. Obviously $S \subseteq S^*$. For (2), let $\varphi_*^s \in L_*$. Then $\varphi \in L$. By Theorem 1, $N_{m,n}(\varphi) \in L$. Since L is closed under (LS) and (US), $\varphi_*^s \in L$. Hence $L_* \subseteq L$. For (3), clearly $\mathsf{IPL} \subseteq L_*$. Let $\varphi_*^s, \varphi_*^s \to \psi_*^t \in L_*$. Then $\varphi, \varphi \to \psi \in L$ and so $\psi \in L$. Hence $\psi_*^t \in L_*$. Clearly L_* is closed under (US). □

Theorem 4. $\mathsf{Ext}(\mathsf{IPL})$ *is lattice isomorphic to* $\mathsf{Ext}(\mathsf{C}_{m,n})$.

Proof. For every $S \in \mathsf{Ext}(\mathsf{IPL})$ and $L \in \mathsf{Ext}(\mathsf{C}_{m,n})$, it is clear that S^* and L_* are uniquely determined. Then $(S^*)_* = S$ and $L = (L_*)^*$. Moreover, $(S_1 \cap S_2)^* = S_1^* \cap S_2^*$ and $(S_1 \oplus S_2)^* = S_1^* \oplus S_2^*$; $(L_1 \cap L_2)_* = L_{1*} \cap L_{2*}$ and $(L_1 \oplus L_2)_* = L_{1*} \oplus L_{2*}$. It follows that $(.)^*$ and $(.)_*$ are lattice isomorphisms. □

Let $\circ$ be the single element frame, i.e., $\circ = (\{w\}, \langle w, w\rangle)$. The *material* $\langle m, n\rangle$*-connexive logic* is defined as the set $\mathsf{K}_{m,n} = \{\varphi \in \mathscr{F} : \circ \models_{m,n} \varphi\}$. Clearly the classical propositional logic CPL is the superintuitionistic logic which is characterized by the frame $\circ$. By Theorem 4, $\mathsf{K}_{m,n} = \mathsf{CPL}^*$.

Corollary 3. *If* $L \in \mathsf{Ext}(\mathsf{C}_{m,n})$ *is consistent, then* $\mathsf{C}_{m,n} \subseteq L \subseteq \mathsf{K}_{m,n}$.

Proof. Let $L \in \mathsf{Ext}(\mathsf{C}_{m,n})$ be consistent. Then L_* is consistent. Hence $\mathsf{IPL} \subseteq L \subseteq \mathsf{CPL}$. By Theorem 4, $\mathsf{C}_{m,n} \subseteq L \subseteq \mathsf{K}_{m,n}$. □

The Hilbert-style axiomatic system $\mathsf{HK}_{m,n}$ for $\mathsf{K}_{m,n}$ is obtained from $\mathsf{HC}_{m,n}$ by replacing all axiom schemata of IPC with all axiom schemata of CPL, or equivalently by adding $\varphi \vee \neg\varphi$ to $\mathsf{HC}_{m,n}$. A formula ψ is an $\langle m, n\rangle$*-consequence* of a set of formulas Γ with respect to the singleton frame $\circ$ (notation: $\Gamma \models_{m,n}^{\circ} \psi$) if $V(\Gamma) \subseteq V(\varphi)$ for every $\langle m, n\rangle$-valuation V in $\circ$. It is quite standard to show the strong completeness of $\mathsf{HK}_{m,n}$, namely, $\Gamma \vdash_{\mathsf{HK}_{m,n}} \psi$ iff $\Gamma \models_{m,n}^{\circ} \psi$.

An $\langle m, n\rangle$-connexive logic L has the *disjunction property* (DP) if $\varphi \vee \psi \in L$ implies $\varphi \in L$ or $\psi \in L$. Obviously, for every $m \geq 1$ and $n \geq 0$, the $\langle m, n\rangle$-connexive logic $\mathsf{K}_{m,n}$ lacks the disjunction property.

Theorem 5. *A logic* $L \in \mathsf{Ext}(\mathsf{C}_{m,n})$ *has the DP iff* $L_* \in \mathsf{Ext}(\mathsf{IPL})$ *has the DP.*

Proof. Let $L \in \mathsf{Ext}(\mathsf{C}_{m,n})$. Assume that L has the DP. Suppose $\varphi_*^s \vee \psi_*^s \in L_*$. Then $\varphi \vee \psi \in L$. Hence $\varphi \in L$ or $\psi \in L$. Then $\varphi_*^s \in L_*^s$ or $\psi_*^s \in L_*$. Assume that L_* has the DP. Suppose $\varphi \vee \psi \in L$. Then $\varphi_*^s \vee \psi_*^s \in \mathsf{IPL}$. By the DP of IPL, $\varphi_*^s \in L_*$ or $\psi_*^s \in L_*$. Then $\varphi \in L$ or $\psi \in L$. □

Corollary 4. *For all $m \geq 1$ and $n \geq 0$, $\mathsf{C}_{m,n}$ has the DP.*

It is well-known CPL is embedded into IPC via Glivenko's double negation translation $\varphi \mapsto \neg\neg\varphi$ where $\neg$ is the intuitionistic negation (cf. [2, p. 46]). This result is extended to intuitionistic and material $\langle m, n\rangle$-connexive logics.

Proposition 5. *For all $m \geq 1$, $n \geq 0$ and $\varphi \in \mathscr{F}$, $\varphi \in \mathsf{K}_{m,n}$ iff $\neg\neg\varphi \in \mathsf{C}_{m,n}$.*

Proof. Assume $\neg\neg\varphi \in \mathsf{C}_{m,n}$. Then $\neg\neg\varphi \in \mathsf{K}_{m,n}$. By $\neg\neg\varphi \to \varphi \in \mathsf{K}_{m,n}$, $\varphi \in \mathsf{K}_{m,n}$. Assume $\varphi \in \mathsf{K}_{m,n}$. By Theorem 4, $\varphi_* \in \mathsf{CPL}$. Then $\neg\neg\varphi_* \in \mathsf{IPL}$. Hence $\neg\neg\varphi_* \in \mathsf{C}_{m,n}$. Clearly $\neg\neg\varphi_* = (\neg\neg\varphi)_*$. Then $N_{m,n}(\neg\neg\varphi) \in \mathsf{C}_{m,n}$. By $\neg\neg\varphi \leftrightarrow N_{m,n}(\neg\neg\varphi) \in \mathsf{C}_{m,n}$, $\neg\neg\varphi \in \mathsf{C}_{m,n}$. □

4 Gentzen Sequent Calculi

In this section, we introduce Gentzen sequent calculi for intuitionistic and material $\langle m, n\rangle$-connexive logics. They are obtained by modifying the construction in [11]. These sequent calculi are based on the G3-style Gentzen sequent calculi for IPL and CPL (cf. e.g. [16]). Let $E[0, 2m+n)$ and $O[0, 2m+n)$ be sets of all even and odd numbers in $[0, 2m+n)$ respectively. A *sequent* is an expression $\Gamma \Rightarrow \psi$ where Γ is a finite (possibly empty) multiset of formulas and ψ is a formula in $\mathscr{F}$. Sequent rules are defined as usual.

Definition 6. *The sequent calculus $\mathsf{G3C}_{m,n}$ consists of the following initial sequents and sequent rules:*

(1) Initial sequents:

$$\text{(Id)}\ \lambda, \Gamma \Rightarrow \lambda, \ \textit{where}\ \lambda \in \mathbb{X}_{m,n}. \quad (\bot)\ \bot, \Gamma \Rightarrow \varphi$$

(2) Sequent rules:

$$\frac{\sim^e\varphi, \sim^e\psi, \Gamma \Rightarrow \chi}{\sim^e(\varphi \wedge \varphi), \Gamma \Rightarrow \chi}(\wedge\Rightarrow) \quad \frac{\Gamma \Rightarrow \sim^e\varphi \quad \Gamma \Rightarrow \sim^e\psi}{\Gamma \Rightarrow \sim^e(\varphi \wedge \psi)}(\Rightarrow\wedge)$$

$$\frac{\sim^e\varphi, \Gamma \Rightarrow \chi \quad \sim^e\psi, \Gamma \Rightarrow \chi}{\sim^e(\varphi \vee \psi), \Gamma \Rightarrow \chi}(\vee\Rightarrow) \quad \frac{\Gamma \Rightarrow \sim^e\varphi_i}{\Gamma \Rightarrow \sim^e(\varphi_1 \vee \varphi_2)}(\Rightarrow\vee)(i = 1, 2)$$

$$\frac{\sim^o\varphi, \Gamma \Rightarrow \chi \quad \sim^o\psi, \Gamma \Rightarrow \chi}{\sim^o(\varphi \wedge \psi), \Gamma \Rightarrow \chi}(\sim\wedge\Rightarrow) \quad \frac{\Gamma \Rightarrow \sim^o\varphi_i}{\Gamma \Rightarrow \sim^o(\varphi_1 \wedge \varphi_2)}(\Rightarrow\sim\wedge)(i = 1, 2)$$

$$\frac{\sim^o\varphi, \sim^o\psi, \Gamma \Rightarrow \chi}{\sim^o(\varphi \vee \varphi), \Gamma \Rightarrow \chi}(\sim\vee\Rightarrow) \quad \frac{\Gamma \Rightarrow \sim^o\varphi \quad \Gamma \Rightarrow \sim^o\psi}{\Gamma \Rightarrow \sim^o(\varphi \vee \psi)}(\Rightarrow\sim\vee)$$

$$\frac{\sim^k(\varphi \to \psi), \Gamma \Rightarrow \varphi \quad \sim^k\psi, \Gamma \Rightarrow \chi}{\sim^k(\varphi \to \psi), \Gamma \Rightarrow \chi}\ (\to\Rightarrow) \quad \frac{\varphi, \Gamma \Rightarrow \sim^k\psi}{\Gamma \Rightarrow \sim^k(\varphi \to \psi)}\ (\Rightarrow\to)$$

$$\frac{\sim^n\varphi, \Gamma \Rightarrow \psi}{\sim^{2m+n}\varphi, \Gamma \Rightarrow \psi}(\sim^{2m+n}\Rightarrow) \quad \frac{\Gamma \Rightarrow \sim^n\varphi}{\Gamma \Rightarrow \sim^{2m+n}\varphi}\ (\Rightarrow\sim^{2m+n})$$

where $e \in E[0, 2m+n)$, $o \in O[0, 2m+n)$ *and* $k \geq 0$.

The formula with connective(s) in the conclusion sequent of a sequent rule is called principal. *Derivations in* $\mathsf{G3C}_{m,n}$ *are denoted by* $\mathcal{D}$, $\mathcal{E}$ *etc. and the* height *of a derivation* $\mathcal{D}$ *is denoted by* $|\mathcal{D}|$. *Let* $\mathsf{G3C}_{m,n} \vdash \Gamma \Rightarrow \psi$ *denote that* $\Gamma \Rightarrow \psi$ *is* derivable *in* $\mathsf{G3C}_{m,n}$. *We use* $\mathsf{G3C}_{m,n} \vdash_h \Gamma \Rightarrow \psi$ *for that there is a derivation* $\mathcal{D}$ *of* $\Gamma \Rightarrow \psi$ *in* $\mathsf{G3C}_{m,n}$ *with* $|\mathcal{D}| \leq h$. *The prefix* $\mathsf{G3C}_{m,n}$ *is omitted if no confusion arises from the context. Admissibility and height-preserving admissibility of a sequent rule in* $\mathsf{G3C}_{m,n}$ *are defined as usual (cf. e.g.* [16]*).*

Lemma 11. *For every formula* $\varphi \in \mathscr{F}$ *and* $k \geq 0$, $\mathsf{G3C}_{m,n} \vdash \sim^k\varphi, \Gamma \Rightarrow \sim^k\varphi$.

Proof. By induction on the complexity $c(\varphi)$. See Appendix A. □

Now we prove the admissibility of structural rules of weakening and contraction in $\mathsf{G3C}_{m,n}$. Then we show the admissibility of the cut rule.

Lemma 12. *The following rule is height-preserving admissible in* $\mathsf{G3C}_{m,n}$:

$$\frac{\Gamma \Rightarrow \psi}{\varphi, \Gamma \Rightarrow \psi}\ (\mathrm{Wk})$$

Proof. Assume $\vdash_h \Gamma \Rightarrow \psi$. We show $\vdash_h \varphi, \Gamma \Rightarrow \psi$ by induction on $h \geq 0$. The case $h = 0$ is trivial. Suppose $h > 0$. Let $\Gamma \Rightarrow \psi$ be obtained by one of the sequent rules and call it (R). We obtain $\varphi, \Gamma \Rightarrow \psi$ by induction hypothesis and the rule (R). □

Lemma 13. *For every* $h \geq 0, e \in E[0, 2m+n)$ *and* $o \in O[0, 2m+n)$*, the following hold in* $\mathsf{G3C}_{m,n}$:

(1) *if* $\vdash_h \sim^e(\varphi \wedge \psi), \Gamma \Rightarrow \chi$, *then* $\vdash_h \sim^e\varphi, \sim^e\psi, \Gamma \Rightarrow \chi$.
(2) *if* $\vdash_h \sim^e(\varphi \vee \psi), \Gamma \Rightarrow \chi$, *then* $\vdash_h \sim^e\varphi, \Gamma \Rightarrow \chi$ *and* $\vdash_h \sim^e\psi, \Gamma \Rightarrow \chi$.
(3) *if* $\vdash_h \sim^o(\varphi \vee \psi), \Gamma \Rightarrow \chi$, *then* $\vdash_h \sim^o\varphi, \sim^o\psi, \Gamma \Rightarrow \chi$.
(4) *if* $\vdash_h \sim^o(\varphi \wedge \psi), \Gamma \Rightarrow \chi$, *then* $\vdash_h \sim^o\varphi, \Gamma \Rightarrow \chi$ *and* $\vdash_h \sim^o\psi, \Gamma \Rightarrow \chi$;
(5) *if* $\vdash_h \sim^{2m+n}\varphi, \Gamma \Rightarrow \psi$, *then* $\vdash_h \sim^n\varphi, \Gamma \Rightarrow \psi$.
(6) *if* $\vdash_h \sim^k(\varphi \to \psi), \Gamma \Rightarrow \chi$, *then* $\vdash_h \sim^k\psi, \Gamma \Rightarrow \chi$.

Proof. We show only (5) and others are shown similarly. Assume $\vdash_h \sim^{2m+n}\varphi, \Gamma \Rightarrow \psi$. The proof proceeds by induction on h. If $h = 0$, then $\sim^{2m+n}\varphi, \Gamma \Rightarrow \psi$ and $\sim^n\varphi, \Gamma \Rightarrow \psi$ are initial. Let $h > 0$ and $\sim^{2m+n}\varphi, \Gamma \Rightarrow \psi$ be obtained by (R). If $\sim^{2m+n}\varphi$ is principal, then $\vdash_{h-1} \sim^n\varphi, \Gamma \Rightarrow \psi$. Suppose $\sim^{2m+n}\varphi$ is not principal in (R). We get $\vdash_h \sim^n\varphi, \Gamma \Rightarrow \psi$ by induction hypothesis and (R). For example, let (R) be $(\to\Rightarrow)$ with premisses $\vdash_{h-1} \sim^k(\chi_1 \to \chi_2), \sim^{2m+n}\varphi, \Sigma \Rightarrow \chi_1$ and $\vdash_{h-1} \sim^k\chi_2, \sim^{2m+n}\varphi, \Sigma \Rightarrow \psi$, and conclusion $\vdash_h \sim^k(\chi_1 \to \chi_2), \sim^{2m+n}\varphi, \Sigma \Rightarrow \psi$. By induction hypothesis, $\vdash_{h-1} \sim^k(\chi_1 \to \chi_2), \sim^n\varphi, \Sigma \Rightarrow \chi_1$ and $\vdash_{h-1} \sim^k\chi_2, \sim^n\varphi, \Sigma \Rightarrow \psi$ By $(\to\Rightarrow)$, $\vdash_h \sim^k(\chi_1 \to \chi_2), \sim^n\varphi, \Sigma \Rightarrow \psi$. □

Lemma 14. *The following rule is height-preserving admissible in* $\mathsf{G3C}_{m,n}$*:*

$$\frac{\varphi, \varphi, \Gamma \Rightarrow \psi}{\varphi, \Gamma \Rightarrow \psi}(\mathrm{Ctr})$$

Proof. Assume $\vdash_h \varphi, \varphi, \Gamma \Rightarrow \psi$. The proof proceeds by induction on h and subinduction on the complexity $c(\varphi)$. See Appendix B. □

Lemma 15. *If* $\mathsf{G3C}_{m,n} \vdash \Gamma \Rightarrow \bot$*, then* $\mathsf{G3C}_{m,n} \vdash \Gamma \Rightarrow \psi$ *for every* $\psi \in \mathscr{F}$*.*

Proof. Assume $\vdash_h \Gamma \Rightarrow \bot$. If $h = 0$, then $\Gamma \Rightarrow \bot$ and $\Gamma \Rightarrow \psi$ are initial sequents. Suppose $h > 0$ and $\Gamma \Rightarrow \bot$ is obtained by a rule (R). Obviously (R) can be only a left rule. We get $\vdash \Gamma \Rightarrow \psi$ by induction hypothesis and (R). For example, let (R) be $(\rightarrow\Rightarrow)$ with premisses $\vdash_{h-1} {\sim}^k(\varphi_1 \rightarrow \varphi_2), \Sigma \Rightarrow \varphi_1$ and $\vdash_{h-1} {\sim}^k\varphi_2, \Sigma \Rightarrow \bot$, and conclusion $\vdash_h {\sim}^k(\varphi_1 \rightarrow \varphi_2), \Sigma \Rightarrow \bot$. By induction hypothesis, $\vdash {\sim}^k\varphi_2, \Sigma \Rightarrow \psi$. By the left premiss and $(\rightarrow\Rightarrow)$, $\vdash_h {\sim}^k(\varphi_1 \rightarrow \varphi_2), \Sigma \Rightarrow \psi$. □

Theorem 6. *The following cut rule is admissible in* $\mathsf{G3C}_{m,n}$*:*

$$\frac{\Gamma \Rightarrow \varphi \quad \varphi, \Delta \Rightarrow \psi}{\Gamma, \Delta \Rightarrow \psi}\ (\mathrm{Cut})$$

Proof. Assume $\vdash_h \Gamma \Rightarrow \varphi$ and $\vdash_j \varphi, \Delta \Rightarrow \psi$. One can show $\vdash \Gamma, \Delta \Rightarrow \psi$ by simultaneous induction on $h + j$ and $c(\varphi)$. See Appendix C. □

Lemma 16. *For every formula* $\varphi \in \mathscr{F}$*, if* $\vdash_{\mathsf{HC}_{m,n}} \varphi$*, then* $\mathsf{G3C}_{m,n} \vdash \Rightarrow \varphi$*.*

Proof. The proof proceeds by induction on the length of a proof of φ in $\mathsf{HC}_{m,n}$. Obviously $\mathsf{G3C}_{m,n} \vdash \Rightarrow \varphi$ if φ is an axiom of IPC. Assume $\vdash \Rightarrow \chi$ and $\vdash \Rightarrow \chi \rightarrow \xi$. Clearly $\vdash \chi, \chi \rightarrow \xi \Rightarrow \xi$. By (Cut), $\vdash \Rightarrow \xi$. Hence $\vdash \Rightarrow \varphi$. □

A sequent $\Gamma \Rightarrow \psi$ is *$\langle m, n\rangle$-valid*, notation $\models_{m,n} \Gamma \Rightarrow \psi$, if $\models_{m,n} \bigwedge\Gamma \rightarrow \psi$. A sequent rule with premisses $\Gamma_i \Rightarrow \psi_i$ for $1 \leq i \leq l$ and conclusion $\Gamma_0 \Rightarrow \psi_0$ *preserves $\langle m, n\rangle$-validity*, if $\models_{m,n} \Gamma_0 \Rightarrow \psi_0$ whenever $\models_{m,n} \Gamma_i \Rightarrow \psi_i$ for all $1 \leq i \leq l$. Then we get the soundness and completeness of $\mathsf{G3C}_{m,n}$.

Theorem 7. $\mathsf{G3C}_{m,n} \vdash \Gamma \Rightarrow \psi$ *iff* $\models_{m,n} \Gamma \Rightarrow \psi$*.*

Proof. Assume $\vdash_h \Gamma \Rightarrow \psi$. The proof proceeds by induction on the height of derivation. Clearly all axioms are valid in $\mathsf{C}_{m,n}$. By Corollary 1, all rules preserve validity in $\mathsf{C}_{m,n}$. Assume $\Gamma \models_{m,n} \psi$. Let $\varphi = \bigwedge\Gamma$. Then $\varphi \rightarrow \psi \in \mathsf{C}_{m,n}$. By Theorem 3, $\vdash_{\mathsf{HC}_{m,n}} \varphi \rightarrow \psi$. By Lemma 16, $\mathsf{G3C}_{m,n} \vdash \Rightarrow \varphi \rightarrow \psi$. Then $\mathsf{G3C}_{m,n} \vdash \varphi \Rightarrow \psi$. Clearly $\mathsf{G3C}_{m,n} \vdash \Gamma \Rightarrow \varphi$. By (Cut), $\mathsf{G3C}_{m,n} \vdash \Gamma \Rightarrow \psi$. □

Definition 7. *The sequent calculus* $\mathsf{G3K}_{m,n}$ *consists of the following initial sequents and sequent rules:*

(1) Initial sequents:

$$(\mathrm{Id})\ \lambda, \Gamma \Rightarrow \Delta, \lambda, \ \textit{where } \lambda \in \mathbb{X}_{m,n}. \quad (\bot)\ \bot, \Gamma \Rightarrow \Delta$$

(2) Sequent rules:

$$\frac{\sim^e\varphi, \sim^e\psi, \Gamma \Rightarrow \Delta}{\sim^e(\varphi \wedge \varphi), \Gamma \Rightarrow \Delta}(\wedge\Rightarrow) \quad \frac{\Gamma \Rightarrow \Delta, \sim^e\varphi \quad \Gamma \Rightarrow \Delta, \sim^e\psi}{\Gamma \Rightarrow \Delta, \sim^e(\varphi \wedge \psi)}(\Rightarrow\wedge)$$

$$\frac{\sim^e\varphi, \Gamma \Rightarrow \Delta \quad \sim^e\psi, \Gamma \Rightarrow \Delta}{\sim^e(\varphi \vee \psi), \Gamma \Rightarrow \Delta}(\vee\Rightarrow) \quad \frac{\Gamma \Rightarrow \Delta, \sim^e\varphi_1, \sim^e\varphi_2}{\Gamma \Rightarrow \Delta, \sim^e(\varphi_1 \vee \varphi_2)}(\Rightarrow\vee)$$

$$\frac{\sim^o\varphi, \Gamma \Rightarrow \Delta \quad \sim^o\psi, \Gamma \Rightarrow \Delta}{\sim^o(\varphi \wedge \psi), \Gamma \Rightarrow \Delta}(\sim\wedge\Rightarrow) \quad \frac{\Gamma \Rightarrow \Delta, \sim^o\varphi_1, \sim^o\varphi_2}{\Gamma \Rightarrow \Delta, \sim^o(\varphi_1 \wedge \varphi_2)}(\Rightarrow\sim\wedge)$$

$$\frac{\sim^o\varphi, \sim^o\psi, \Gamma \Rightarrow \Delta}{\sim^o(\varphi \vee \varphi), \Gamma \Rightarrow \Delta}(\sim\vee\Rightarrow) \quad \frac{\Gamma \Rightarrow \Delta, \sim^o\varphi \quad \Gamma \Rightarrow \Delta, \sim^o\psi}{\Gamma \Rightarrow \Delta, \sim^o(\varphi \vee \psi)}(\Rightarrow\sim\vee)$$

$$\frac{\Gamma \Rightarrow \Delta, \varphi \quad \sim^k\psi, \Gamma \Rightarrow \Delta}{\sim^k(\varphi \to \psi), \Gamma \Rightarrow \Delta}(\to\Rightarrow) \quad \frac{\varphi, \Gamma \Rightarrow \Delta, \sim^k\psi}{\Gamma \Rightarrow \Delta, \sim^k(\varphi \to \psi)}(\Rightarrow\to)$$

$$\frac{\sim^n\varphi, \Gamma \Rightarrow \Delta}{\sim^{2m+n}\varphi, \Gamma \Rightarrow \Delta}(\sim^{2m+n}\Rightarrow) \quad \frac{\Gamma \Rightarrow \Delta, \sim^n\varphi}{\Gamma \Rightarrow \Delta, \sim^{2m+n}\varphi}(\Rightarrow\sim^{2m+n})$$

where $e \in E[0, 2m+n)$, $o \in O[0, 2m+n)$ *and* $k \geq 0$.

Let $\mathsf{G3K}_{m,n} \vdash \Gamma \Rightarrow \Delta$ *stand for that* $\Gamma \Rightarrow \Delta$ *is derivable in* $\mathsf{G3K}_{m,n}$. *Derivability and related notions are defined as in Definition 6.*

Lemma 17. *For every* $\varphi \in \mathscr{F}$ *and* $k \geq 0$, $\mathsf{G3K}_{m,n} \vdash \sim^k\varphi, \Gamma \Rightarrow \Delta, \sim^k\varphi$.

Proof. The proof is similar to Lemma 11. □

Lemma 18 (Invertibility). *For every* $h \geq 0, e \in E[0, 2m+n)$ *and* $o \in O[0, 2m+n)$, *the following hold in* $\mathsf{G3K}_{m,n}$:

(1) *if* $\vdash_h \sim^e(\varphi \wedge \psi), \Gamma \Rightarrow \Delta$, *then* $\vdash_h \sim^e\varphi, \sim^e\psi, \Gamma \Rightarrow \Delta$.
(2) *if* $\vdash_h \Gamma \Rightarrow \Delta, \sim^e(\varphi \wedge \psi)$; *then* $\vdash_h \Gamma \Rightarrow \Delta, \sim^e\varphi$ *and* $\vdash_h \Gamma \Rightarrow \Delta, \sim^e\psi$.
(3) *if* $\vdash_h \sim^e(\varphi \vee \psi), \Gamma \Rightarrow \Delta$, *then* $\vdash_h \sim^e\varphi, \Gamma \Rightarrow \Delta$ *and* $\vdash_h \sim^e\psi, \Gamma \Rightarrow \Delta$.
(4) *if* $\vdash_h \Gamma \Rightarrow \Delta, \sim^e(\varphi \vee \psi)$, *then* $\vdash_h \Gamma \Rightarrow \Delta, \sim^e\varphi, \sim^e\psi$.
(5) *if* $\vdash_h \sim^o(\varphi \vee \psi), \Gamma \Rightarrow \Delta$, *then* $\vdash_h \sim^o\varphi, \Gamma \Rightarrow \Delta$ *and* $\vdash_h \sim^o\psi, \Gamma \Rightarrow \Delta$.
(6) *if* $\vdash_h \Gamma \Rightarrow \Delta, \sim^o(\varphi \wedge \psi)$, *then* $\vdash_h \Gamma \Rightarrow \Delta, \sim^o\varphi, \sim^o\psi$.
(7) *if* $\vdash_h \sim^o(\varphi \vee \psi)\Gamma \Rightarrow \Delta$, *then* $\vdash_h \sim^o\varphi, \sim^o\psi, \Gamma \Rightarrow \Delta$.
(8) *if* $\vdash_h \Gamma \Rightarrow \Delta, \sim^o(\varphi \vee \psi)$, *then* $\vdash_h \Gamma \Rightarrow \Delta, \sim^o\varphi$ *and* $\vdash_h \Gamma \Rightarrow \Delta, \sim^o\psi$;
(9) *if* $\vdash_h \sim^{2m+n}\varphi, \Gamma \Rightarrow \Delta$, *then* $\vdash_h \sim^n\varphi, \Gamma \Rightarrow \Delta$.
(10) *if* $\vdash_h \Gamma \Rightarrow \Delta, \sim^{2m+n}\varphi$, *then* $\vdash_h \Gamma \Rightarrow \Delta, \sim^n\varphi$.
(11) *if* $\vdash_h \sim^k(\varphi \to \psi), \Gamma \Rightarrow \Delta$, *then* $\vdash_h \Gamma \Rightarrow \Delta, \varphi$ *and* $\vdash_h \sim^k\psi, \Gamma \Rightarrow \Delta$.
(12) *if* $\vdash_h \Gamma \Rightarrow \Delta, \sim^k(\varphi \to \psi)$, *then* $\vdash_h \varphi, \Gamma \Rightarrow \Delta, \sim^k\psi$.

Proof. The proof is similar to Lemma 13. □

Lemma 19. *The following rules are height-preserving admissible in* $\mathsf{G3K}_{m,n}$:

$$\frac{\Gamma \Rightarrow \Delta}{\varphi, \Gamma \Rightarrow \Delta}(\mathrm{Wk}\Rightarrow) \quad \frac{\Gamma \Rightarrow \Delta}{\Gamma \Rightarrow \Delta, \varphi}(\Rightarrow\mathrm{Wk})$$

Proof. Assume $\vdash_h \Gamma \Rightarrow \Delta$. The proof proceeds by simultaneous induction on h. The case $h = 0$ is trivial. For $h > 0$, $\Gamma \Rightarrow \Delta$ is obtained by a rule (R). By induction hypothesis and (R), we obtain the conclusions. □

Lemma 20. *The following rules are height-preserving admissible in* $\mathsf{G3K}_{m,n}$*:*

$$\frac{\varphi, \varphi, \Gamma \Rightarrow \Delta}{\varphi, \Gamma \Rightarrow \Delta}(\mathrm{Ctr}\Rightarrow) \quad \frac{\Gamma \Rightarrow \Delta, \varphi, \varphi}{\Gamma \Rightarrow \Delta, \varphi}(\Rightarrow\mathrm{Ctr})$$

Proof. The proof is similar to Lemma 14. Note that the invertibility of rules in Lemma 18 is used in the proof. Details are omitted. □

Lemma 21. *If* $\mathsf{G3K}_{m,n} \vdash \Gamma \Rightarrow \Sigma, \bot$, *then* $\mathsf{G3K}_{m,n} \vdash \Gamma \Rightarrow \Sigma, \Theta$.

Proof. The proof is quite similar to the proof of Lemma 15. □

Theorem 8. *The following cut rule is admissible in* $\mathsf{G3K}_{m,n}$*:*

$$\frac{\Gamma \Rightarrow \Sigma, \varphi \quad \varphi, \Delta \Rightarrow \Theta}{\Gamma, \Delta \Rightarrow \Sigma, \Theta}(\mathrm{Cut})$$

Proof. The proof is similar to Theorem 6. □

Lemma 22. *For every formula* $\varphi \in \mathscr{F}$, *if* $\vdash_{\mathsf{HK}_{m,n}} \varphi$, *then* $\mathsf{G3K}_{m,n} \vdash \Rightarrow \varphi$.

Proof. The proof is quite similar to the proof of Lemma 16. □

A sequent $\Gamma \Rightarrow \Delta$ is $\langle m, n\rangle$*-valid* in $\circ$ (notation: $\circ \models_{m,n} \Gamma \Rightarrow \Delta$) if $\bigwedge \Gamma \rightarrow \bigvee \Delta \in \mathsf{K}_{m,n}$. The notion of preserving $\langle m, n\rangle$-validity of a sequent rule is defined naturally. We get the soundness and completeness of $\mathsf{G3K}_{m,n}$.

Theorem 9. $\mathsf{G3K}_{m,n} \vdash \Gamma \Rightarrow \Delta$ *iff* $\circ \models_{m,n} \Gamma \Rightarrow \Delta$.

Proof. Assume $\vdash_h \Gamma \Rightarrow \Delta$. We get $\circ \models_{m,n} \Gamma \Rightarrow \Delta$ by induction on the height of derivation. The proof is similar to Theorem 7. Assume $\circ \models_{m,n} \Gamma \Rightarrow \Delta$. Let $\varphi = \bigwedge \Gamma$ and $\psi = \bigvee \Delta$. Then $\vdash_{\mathsf{HK}_{m,n}} \varphi \rightarrow \psi$. By Lemma 22, $\vdash \Rightarrow \varphi \rightarrow \psi$. Then $\vdash \varphi \Rightarrow \psi$. Clearly $\vdash \Gamma \Rightarrow \varphi$ and $\vdash \psi \Rightarrow \Delta$. By (Cut), $\vdash \Gamma \Rightarrow \Delta$. □

5 Concluding Remarks

The present work makes contributions to the study of intuitionistic connexive logics. Inspired by Wansing's work [24], we develop infinitely many intuitionistic connexive logics. For each superintuitionistic logic L, we introduce $\langle m, n\rangle$-connexive logics by adding a set of connexive principles. The semantics for these connexive logics is given by extending the domain of a valuation from the set of all propositional variables to the set of all literals. Finally we provide G3-style sequent calculi for the intuitionistic and classical $\langle m, n\rangle$-connexive logics. The road to connexivity we take in this work can be extended to many other logics. At least one can obtain first-order and modal extensions of them as in [24].

Appendix

A Proof of Lemma 11

Proof. Assume $c(\varphi) = 0$. Suppose $\varphi = p$. If $k < 2m + n$, then $\sim^k p, \Gamma \Rightarrow \sim^k p$ is an instance of (Id). Suppose $k \geq 2m + n$. There exist $l \in [0, 2m + n)$ and $r \geq 1$ with $k = 2rm + l$. Obviously $\vdash \sim^l p, \Gamma \Rightarrow \sim^l p$. By $(\sim^{2m+n}\Rightarrow)$ and $(\Rightarrow\sim^{2m+n})$, $\vdash \sim^k p, \Gamma \Rightarrow \sim^k p$. Suppose $\varphi = \bot$. If $k < 2m + n$, then $\sim^k \bot, \Gamma \Rightarrow \sim^k \bot$ is an instance of $(\bot\Rightarrow)$. Suppose $k \geq 2m + n$. There exist $l \in [0, 2m + n)$ and $r \geq 1$ with $k = 2rm + l$. Then $\vdash \sim^l \bot, \Gamma \Rightarrow \sim^l \bot$. By $(\sim^{2m+n}\Rightarrow)$ and $(\Rightarrow\sim^{2m+n})$, $\vdash \sim^k \bot, \Gamma \Rightarrow \sim^k \bot$. Assume $c(\varphi) > 0$. Suppose $k \geq 2m + n$. There exist $l \in [0, 2m + n)$ and $r \geq 1$ with $k = 2rm + l$. Clearly $c(\sim^l \varphi) < c(\sim^k \varphi)$. By induction hypothesis, $\vdash \sim^l \varphi, \Gamma \Rightarrow \sim^l \varphi$. By $(\sim^{2m+n}\Rightarrow)$ and $(\Rightarrow\sim^{2m+n})$, $\vdash \sim^k \varphi, \Gamma \Rightarrow \sim^k \varphi$. Suppose $k < 2m + n$. Assume $\varphi = \varphi_1 \wedge \varphi_2$. Let $\varphi = \sim^e(\varphi_1 \wedge \varphi_2)$ with $e \in E[0, 2m + n)$. Clearly $c(\sim^e \varphi_1) < c(\sim^e(\varphi_1 \wedge \varphi_2))$ and $c(\sim^e \varphi_2) < c(\sim^e(\varphi_1 \wedge \varphi_2))$. By induction hypothesis, $\vdash \sim^e \varphi_1, \sim^e \varphi_2, \Gamma \Rightarrow \sim^e \varphi_1$ and $\vdash \sim^e \varphi_1, \sim^e \varphi_2, \Gamma \Rightarrow \sim^e \varphi_2$. By $(\Rightarrow\wedge)$, $\vdash \sim^e \varphi_1, \sim^e \varphi_2, \Gamma \Rightarrow \sim^e(\varphi_1 \wedge \varphi_2)$. By $(\wedge\Rightarrow)$, $\vdash \sim^e(\varphi_1 \wedge \varphi_2), \Gamma \Rightarrow \sim^e(\varphi_1 \wedge \varphi_2)$. Now let $\varphi = \sim^o(\varphi_1 \wedge \varphi_2)$ with $o \in O[0, 2m + n)$. Clearly, $c(\sim^o \varphi_1) < c(\sim^o(\varphi_1 \wedge \varphi_2))$ and $c(\sim^o \varphi_2) < c(\sim^o(\varphi_1 \wedge \varphi_2))$. By induction hypothesis, $\vdash \sim^o \varphi_1, \Gamma \Rightarrow \sim^o \varphi_1$ and $\vdash \sim^o \varphi_2, \Gamma \Rightarrow \sim^o \varphi_2$. By $(\Rightarrow\sim\wedge)$, $\vdash \sim^o \varphi_1, \Gamma \Rightarrow \sim^o(\varphi_1 \wedge \varphi_2)$ and $\vdash \sim^o \varphi_2, \Gamma \Rightarrow \sim^o(\varphi_1 \wedge \varphi_2)$. By $(\sim\wedge\Rightarrow)$, $\vdash \sim^o(\varphi_1 \wedge \varphi_2), \Gamma \Rightarrow \sim^o(\varphi_1 \wedge \varphi_2)$. The case $\varphi = \varphi_1 \vee \varphi_2$ is shown similarly. Suppose $\varphi = \varphi_1 \rightarrow \varphi_2$. Clearly $c(\varphi_1) < c(\sim^k(\varphi_1 \rightarrow \varphi_2))$ and $c(\varphi_2) < c(\sim^k(\varphi_1 \rightarrow \varphi_2))$. By induction hypothesis, $\vdash \varphi_1, \sim^k(\varphi_1 \rightarrow \varphi_2), \Gamma \Rightarrow \varphi_1$ and $\vdash \varphi_1, \sim^k \varphi_2, \Gamma \Rightarrow \sim^k \varphi_2$. By $(\rightarrow\Rightarrow)$, $\vdash \varphi_1, \sim^k(\varphi_1 \rightarrow \varphi_2), \Gamma \Rightarrow \sim^k \varphi_2$. By $(\Rightarrow\rightarrow)$, $\vdash \sim^k(\varphi_1 \rightarrow \varphi_2), \Gamma \Rightarrow \sim^k(\varphi_1 \rightarrow \varphi_2)$. □

B Proof of Lemma 14

Proof. Assume $\vdash_h \varphi, \varphi, \Gamma \Rightarrow \psi$. The proof proceeds by induction on h and subinduction on the complexity $c(\varphi)$. The case $h = 0$ is trivial. Suppose $h > 0$ and $\varphi, \varphi, \Gamma \Rightarrow \psi$ is obtained by a rule (R). Suppose φ is not principal in (R). Then $\vdash \varphi, \Gamma \Rightarrow \psi$ by induction hypothesis and (R). For example, let (R) be $(\Rightarrow \wedge)$ with premisses $\vdash_{h-1} \varphi, \varphi, \Gamma \Rightarrow \sim^e \psi_1$ and $\vdash_{h-1} \varphi, \varphi, \Gamma \Rightarrow \sim^e \psi_2$, and conclusion $\vdash_h \varphi, \varphi, \Gamma \Rightarrow \sim^e(\psi_1 \wedge \psi_2)$. By induction hypothesis, $\vdash_{h-1} \varphi, \Gamma \Rightarrow \sim^e \psi_1$ and $\vdash_{h-1} \varphi, \Gamma \Rightarrow \sim^e \psi_2$. By $(\Rightarrow \wedge)$, $\vdash_h \varphi, \Gamma \Rightarrow \sim^e(\psi_1 \wedge \psi_2)$. Other cases are shown similarly. Suppose φ is principal in (R). Assume $\varphi = \sim^e(\varphi_1 \wedge \varphi_2)$. Let (R) end with premiss $\vdash_{h-1} \sim^e \varphi_1, \sim^e \varphi_2, \sim^e(\varphi_1 \wedge \varphi_2), \Gamma \Rightarrow \psi$ and conclusion $\vdash_h \sim^e(\varphi_1 \wedge \varphi_2), \sim^e(\varphi_1 \wedge \varphi_2), \Gamma \Rightarrow \psi$. By Lemma 13 (1), $\vdash_{h-1} \sim^e \varphi_1, \sim^e \varphi_2, \sim^e \varphi_1, \sim^e \varphi_2, \Gamma \Rightarrow \psi$. By induction hypothesis, $\vdash_{h-1} \sim^e \varphi_1, \sim^e \varphi_2, \Gamma \Rightarrow \psi$. By $(\wedge\Rightarrow)$, $\vdash_h \sim^e(\varphi_1 \wedge \varphi_2), \Gamma \Rightarrow \psi$. Assume $\varphi = \sim^o(\varphi_1 \wedge \varphi_2)$. Let (R) end with premisses $\vdash \sim^o \varphi_1, \sim^o(\varphi_1 \wedge \varphi_2), \Gamma \Rightarrow \psi$ and $\vdash \sim^o \varphi_2, \sim^o(\varphi_1 \wedge \varphi_2), \Gamma \Rightarrow \psi$, and conclusion $\vdash_h \sim^o(\varphi_1 \wedge \varphi_2), \sim^o(\varphi_1 \wedge \varphi_2), \Gamma \Rightarrow \psi$. By Lemma 13 (4), $\vdash \sim^o \varphi_1, \sim^o \varphi_1, \Gamma \Rightarrow \psi$ and $\vdash \sim^o \varphi_2, \sim^o \varphi_2, \Gamma \Rightarrow \psi$. By induction hypothesis, $\vdash \sim^o \varphi_1, \Gamma \Rightarrow \psi$ and $\vdash \sim^o \varphi_2, \Gamma \Rightarrow \psi$. By $(\sim\wedge\Rightarrow)$, $\vdash_h \sim^o(\varphi_1 \wedge \varphi_2), \Gamma \Rightarrow \psi$. Assume $\varphi = \sim^e(\varphi_1 \vee \varphi_2)$ or $\sim^o(\varphi_1 \vee \varphi_2)$. The proof is similar to previous cases. Assume $\varphi = \sim^{2m+n}\chi$. Let (R) end with premiss $\vdash \sim^n\chi, \sim^{2m+n}\chi, \Gamma \Rightarrow \psi$ and conclusion $\vdash_h \sim^{2m+n}\chi, \sim^{2m+n}\chi, \Gamma \Rightarrow \psi$. By Lemma 13 (5), $\vdash \sim^n\chi, \sim^n\chi, \Gamma \Rightarrow \psi$. By induction hypothesis, $\vdash \sim^n\chi, \Gamma \Rightarrow \psi$. By $(\sim^{2m+n}\Rightarrow)$, $\vdash_h \sim^{2m+n}\chi, \Gamma \Rightarrow \psi$. Assume $\varphi = \sim^k(\varphi_1 \to \varphi_2)$. Let (R) end with premisses $\vdash \sim^k(\varphi_1 \to \varphi_2), \sim^k(\varphi_1 \to \varphi_2), \Gamma \Rightarrow \varphi_1$ and $\vdash \sim^k\varphi_2, \Gamma \Rightarrow \psi$, and conclusion $\vdash_h \sim^k(\varphi_1 \to \varphi_2), \sim^k(\varphi_1 \to \varphi_2), \Gamma \Rightarrow \psi$. By induction hypothesis, $\vdash \sim^k(\varphi_1 \to \varphi_2), \Gamma \Rightarrow \varphi_1$. By premiss and $(\to\Rightarrow)$, $\vdash_h \sim^k(\varphi_1 \to \varphi_2), \Gamma \Rightarrow \psi$. □

C Proof of Theorem 6

Proof. Assume $\vdash_h \Gamma \Rightarrow \varphi$ and $\vdash_j \varphi, \Delta \Rightarrow \psi$. We show $\vdash \Gamma, \Delta \Rightarrow \psi$ by simultaneous induction on the cut height $h + j$ and the complexity $c(\varphi)$. Assume $h = 0$ or $j = 0$. Suppose $h = 0$. Suppose $\Gamma \Rightarrow \varphi$ is an instance of (Id). Then $\varphi \in \Gamma$. By $\vdash \varphi, \Delta \Rightarrow \psi$ and (Wk), $\vdash \Gamma, \Delta \Rightarrow \psi$. If $\bot \in \Gamma$, then $\vdash \Gamma, \Delta \Rightarrow \psi$. Suppose $j = 0$. Let $\varphi, \Delta \Rightarrow \psi$ be an instance of (Id). If $\varphi = \psi$, by $\vdash \Gamma \Rightarrow \varphi$ and (Wk), $\vdash \Gamma, \Delta \Rightarrow \varphi$. Let $\varphi, \Delta \Rightarrow \psi$ be an instance of $(\bot)$. If $\bot \in \Delta$, then $\vdash \Gamma, \Delta \Rightarrow \psi$. If $\varphi = \bot$, by $\vdash \Gamma \Rightarrow \bot$, Lemma 15 and (Wk), $\vdash \Gamma, \Delta \Rightarrow \psi$. Now assume $h, j > 0$. Let the premisses of (Cut) be obtained by the rules (R1) and (R2) respectively. Suppose φ is not principal in (R1). Then (R1) is a left rule. We apply (Cut) to the premiss(es) of (R1) and $\varphi, \Delta \Rightarrow \psi$, and then apply (R1). Suppose φ is not principal in (R2). We apply (Cut) to $\Gamma \Rightarrow \varphi$ and the premiss(es) of (R2), and then apply (R2). Suppose φ is principal in both (R1) and (R2). The proof proceeds by induction on $c(\varphi)$. We have the following cases:

(1) $\varphi = \sim^e(\varphi_1 \wedge \varphi_2)$ and the derivations end with

$$\frac{\Gamma \Rightarrow \sim^e\varphi_1 \quad \Gamma \Rightarrow \sim^e\varphi_2}{\Gamma \Rightarrow \sim^e(\varphi_1 \wedge \varphi_2)}(\Rightarrow\wedge) \quad \frac{\sim^e\varphi_1, \sim^e\varphi_2, \Delta \Rightarrow \psi}{\sim^e(\varphi_1 \wedge \varphi_2), \Delta \Rightarrow \psi}(\wedge\Rightarrow)$$

By applying (Cut) to sequents with cut formula of less complexity, we have

$$\cfrac{\cfrac{\Gamma \Rightarrow \sim^e\varphi_1 \quad \cfrac{\Gamma \Rightarrow \sim^e\varphi_2 \quad \sim^e\varphi_1, \sim^e\varphi_2, \Delta \Rightarrow \psi}{\sim^e\varphi_1, \Gamma, \Delta \Rightarrow \psi}\text{(Cut)}}{\Gamma, \Gamma, \Delta \Rightarrow \psi}\text{(Cut)}}{\Gamma, \Delta \Rightarrow \psi}\text{(Ctr)}$$

(2) $\varphi = \sim^e(\varphi_1 \vee \varphi_2)$ and the derivations end with

$$\frac{\Gamma \Rightarrow \sim^e\varphi_i}{\Gamma \Rightarrow \sim^e(\varphi_1 \vee \varphi_2)}(\Rightarrow\vee)(i = 1, 2) \quad \frac{\sim^e\varphi_1, \sim^e\varphi_2, \Delta \Rightarrow \psi}{\sim^e(\varphi_1 \vee \varphi_2), \Delta \Rightarrow \psi}(\vee\Rightarrow)$$

By applying (Cut) to sequents with cut formula of less complexity, and by (Ctr), we obtain $\Gamma, \Delta \Rightarrow \psi$.

(3) $\varphi = \sim^o(\varphi_1 \wedge \varphi_2)$ or $\sim^o(\varphi_1 \vee \varphi_2)$. The proof is similar to (1) or (2).

(4) $\varphi = \sim^k(\varphi_1 \to \varphi_2)$ and the derivations end with

$$\frac{\varphi_1, \Gamma \Rightarrow \sim^k\varphi_2}{\Gamma \Rightarrow \sim^k(\varphi_1 \to \varphi_2)}(\Rightarrow\to) \quad \frac{\sim^k(\varphi_1 \to \varphi_2), \Delta \Rightarrow \varphi_1 \quad \sim^k\varphi_2, \Delta \Rightarrow \psi}{\sim^k(\varphi_1 \to \varphi_2), \Delta \Rightarrow \psi}(\to\Rightarrow)$$

By applying (Cut) to sequents with cut formula of less complexity, we have $\vdash \sim^k(\varphi_1 \to \varphi_2), \Gamma, \Delta, \Delta \Rightarrow \psi$. By (Ctr), $\vdash \sim^k(\varphi_1 \to \varphi_2), \Gamma, \Delta \Rightarrow \psi$.

(5) $\varphi = \sim^{2m+n}\psi$ and the derivations end with

$$\frac{\Gamma \Rightarrow \sim^n\psi}{\Gamma \Rightarrow \sim^{2m+n}\psi}(\Rightarrow\sim^{2m+n}) \quad \frac{\sim^n\psi, \Delta \Rightarrow \chi}{\sim^{2m+n}\psi, \Delta \Rightarrow \chi}(\sim^{2m+n}\Rightarrow)$$

By applying (Cut) to premisses, we get $\vdash \Gamma, \Delta \Rightarrow \chi$. □

References

1. Berman, J.: Distributive lattices with an additional unary operation. Aequationes Math. **16**, 165–171 (1977)
2. Chagrov, A., Zakharyaschev, M.: Modal Logic. Clarendon Press, Oxford (1997)
3. Cornejo, J.M., Sankappanavar, H.P.: A logic for dually hemimorphic semi-Heyting algebras and its axiomatic extensions. Bull. Section Logic (2022, to appear)
4. Fazio, D., Ledda, A., Paoli, F.: Intuitionistic logic is a connexive logic. arXiv:2208.14715v1 [math.LO] (2022)
5. Humberstone, L.: Contra-classical logics. Australas. J. Philos. **78**(4), 438–474 (2000)
6. Kamide, N.: Paraconsistent double negations as classical and intuitionistic negations. Stud. Logica **105**(6), 1167–1191 (2017)
7. Kamide, N.: Kripke-completeness and cut-elimination theorems for intuitionistic paradefinite logics with and without quasi-explosion. J. Philos. Log. **49**(6), 1185–1212 (2020)
8. Kamide, N., Wansing, H.: Connexive modal logic based on positive S4. Logic without Frontiers. Festschrift for Walter Alexandre Carnielli on the Occasion of His 60th Birthday, pp. 389–409. College Publications, London (2011)
9. Kamide, N., Wansing, H.: Proof theory of Nelson's paraconsistent logic: a uniform perspective. Theor. Comput. Sci. **415**, 1–38 (2012)
10. Kamide, N., Wansing, H.: Proof Theory of N4-Related Paraconsistent Logics. College Publications, London (2015)
11. Ma, M., Lin, Y.: Countably many weakenings of Belnap-Dunn logic. Stud. Logica **108**, 163–198 (2020)
12. Malinowski, J., Palczewski, R.: Relating semantics for connexive logic. In: Giordani, A., Malinowski, J. (eds.) Logic in High Definition. TL, vol. 56, pp. 49–65. Springer, Cham (2021). https://doi.org/10.1007/978-3-030-53487-5_4
13. McCall, S.: Non-classical Propositional Calculi. Ph.D thesis, University of Oxford (1964)
14. McCall, S.: Connexive implication. J. Symbolic Log. **31**(3), 415–433 (1966)
15. McCall, S.: A history of connexivity. In: Gabbay, D.M., Pelletier, F.J., Woods, J. (eds.) Handbook of the History of Logic, vol. 11, pp. 415–449. Elsevier (2012)
16. Negri, S., von Plato, J.: Structural Proof Theory. Cambridge University Press (2001)
17. Odinstov, S.: Constructive Negations and Paraconsistency. Springer, Heidelberg (2008). https://doi.org/10.1007/978-1-4020-6867-6
18. Omori, H.: From paraconsistent logic to dialetheic logic. In: Andreas, H., Verdée, P. (eds.) Logical Studies of Paraconsistent Reasoning in Science and Mathematics. TL, vol. 45, pp. 111–134. Springer, Cham (2016). https://doi.org/10.1007/978-3-319-40220-8_8

19. Omori, H., Wansing, H.: On contra-classical variants of Nelson logic N4 and its classical extension. Rev. Symbolic Log. **11**(4), 805–820 (2018)
20. Omori, H., Wansing, H.: 40 years of FDE: an introductory overview. Stud. Logica **105**(6), 1021–1049 (2017)
21. Omori, H., Wansing, H.: Connexive logics: an overview and current trends. Logic Log. Philos. **28**(3), 371–387 (2019)
22. Priest, G.: Negation as cancellation, and connexive logic. Topoi **18**(2), 141–148 (1999)
23. Routley, R.: Semantics for connexive logics. I. Studia Logica **37**(4), 393–412 (1978)
24. Wansing, H.: Connexive modal logic. In: Schmidt, R., Pratt-Hartmann, I., Reynolds, M., Wansing, H. (eds.) Advances in Modal Logic, vol. 5, pp. 367–383. King's College Publications, London (2004)
25. Wansing, H.: Negation. In: Goble, L. (eds.) The Blackwell Guide to Philosophical Logic, pp. 415–436. Blackwell Publishers Ltd. (2001)
26. Wansing, H., Skurt, D.: Negation as cancellation, connexive logic, and qLPm. Australas. J. Log. **15**(2), 476–488 (2018)

Relational Semantics for Normal Topological Quasi-Boolean Logic

Hao Wu[1] and Minghui Ma[2(✉)]

[1] Institute of Logic and Cognition, Sun Yat-Sen University, Guangzhou, China
wuhao43@mail2.sysu.edu.cn

[2] Department of Philosophy, Sun Yat-Sen University, Guangzhou, China
mamh6@mail.sysu.edu.cn

Abstract. This work introduces modal logics for varieties of normal topological quasi-Boolean algebras. Relational semantics for these modal logics using involutive frames are established. A discrete duality is given for involutive frames and normal topological quasi-Boolean algebras. Some results on Kripke-completeness and finite model property are given.

Keywords: Relational semantics · Quasi-Boolean algebra · Topological quasi-Boolean algebra · Finite model property

1 Introduction

A topological quasi-Boolean algebra (tqBa) is a quasi-Boolean algebra (qBa, also known as De Morgan algebra) with an interior operator (cf. e.g. [20]). The Belnap-Dunn four-valued logic BD for quasi-Boolean algebras has been well-developed in the literature (cf. e.g. [2,11,14]). The logic for tqBas is exactly the quasi-Boolean counterpart of modal logic S4. In the setting of the study on quasi-Boolean algebras with operators (cf. e.g. [15,18,19]), one can find the extension of BD with S4 axioms. In the study of algebras from rough set theory (cf. e.g. [1]), classes of pre-rough algebras which are subvarieties of tqBa are developed in [21]. In a recent work [16], residuated pre-rough algebras are investigated.

Celani [6] introduced the variety $\mathcal{CMD}_\Box$ of classical De Morgan algebras which is indeed a generalization of the variety $\mathcal{TMA}$ of tetravalent modal algebras (cf. e.g. [13,17]). A tqBa $(A, \Box)$ is a *classical De Morgan algebra* if it satisfies the equation $\Box x \vee {\sim}\Box x = 1$. The De Morgan dual $\Diamond$ of the interior operator $\Box$ is a closure operator in a classical De Morgan algebra. However, the interaction axioms $\Box x \wedge \Diamond y \leq \Diamond(x \wedge y)$ and $\Box(x \vee y) \leq \Box x \vee \Diamond y$ between modal operators are not well-investigated in the setting of tqBas. These interaction axioms are used to define *positive modal algebra* in [12], and the representation of distributive modal algebras and Priestley duality for positive modal algebras have been

This work was supported by Chinese National Funding of Social Sciences (Grant no. 18ZDA033).

M. Banerjee and A. V. Sreejith (Eds.): ICLA 2023, LNCS 13963, pp. 207–221, 2023.
https://doi.org/10.1007/978-3-031-26689-8_15

explored in [7,8]. In the present paper, we consider topological quasi-Boolean algebras with interaction axioms.

Normal topological quasi-Boolean logics are defined as consequence theories of normal tqBas. The present work gives a relational semantics for these logics by introducing *involutive* frames and models. Białynicki-Birula and Rasiowa [5] proposed a set representation of quasi-Boolean algebras using an involution of a set X for the interpretation of quasi-complementation (or De Morga negation). In the present paper, we introduce involutive frames which are defined as pre-ordered sets with an involution. Then a discrete duality for normal tqBas is developed. Furthermore, using the canonical involutive frame, we get completeness results for the minimal normal topological quasi-Boolean logic wS4 as well as its extensions wS4C and wS5. Finally, we extend the filtration method to our setting and show the finite model property for wS4, wS4C and wS5.

2 Normal Topological Quasi-Boolean Logic

An algebra $(A, \wedge, \vee, \sim, 0, 1)$ is a *quasi-Boolean algebra* (qBa, also known as De Morgan algebra) if $(A, \wedge, \vee, 0, 1)$ is a bounded distributive lattice satisfying $\sim 0 = 1$, $\sim\sim x = x$ and $\sim(x \vee y) = \sim x \wedge \sim y$ for all $x, y \in A$. The lattice order on A is denoted by $\leq_A$ or simply by $\leq$. A unary operator $\Box : A \to A$ on a qBa is an *interior operator* if $\Box 1 = 1$ and for all $x, y \in A$, $\Box(x \wedge y) = \Box x \wedge \Box y$, $\Box x \leq x$ and $\Box\Box x = \Box x$. An operator $\Diamond : A \to A$ is a *closure operator* if $\Diamond 0 = 0$ and for all $x, y \in A$, $\Diamond(x \vee y) = \Diamond x \vee \Diamond y$, $x \leq \Diamond x$ and $\Diamond\Diamond x = \Diamond x$. An algebra $(A, \wedge, \vee, \sim, 0, 1, \Box)$ is a *topological quasi-Boolean algebra* (tqBa) if $(A, \wedge, \vee, \sim, 0, 1)$ is a qBa and $\Box$ is an interior operator on A. We use $(\mathcal{A}, \Box)$ for a tqBa where $\mathcal{A}$ is supposed to be a qBa. A tqBa $(\mathcal{A}, \Box)$ is *normal* (ntqBa) if the following condition holds for all $x, y \in A$,

$$\Box x \wedge \Diamond y \leq \Diamond(x \wedge y) \tag{N}$$

where $\Diamond : A \to A$ is defined by $\Diamond x := \sim\Box\sim x$. Clearly $\Diamond$ is a closure operator on A. Let NtqBa be the variety of all normal tqBas.

Lemma 1. *Let $(\mathcal{A}, \Box)$ be a tqBa. For all $x, y \in A$, (1) $x \leq y$ if and only if $\sim y \leq \sim x$; (2) if $x \leq y$, then $\Box x \leq \Box y$ and $\Diamond x \leq \Diamond y$; (3) $\Diamond\Box\Diamond x \leq \Diamond x$, $\Box\Diamond x = \Box\Diamond\Box\Diamond x$ and $\Diamond\Box x = \Diamond\Box\Diamond\Box x$; (4) $\Box(x \vee y) \leq \Diamond x \vee \Box y$.*

Proof. Clearly (1) and (2) hold. For (3), $\Diamond\Box\Diamond x \leq \Diamond\Diamond\Diamond x = \Diamond x$. Clearly $\Box\Diamond x \leq \Diamond\Box\Diamond x$. By (2), $\Box\Box\Diamond x \leq \Box\Diamond\Box\Diamond x$. Then $\Box\Diamond x = \Box\Box\Diamond x \leq \Box\Diamond\Box\Diamond x$. Clearly $\Box\Diamond\Box\Diamond x \leq \Box\Diamond x$. Then $\Box\Diamond x = \Box\Diamond\Box\Diamond x$. We have $\Box\Diamond\sim x = \Box\Diamond\Box\Diamond\sim x$. Then $\Diamond\Box x = \Diamond\Box\Diamond\Box x$. For (4), by (N), $\Box\sim x \wedge \Diamond\sim y \leq \Diamond(\sim x \wedge \sim y)$. By (1), $\Box(x \vee y) = \sim\Diamond(\sim x \wedge \sim y) \leq \sim(\Box\sim x \wedge \Diamond\sim y) = \Diamond x \vee \Box y$. □

Let $\mathbb{P} = \{p_i : i \in \omega\}$ be a denumerable set of propositional variables. The set of all formulas For is defined inductively as follows:

$$\mathsf{For} \ni \varphi ::= p \mid \bot \mid \sim\varphi \mid (\varphi_1 \wedge \varphi_2) \mid \Box\varphi$$

where $p \in \mathbb{P}$. A propositional variable or $\bot$ is called *atomic*. We use abbreviations $\top := \sim\bot$, $\varphi_1 \vee \varphi_2 := \sim(\sim\varphi_1 \wedge \sim\varphi_2)$ and $\Diamond\varphi := \sim\Box\sim\varphi$. A *substitution* σ is a function $\sigma : \mathbb{P} \to \mathsf{For}$. Let φ^σ denote the formula obtained from φ by using substitution σ. Let $sub(\varphi)$ be the set of all subformulas of φ.

A *consequence* is an expression $\varphi \vdash \psi$ where $\varphi, \psi \in \mathsf{For}$. Let s, t etc. with or without subscripts denote consequences. The set of all consequences in For is denoted by $C(\mathsf{For})$. A *consequence rule* is an expression

$$\frac{s_1 \ldots s_n}{s_0}(R)$$

where $s_1, \ldots, s_n$ are *premisses* and s_0 is the *conclusion* of (R).

Definition 1. Let A be a ntqBa. A *valuation* is a function $V : \mathbb{P} \to A$. A valuation V in A is extended to a function $V : \mathsf{For} \to A$ as follows:

$$V(\bot) = 0,\ V(\sim\varphi) = \sim V(\varphi),\ V(\varphi \wedge \psi) = V(\varphi) \wedge V(\psi),\ V(\Box\varphi) = \Box V(\varphi).$$

A *consequence* $\varphi \vdash \psi$ is *valid* in A (notation: $\varphi \models_A \psi$) if $V(\varphi) \leq V(\psi)$ for all valuations in A. The *consequence theory* of a class of ntqBas $\mathcal{K}$ is defined as $\mathsf{Th}(\mathcal{K}) = \{\varphi \vdash \psi : \varphi \models_{\mathcal{K}} \psi\}$ where $\varphi \models_{\mathcal{K}} \psi$ means $\varphi \models_A \psi$ for all $A \in \mathcal{K}$.

Definition 2. A *normal toplogical quasi-Boolean logic* (ntqBl) is a set L of consequences in For such that the following conditions hold:

(1) L contains all instances of the following axiom schemes:

$$(\mathrm{Id})\ \varphi \vdash \varphi \quad (\bot)\ \bot \vdash \varphi \quad (\top)\ \varphi \vdash \top \quad (\mathrm{D})\ \varphi \wedge (\psi \vee \chi) \vdash (\varphi \wedge \psi) \vee (\varphi \wedge \chi)$$

$$(\Box\wedge)\ \Box\varphi \wedge \Box\psi \vdash \Box(\varphi \wedge \psi) \quad (\mathrm{N})\ \Box\varphi \wedge \Diamond\psi \vdash \Diamond(\varphi \wedge \psi)$$

$$(\Box\top)\ \top \vdash \Box\top \quad (\mathrm{T})\ \Box\varphi \vdash \varphi \quad (4)\ \Box\varphi \vdash \Box\Box\psi$$

(2) L is closed under the following rules:

$$\frac{\varphi \vdash \chi}{\varphi \wedge \psi \vdash \chi}(\wedge\vdash_1) \quad \frac{\psi \vdash \chi}{\varphi \wedge \psi \vdash \chi}(\wedge\vdash_2) \quad \frac{\varphi \vdash \psi \quad \varphi \vdash \chi}{\varphi \vdash \psi \wedge \chi}(\vdash\wedge)$$

$$\frac{\varphi \vdash \psi}{\sim\psi \vdash \sim\varphi}(\mathrm{CP}) \quad \frac{\varphi \vdash \psi}{\sim\sim\varphi \vdash \psi}(\sim\sim\vdash) \quad \frac{\varphi \vdash \psi}{\varphi \vdash \sim\sim\psi}(\vdash\sim\sim)$$

$$\frac{\varphi \vdash \psi}{\Box\varphi \vdash \Box\psi}(\Box) \quad \frac{\varphi \vdash \chi \quad \chi \vdash \psi}{\varphi \vdash \psi}(\mathrm{Cut})$$

(3) L is closed under uniform substitution, i.e., if $\varphi \vdash \psi \in L$, then $\varphi^\sigma \vdash \psi^\sigma \in L$ for every substitution σ.

A consequence $\varphi \vdash \psi$ is *derivable* in L (notation: $\varphi \vdash_L \psi$) if $\varphi \vdash \psi \in L$. A consequence rule (R) with premisses $s_1, \ldots, s_n$ and conclusion s_0 is *admissible* in L if $s_0 \in L$ whenever $s_i \in L$ for $1 \leq i \leq n$. The index L is omitted if no confusion arises. A formula φ is *L-equivalent* to ψ (notation: $\varphi \equiv_L \psi$) if $\varphi \vdash_L \psi$ and $\psi \vdash_L \varphi$. The smallest ntqBl is denoted by wS4.

Fact 1. *For every ntqBl L, the following hold:*

(1) $\varphi \equiv_L \sim\sim\varphi$.
(2) $\sim(\varphi \wedge \psi) \equiv_L \sim\varphi \vee \sim\psi$ *and* $\sim(\varphi \vee \psi) \equiv_L \sim\varphi \wedge \sim\psi$.
(3) $\sim\Box\varphi \equiv_L \Diamond\sim\varphi$ *and* $\sim\Diamond\varphi \equiv_L \Box\sim\varphi$.
(4) $\Box(\varphi \wedge \psi) \equiv_L \Box\varphi \wedge \Box\psi$ *and* $\Diamond(\varphi \vee \psi) \equiv_L \Diamond\varphi \vee \Diamond\psi$.
(5) $\Box\top \equiv_L \top$ *and* $\Diamond\bot \equiv_L \bot$.
(6) $\varphi \vdash_L \Diamond\varphi$, $\Box\varphi \equiv_L \Box\Box\varphi$ *and* $\Diamond\varphi \equiv_L \Diamond\Diamond\varphi$.
(7) $\Box\Diamond\varphi \equiv_L \Box\Diamond\Box\Diamond\varphi$ *and* $\Diamond\Box\varphi \equiv_L \Diamond\Box\Diamond\Box\varphi$.
(8) $\Box(\varphi \vee \psi) \vdash_L \Box\varphi \vee \Diamond\psi$.

The following rules are admissible in L:

$$\frac{\varphi \vdash \chi \quad \psi \vdash \chi}{\varphi \vee \psi \vdash \chi}\,(\vee\vdash) \qquad \frac{\varphi \vdash \psi}{\varphi \vdash \psi \vee \chi}\,(\vdash\vee_1) \qquad \frac{\varphi \vdash \chi}{\varphi \vdash \psi \vee \chi}\,(\vdash\vee_2) \qquad \frac{\varphi \vdash \psi}{\Diamond\varphi \vdash \Diamond\psi}\,(\Diamond)$$

Let L be a ntqBl. For every set of consequences Σ, let $L \oplus \Sigma$ be the smallest ntqBl containing $L \cup \Sigma$. If $\Sigma = \{s_1, \ldots, s_n\}$, we write $L \oplus s_1 \oplus \ldots \oplus s_n$. Let $\mathsf{NExt}(L)$ be the set of all ntqBls containing L. Obviously $\mathsf{NExt}(L)$ is a bounded lattice with respect to the operations $\cap$ and $\oplus$, and it is also closed under arbitrary intersections. For a set of formulas Σ, let $\mathsf{Alg}(\Sigma)$ be the variety of ntqBas validating all consequences in Σ. Clearly $L_1 \subseteq L_2$ implies $\mathsf{Alg}(L_2) \subseteq \mathsf{Alg}(L_1)$. By Fact 1, the equivalence relation $\equiv_L$ on L is a congruence relation on the algebra of formulas. Thus the *Lindenbaum-Tarski algebra* A^L for L is well-defined such that $L = \mathsf{Th}(A^L)$. Clearly $A^L \models L$. Then $L = \mathsf{Th}(\mathsf{Alg}(L))$.

3 Involutive Frames and Discrete Duality

In this section, we introduce involutive frames and show a discrete duality between normal topological quasi-Boolean algebras and involutive frames. We recall some basic notions from [10,20]. Some basic facts about filters and ideals shall be used without mentioning a reference.

Definition 3. An *involutive frame* is a structure $\mathfrak{F} = (W, g, R)$ such that

(F1) $g : W \to W$ is involutive, i.e., $g(g(w)) = w$ for all $w \in W$.
(F2) R is a reflexive and transitive relation on W.
(F3) for all $w, u \in W$, if wRu, then $g(w)Rg(u)$.

Given an involutive frame $\mathfrak{F}$, for all $w \in W$ and $X \subseteq W$, let $R(w) = \{u \in W : wRu\}$ and $R(X) = \bigcup_{w\in X} R(w)$. Operations $\sim_g$ and $\Box_R$ on the powerset $\mathcal{P}(W)$ are defined by setting $\sim_g X = \overline{g(X)}$ and $\Box_R X = \{w \in W : R(w) \subseteq X\}$ where $-$ is the complement in W. Let $\Diamond_R X = \sim_g\Box_R\sim_g X$. The *complex algebra* of an involutive frame $\mathfrak{F}$ is defined as $\mathfrak{F}^+ = (\mathcal{P}(W), \cup, \cap, \sim_g, \Box_R, \varnothing, W)$.

Note that every involution $g : W \to W$ is one-one and onto, so $g^{-1}(w) = g(w)$ for all $w \in W$. Hence for every subset $X \subseteq W$ in an involutive frame, $w \in g(X)$ if and only if $g(w) \in X$. The condition (F3) in Definition 3 guarantees the correctness of the normality axiom (N). Moreover, for every $X \subseteq W$ in an involutive frame, we have $\Diamond_R X = \{w \in W : R(g(w)) \cap g(X) \neq \varnothing\}$.

Let A be a ntqBa. Let $\mathcal{F}(A)$ and $\mathcal{I}(A)$ be sets of all filters and ideals in A respectively. Let $\mathcal{F}_{\mathrm{p}}(A)$ and $\mathcal{I}_{\mathrm{p}}(A)$ be sets of all prime filters and prime ideals in A respectively. For every subset $B \subseteq A$, let $\overline{B} = A \backslash B$ and $\sim B = \{\sim x : x \in B\}$. The filter and ideal generated by B in A are denoted by $[B)_A$ and $(B]_A$ respectively. Recall that $\Box^{-1}(B) = \{x \in A : \Box x \in B\}$ and $\Diamond^{-1}(B) = \{x \in A : \Diamond x \in B\}$.

Lemma 2. *Let $(A, \Box)$ be a ntqBa and $F \in \mathcal{F}_{\mathrm{p}}(A)$. Then (1) $\sim x \in F$ if and only if $x \in \sim F$; and (2) $\sim x \in F$ if and only if $x \notin g_A(F)$.*

Proof. For (1), assume $\sim x \in F$. Then $\sim\sim x \in \sim F$. By $\sim\sim x = x$, we have $x \in \sim F$. Assume $x \in \sim F$. Then $x = \sim y$ for some $y \in F$. Then $\sim x = \sim\sim y = y \in F$. Clearly (2) follows from (1) immediately. □

Lemma 3. *Let $(A, \Box)$ be a ntqBa, $F \in \mathcal{F}(A)$ and $I \in \mathcal{I}(A)$. Then (1) $F \in \mathcal{F}_{\mathrm{p}}(A)$ if and only if $\overline{F} \in \mathcal{I}_{\mathrm{p}}(A)$; (2) $F \in \mathcal{F}_{\mathrm{p}}(A)$ if and only if $\sim F \in \mathcal{I}_{\mathrm{p}}(A)$; and (3) if $F \cap I = \varnothing$, there exists $G \in \mathcal{F}_{\mathrm{p}}(A)$ with $F \subseteq G$ and $I \subseteq \overline{G}$.*

Proof. (1) holds obviously. For (2), assume $F \in \mathcal{F}_{\mathrm{p}}(A)$. Suppose $x, y \in \sim F$. Then $\sim x, \sim y \in F$ and so $\sim x \wedge \sim y = \sim(x \vee y) \in F$. Then $x \vee y \in \sim F$. Suppose $x \in \sim F$ and $y \leq x$. Then $\sim x \leq \sim y$ and so $\sim y \in F$. Hence $y \in \sim F$. Suppose $x \wedge y \in \sim F$. Then $\sim(x \wedge y) \in F$ and so $\sim x \vee \sim y \in F$. Then $\sim x \in F$ or $\sim y \subset F$. Then $x \in \sim F$ or $y \in \sim F$. Hence $\sim F \in \mathcal{I}_{\mathrm{p}}(A)$. The other direction is similar. The item (3) is shown by Zorn's Lemma (cf. e.g. [10]). □

Definition 4. Let $(A, \Box)$ be a ntqBa. The *dual frame* of A is defined as the structure $A_+ = (\mathcal{F}_{\mathrm{p}}(A), g_A, R_A)$ where (1) $g_A(F) = \overline{\sim F}$ and (2) FR_AG if and only if $\Box^{-1}(F) \subseteq G$ and $\Diamond^{-1}(\overline{F}) \subseteq \overline{G}$.

Note that the function g_A in Definition 4 is well-defined since $g_A(F) \in \mathcal{F}_{\mathrm{p}}(A)$ by Lemma 3 (1) and (2).

Proposition 1. *Let $\mathfrak{F} = (W, g, R)$ be an involutive frame and $(A, \Box)$ a ntqBa. Then (1)$\mathfrak{F}^+$ is a ntqBa, and (2)A_+ is an involutive frame.*

Proof.(1) Clearly $\sim_g\varnothing = W$. Let $X, Y \subseteq W$. Then $\sim_g\sim_g X = \sim_g\overline{g(X)} = \overline{g(\overline{g(X)})}$. Suppose $x \in X$. If $x \in g(\overline{g(X)})$, then $g(x) \notin g(X)$ which contradicts $x \in X$. Hence $X \subseteq \sim_g\sim_g X$. Suppose $x \notin g(\overline{g(X)})$. Then $g(x) \in g(X)$ and so $x = g^{-1}g(x) \in g^{-1}g(X) = X$. Moreover, $\sim_g(X \cup Y) = \overline{g(X \cup Y)} = \overline{g(X) \cup g(Y)} = \overline{g(X)} \cap \overline{g(Y)} = \sim_g X \cap \sim_g Y$. Clearly $\Box_R(X \cap Y) = \Box_R X \cap \Box_R Y$, $\Box_R X \subseteq X$ and $\Box_R X \subseteq \Box_R\Box_R X$. Hence $\mathfrak{F}^+$ is a tqBa. Now we show the condition (N). Assume $w \in \Box_R X \cap \Diamond_R Y$. Then $R(w) \subseteq X$ and $R(g(w)) \cap g(Y) \neq \varnothing$. Let $u \in R(g(w))$ and $u \in g(Y)$. By (F3), $g(u) \in R(w)$ and so $g(u) \in X$. Clearly $g(u) \in Y$. Then $g(u) \in X \cap Y$ and so $u \in g(X \cap Y)$. Hence $R(g(w)) \cap g(X \cap Y) \neq \varnothing$, i.e., $w \in \Diamond_R(X \cap Y)$. It follows that $\mathfrak{F}^+$ is a ntqBa.

(2) We show that (F1)–(F3) hold for A_+ and hence A_+ is involutive.

(F1) Let $F \in \mathcal{F}_{\mathrm{p}}(A)$. Suppose $x \in F$ and $x \in {\sim}(\overline{{\sim}F})$. Then $x = {\sim}y$ for some $y \notin {\sim}F$. Then ${\sim}x = y \in {\sim}F$ which is impossible. Then $F \subseteq g_A(g_A(F))$. Suppose $x \in g_A(g_A(F))$ and $x \notin F$. Then $x \notin {\sim}(\overline{{\sim}F})$ and ${\sim}x \notin {\sim}F$. Then ${\sim}x \in \overline{{\sim}F}$ and so $x \in {\sim}(\overline{{\sim}F})$ which is impossible. Then $g_A(g_A(F)) \subseteq F$.

(F2) Let $F \in \mathcal{F}_{\mathrm{p}}(A)$. If $\Box x \in F$, then $\Box x \leq x \in F$. If $\Diamond x \in \overline{F}$, then $x \leq \Diamond x$ and so $x \in \overline{F}$. Hence FR_AF. Assume FR_AG and GR_AH. Suppose $\Box x \in F$. Then $\Box x \leq \Box\Box x \in F$. Then $\Box x \in G$ and so $x \in H$. Suppose $\Diamond x \in \overline{F}$. Then $\Diamond\Diamond x \leq \Diamond x$ and so $\Diamond\Diamond x \in \overline{F}$. Then $\Diamond x \in \overline{G}$ and so $x \in \overline{H}$. Hence FR_AH.

(F3) Assume FR_AG. Suppose $\Box x \in g_A(F)$. By Lemma 2 (2), ${\sim}\Box x \notin F$. Then ${\sim}x \notin G$ and so $x \in g_A(G)$. Suppose $\Diamond x \in \overline{g_A(F)}$. Then $\Diamond x \in {\sim}F$. Then $\Diamond x = {\sim}y$ for some $y \in F$. Then $y = {\sim}{\sim}y = {\sim}\Diamond x = \Box{\sim}x \in F$. Hence ${\sim}x \in G$ and so $x \in {\sim}G$. Then $x \notin g_A(G)$ and so $x \in \overline{g_A(G)}$. □

Lemma 4. *Let* $(A, \Box)$ *be a ntqBa and* $F \in \mathcal{F}_{\mathrm{p}}(A)$. *Then* (1) $\Box x \in F$ *if and only if for all* $G \in \mathcal{F}_{\mathrm{p}}(A)$, FR_AG *implies* $x \in G$; *and* (2) $\Diamond x \in F$ *if and only if there exists* $G \in \mathcal{F}_{\mathrm{p}}(A)$ *with* $g_A(F)R_AG$ *and* $x \in g_A(G)$.

Proof. For (1), assume $\Box x \in F$. Suppose FR_AG. Then $x \in G$. Assume $\Box x \notin F$. Then $\Box^{-1}(F)$ is a filter and $x \notin \Box^{-1}(F)$. Then $\overline{\Box^{-1}(F)}$ is an ideal. By Lemma 3 (3), let $G \in \mathcal{F}_{\mathrm{p}}(A)$, $\Box^{-1}(F) \subseteq G$ and $\overline{\Box^{-1}(F)} \subseteq \overline{G}$. Then $x \notin G$. If $\Box y \in F$, then $y \in G$. Suppose $\Diamond y \in \overline{F}$ and $y \notin \overline{G}$. Then $y \notin \overline{\Box^{-1}(F)}$ and so $y \in \Box^{-1}(F)$. Then $\Box y \in F$ and so $\Box y \leq \Diamond y \in F$ which contradicts $\Diamond y \in \overline{F}$. Then $y \in \overline{G}$. Hence FR_AG. For (2), assume $g_A(F)R_AG$ and $x \in g_A(G)$. Then $x \notin {\sim}G$. Then ${\sim}x \in G$ and so $\Diamond{\sim}x \in g_A(F)$. Then $\Diamond{\sim}x \in {\sim}F$. Hence $\Box x \in F$ and so $\Box x \leq \Diamond x \in F$. Assume $\Diamond x \in F$. By Lemma 2 (2), $\Box{\sim}x \notin g_A(F)$. By (1), let $G \in \mathcal{F}_{\mathrm{p}}(A)$, $g_A(F)R_AG$ and ${\sim}x \notin G$. Then $x \notin {\sim}G$ and so $x \in g_A(G)$. □

Let $\mathfrak{F} = (W, g, R)$ and $\mathfrak{G} = (T, h, S)$ be involutive frames. A function $\alpha : W \to T$ is called an *embedding* if (i) α is one-one; (ii) $\alpha(g(w)) = h(\alpha(w))$ for all $w \in W$; and (iii) wRu if and only if $\alpha(w)S\alpha(u)$. An involutive frame $\mathfrak{F}$ is *embedded* into $\mathfrak{G}$ if there exists an embedding from $\mathfrak{F}$ to $\mathfrak{G}$. A ntqBa A is *embedded* into a ntqBa B if there exists a one-one homomorphism from A to B.

Theorem 2. *Let* $\mathfrak{F} = (W, g, R)$ *be an involutive frame and* $(A, \Box)$ *a ntqBa. Then* (1) $\mathfrak{F}$ *is embedded into* $(\mathfrak{F}^+)_+$; *and* (2) A *is embedded into* $(A_+)^+$.

Proof.(1) Let $\alpha : W \to \mathcal{F}_{\mathrm{p}}(\mathfrak{F}^+)$ be the function with $\alpha(w) = \{X \subseteq W : w \in X\}$ for all $w \in W$. Clearly each $\alpha(w) \in \mathcal{F}_{\mathrm{p}}(\mathfrak{F}^+)$. Now we show α is an embedding. If $w \neq u$, then $\alpha(w) \neq \alpha(u)$. Hence α is one-one. Now we show $\alpha(g(w)) = g_{\mathfrak{F}^+}(\alpha(w)) = {\sim}_g\alpha(w)$. Assume $X \in \alpha(g(w))$. Then $g(w) \in X$. Suppose $X \in {\sim}_g\alpha(w)$. Let $X = {\sim}_gY = \overline{g(Y)}$ for some $Y \in \alpha(w)$. Then $g(w) \notin g(Y)$ and $w \in Y$ which is impossible. Then $X \notin {\sim}_g\alpha(w)$. Hence $\alpha(g(w)) \subseteq g_{\mathfrak{F}^+}(\alpha(w))$. Assume $X \notin \alpha(g(w))$. Then $g(w) \notin X$ and so $w \notin g(X)$. Then $w \in {\sim}_gX$. Suppose $X \notin {\sim}_g\alpha(w)$. Then ${\sim}_gX \notin \alpha(w)$ and so $w \notin {\sim}_gX$ which yields a contradiction. Then $X \in \sim_g \alpha(w)$. Hence $g_{\mathfrak{F}^+}(\alpha(w)) \subseteq \alpha(g(w))$. It follows that $\alpha(g(w)) = g_{\mathfrak{F}^+}(\alpha(w))$. Now we show wRu if and only if $\alpha(w) R_{\mathfrak{F}^+} \alpha(u)$. Assume wRu. Suppose $\Box_R X \in \alpha(w)$. Then $w \in \Box_R X$ and so $R(w) \subseteq X$. By wRu, we have $R(u) \subseteq R(w)$. Then $R(u) \subseteq X$ and so $u \in \Box_R X$. Then $\Box_R X \in \alpha(u)$. Since $\Box_R X \subseteq X$, we have $X \in \alpha(u)$. Suppose $\Diamond_R X \notin \alpha(w)$. Then $w \notin \Diamond_R X$ and so $g(w) \notin g(\Diamond_R X)$. Then $g(w) \in \sim_g(\Diamond_R X)$. Then $\Diamond_R X \in \sim_g \alpha(g(w))$ and so $\sim_g(\Diamond_R X) \in \alpha(g(w))$. Then $\Box_R {\sim_g} X \in \alpha(g(w))$ and so $g(w) \in \Box_R {\sim_g} X$. By wRu, we have $g(w) R g(u)$. Then $g(u) \in \sim_g X$ and so $g(u) \notin g(X)$. Then $u \notin X$ and so $X \notin \alpha(u)$. Hence $\alpha(w) R_{\mathfrak{F}^+} \alpha(u)$. Assume $\alpha(w) R_{\mathfrak{F}^+} \alpha(u)$. Clearly $\Box_R R(w) = R(w) \in \alpha(w)$. Then $R(w) \in \alpha(u)$. Hence $u \in R(w)$.

(2) Let $\beta : A \to \mathcal{P}(\mathfrak{F}_p(A))$ be the function with $\beta(x) = \{F \in \mathfrak{F}_p(A) : x \in F\}$. If $x \neq y$, then $x \nleq y$ or $y \nleq x$. In each case there exists a prime filter in A containing one of x and y but not the other. Hence β is one-one. It suffices to show β is a homomorphism. First, we show $\beta(\sim x) = \sim_{g_A} \beta(x) = \overline{g_A(\beta(x))}$. Assume $F \in \beta(\sim x)$, i.e., $\sim x \in F$. Suppose $F \in g_A(\beta(x))$. Then $g_A(F) \in \beta(x)$, i.e., $x \in g_A(F) = \overline{\sim F}$. By $\sim x \in F$, we have $x \in \sim F$ which is impossible. Then $\beta(\sim x) \subseteq \sim_{g_A} \beta(x)$. Assume $F \notin g_A(\beta(x))$. Then $g_A(F) \notin \beta(x)$, i.e., $x \notin g_A(F) = \overline{\sim F}$. Suppose $F \notin \beta(\sim x)$. Then $\sim x \notin F$ and so $x \notin \sim F$ which is impossible. Then $\sim_{g_A} \beta(x) \subseteq \beta(x')$. Hence $\beta(\sim x) = \sim_{g_A} \beta(x)$. Obviously $\beta(x \wedge y) = \beta(x) \cap \beta(y)$, $\beta(x \vee y) = \beta(x) \cup \beta(y)$, $\beta(0) = \varnothing$ and $\beta(1) = \mathcal{P}(\mathcal{F}_p(A))$. Finally, we show $\beta(\Box x) = \Box_{R_A}(\beta(x))$. Assume $\Box x \in F$. Suppose $F R_A G$. Then $x \in G$ and so $G \in \beta(x)$. Hence $F \in \Box_{R_A}(\beta(x))$. Assume $F \in \Box_{R_A}(\beta(x))$. Suppose $\Box x \notin F$. By Lemma 4, there exists $G \in \mathfrak{F}_p(A)$ with $F R_A G$ and $x \notin G$. Then $G \in \beta(x)$ and so $x \in G$ which yields a contradiction. Hence $\beta(\Box x) = \Box_{R_A}(\beta(x))$. □

4 Some Completeness Results

In this section, we introduce the canonical involutive model for every ntqBl and show some completeness results. A *valuation* in an involutive frame $\mathfrak{F} = (W, g, R)$ is a function $V : \mathbb{P} \to \mathcal{P}(W)$. An *involutive model* is a tuple $\mathfrak{M} = (W, g, R, V)$ where (W, g, R) is an involutive frame and V is a valuation in it.

Definition 5. Let $\mathfrak{M} = (W, g, R, V)$ be an involutive model. The *truth set* $V(\varphi)$ of a formula φ in $\mathfrak{M}$ is defined as follows:

$$V(\bot) = \varnothing, V(\sim\varphi) = \sim_g V(\varphi), V(\varphi \wedge \psi) = V(\varphi) \cap V(\psi), V(\Box\varphi) = \Box_R V(\varphi).$$

We write $\mathfrak{M}, w \models \varphi$ if $w \in V(\varphi)$. A consequence $\varphi \vdash \psi$ is *true* in $\mathfrak{M}$ if $V(\varphi) \subseteq V(\psi)$. Let $\mathfrak{M} \models s$ stand for that a consequence s is true in $\mathfrak{M}$. Given a set of consequences T, we write $\mathfrak{M} \models T$ if $\mathfrak{M} \models s$ for all $s \in T$. A consequence $\varphi \vdash \psi$ is *valid* in an involutive frame $\mathfrak{F}$ (notation: $\varphi \models_{\mathfrak{F}} \psi$) if $V(\varphi) \subseteq V(\psi)$ for all valuations V in $\mathfrak{F}$. Let $\mathfrak{F} \models s$ stand for that a consequence s is valid in $\mathfrak{F}$. We write $\mathfrak{F} \models T$ if $\mathfrak{F} \models s$ for all $s \in T$. A consequence $\varphi \vdash \psi$ is *valid* in a class of involutive frames $\mathcal{S}$ (notation: $\varphi \models_{\mathcal{S}} \psi$) if $\varphi \models_{\mathfrak{F}} \psi$ for all $\mathfrak{F} \in \mathcal{S}$. The *Kripke-theory* of a class of involutive frames $\mathcal{S}$ is defined as the set of consequences $\mathsf{Th}_K(\mathcal{S}) = \{\varphi \vdash \psi : \varphi \models_{\mathcal{S}} \psi\}$. A formula φ is *$\mathcal{S}$-equivalent* to ψ (notation: $\varphi \equiv_{\mathcal{S}} \psi$) if $\varphi \models_{\mathcal{S}} \psi$ and $\psi \models_{\mathcal{S}} \varphi$. For a set of consequences Σ, let $\mathsf{Fr}(\Sigma) = \{\mathfrak{F} : \mathfrak{F} \models \Sigma\}$. A ntqBl L is *Kripke-complete* if $L = \mathsf{Th}_K(\mathsf{Fr}(L))$.

Let $\mathfrak{M} = (W, g, R, V)$ be an involutive model and $w \in W$. We write $w \models \varphi$ for $\mathfrak{M}, w \models \varphi$ if no confusion arises. By Definition 5, we have the following statements: (1) $w \models {\sim}\varphi$ if and only if $g(w) \not\models \varphi$; (2) $w \models \Box\varphi$ if and only if $u \models \varphi$ for all $u \in R(w)$; (3) $w \models \Diamond\varphi$ if and only if there exists $u \in W$ with $g(w)Ru$ and $g(u) \models \varphi$. Note that $\Diamond\varphi \equiv_{\mathcal{S}} {\sim}\Box{\sim}\varphi$ for every class of involutive frames $\mathcal{S}$.

Lemma 5. *If $\varphi \vdash_{\mathsf{wS4}} \psi$, then $\varphi \models_{\mathfrak{F}} \psi$ for every involutive frame $\mathfrak{F}$.*

Proof. Let $\mathfrak{F} = (W, g, R)$. Assume $\varphi \vdash_{\mathsf{wS4}} \psi$. All axioms schemes are valid and all rules preserve validity in $\mathfrak{F}$. Here we show only the validity of (N). Let V be any valuation in $\mathfrak{F}$ and $w \in W$. Assume $w \models \Box\varphi \wedge \Diamond\psi$. Then there exists $u \in W$ with $g(w)Ru$ and $g(u) \models \psi$. By $g(w)Ru$ and the condition (F3), $g(g(w))Rg(u)$ and so $wRg(u)$. By $w \models \Box\varphi$, $g(u) \models \varphi$. Then $g(u) \models \varphi \wedge \psi$. Hence $w \models \Diamond(\varphi \wedge \psi)$. □

Let Θ be a set of formulas. Let $\overline{\Theta} = \mathsf{For} \setminus \Theta$; ${\sim}\Theta = \{{\sim}\varphi : \varphi \in \Theta\}$; $\Box^{-1}(\Theta) = \{\varphi : \Box\varphi \in \Theta\}$ and $\Diamond^{-1}(\Theta) = \{\varphi : \Diamond\varphi \in \Theta\}$. Let $\mathcal{P}_{<\omega}(\Theta)$ be the set of all finite subsets of Θ. For each $\Delta \in \mathcal{P}_{<\omega}(\Theta)$, let $\bigvee \Delta$ and $\bigwedge \Delta$ be the disjunction and conjunction of formulas in Δ respectively. In particular, $\bigvee \varnothing = \bot$ and $\bigwedge \varnothing = \top$. Let L be a ntqBl. A formula φ is *L-derivable* from Θ (notation: $\Theta \vdash_L \varphi$) if there exists $\Delta \in \mathcal{P}_{<\omega}(\Theta)$ with $\bigwedge \Delta \vdash_L \varphi$. A set Θ is *L-consistent* if $\Theta \not\vdash_L \bot$. A set Θ is an *L-filter* if Θ is closed under $\vdash_L$, i.e., if $\Theta \vdash_L \varphi$, then $\varphi \in \Theta$. The L-filter *generated* by a set of formulas Θ is denoted by $[\Theta)_L$. An *L-ideal* is a set of formulas Σ such that $\psi \in \Sigma$ whenever $\psi \vdash_L \bigvee \Delta$ for some $\Delta \in \mathcal{P}_{<\omega}(\Sigma)$. The L-ideal *generated* by a set of formulas Θ is denoted by $(\Theta]_L$. If $\Theta = \{\varphi\}$, we write $[\varphi)_L$ and $(\varphi]_L$. An L-filter or L-ideal is *proper* if it is not equal to For. A proper L-filter Θ is *prime* if $\varphi \vee \psi \in \Theta$ implies $\varphi \in \Theta$ or $\psi \in \Theta$. A proper L-ideal Σ is *prime* if $\varphi \wedge \psi \in \Sigma$ implies $\varphi \in \Sigma$ or $\psi \in \Sigma$. Let $\mathcal{F}(L)$ and $\mathcal{I}(L)$ be sets of all L-filters and L-ideals respectively. Let $\mathcal{F}_{\mathrm{p}}(L)$ and $\mathcal{I}_{\mathrm{p}}(L)$ be sets of all prime L-filters and prime L-ideals respectively.

Lemma 6. *Let $\Theta \in \mathcal{F}(L)$ and $\Sigma \in \mathcal{I}(L)$. The following hold:*

(1) *if $\varphi \notin \Theta$, there exists $\Delta \in \mathcal{F}_{\mathrm{p}}(L)$ with $\Theta \subseteq \Delta$ and $\varphi \notin \Delta$.*
(2) *if $\varphi \not\vdash_L \psi$, there exists $\Delta \in \mathcal{F}_{\mathrm{p}}(L)$ with $\varphi \in \Delta$ and $\psi \notin \Delta$.*
(3) *$\Sigma \in \mathcal{I}_{\mathrm{p}}(L)$ if and only if $\overline{\Sigma} \in \mathcal{F}_{\mathrm{p}}(L)$.*
(4) *$\Theta \in \mathcal{F}_{\mathrm{p}}(L)$ if and only if $({\sim}\Theta]_L \in \mathcal{I}_{\mathrm{p}}(L)$.*
(5) *if $\Theta \cap \Sigma = \varnothing$, there exists $\Omega \in \mathcal{F}_{\mathrm{p}}(L)$ with $\Theta \subseteq \Omega$ and $\Sigma \subseteq \overline{\Omega}$.*

Proof. Items (1) and (5) are obtained by Zorn's lemma. For (2), assume $\varphi \not\vdash_L \psi$. Then $\psi \notin [\varphi)_L$. By (1), there exists $\Delta \in \mathcal{F}_{\mathrm{p}}(L)$ with $[\varphi)_L \subseteq \Delta$ and $\psi \notin \Delta$. For

(3), assume $\Sigma \in \mathcal{I}_{\mathrm{p}}(L)$. Clearly $\bot \notin \overline{\Sigma}$. Suppose $\overline{\Sigma} \vdash_L \varphi$. If $\top \vdash_L \varphi$, then $\varphi \notin \Sigma$. Let $\bigwedge \Gamma \vdash_L \varphi$ for some $\varnothing \neq \Gamma \in \mathcal{P}_{<\omega}(\overline{\Sigma})$. Suppose $\varphi \in \Sigma$. Then $\bigwedge \Gamma \in \Sigma$. Hence $\psi \in \Sigma$ for some $\psi \in \Gamma$ which contradicts $\Gamma \subseteq \overline{\Sigma}$. Then $\overline{\Sigma}$ is a proper L-filter. Suppose $\psi \vee \chi \notin \Sigma$. Then $\psi \notin \Sigma$ or $\chi \notin \Sigma$. Hence $\overline{\Sigma} \in \mathcal{F}_{\mathrm{p}}(L)$. The other direction is shown similarly. For (4), assume $\Theta \in \mathcal{F}_{\mathrm{p}}(L)$. Clearly ${\sim}\bot \notin ({\sim}\Theta]_L$. If $\psi \vdash_L \bot$, then ${\sim}\psi \in \Theta$ and so $\psi \in ({\sim}\Theta]_L$. Suppose $\psi \vdash_L \bigvee {\sim}\Delta$ for some $\varnothing \neq \Delta \in \mathcal{P}_{<\omega}(\Theta)$. By (CP), $\bigwedge \Delta \vdash_L {\sim}\psi$ and so ${\sim}\psi \in \Theta$. Then $\psi \in ({\sim}\Theta]_L$. Hence $({\sim}\Theta]_L$ is a proper L-ideal. Suppose $\varphi \wedge \psi \in ({\sim}\Theta]_L$. Then $\varphi \wedge \psi \vdash_L \bigvee {\sim}\Delta$ for some $\Delta \in \mathcal{P}_{<\omega}(\Theta)$. By (CP), $\bigwedge \Delta \vdash_L {\sim}(\varphi \wedge \psi)$ and so ${\sim}(\varphi \wedge \psi) \in \Theta$. By ${\sim}(\varphi \wedge \psi) \vdash_L {\sim}\varphi \vee {\sim}\psi$, we have ${\sim}\varphi \vee {\sim}\psi \in \Theta$. Then ${\sim}\varphi \in \Theta$ or ${\sim}\psi \in \Theta$. Then $\varphi \in ({\sim}\Theta]_L$ or $\psi \in ({\sim}\Theta]_L$. Hence $({\sim}\Theta]_L \in \mathcal{I}_{\mathrm{p}}(L)$. □

Definition 6. Let L be a ntqBl. The *canonical involutive model* for L is defined as the structure $\mathfrak{M}^L = (\mathcal{F}_{\mathrm{p}}(L), g^L, R^L, V^L)$ where

(1) $g^L(\Theta) = \overline{({\sim}\Theta]_L}$.
(2) $\Theta R^L \Sigma$ if and only if $\Box^{-1}(\Theta) \subseteq \Sigma$ and $\Diamond^{-1}(\overline{\Theta}) \subseteq \overline{\Sigma}$.
(3) $V^L(p) = \{\Theta \in \mathcal{F}_{\mathrm{p}}(L) : p \in \Theta\}$ for each $p \in \mathbb{P}$.

The structure $\mathfrak{F}^L = (\mathcal{F}_{\mathrm{p}}(L), g^L, R^L)$ is the *canonical involutive frame* for L.

Lemma 7. *Let $\Theta, \Sigma \in \mathcal{F}_{\mathrm{p}}(L)$. Then (1) ${\sim}\varphi \in \Theta$ if and only if $\varphi \notin g^L(\Theta)$; (2) g^L is an involution on $\mathcal{F}_{\mathrm{p}}(L)$; and (3) if $\Theta R^L \Sigma$, then $g^L(\Theta) R^L g^L(\Sigma)$.*

Proof.(1) Assume ${\sim}\varphi \in \Theta$. Then $\varphi \in ({\sim}\Theta]_L$ and so $\varphi \notin g^L(\Theta)$. Assume $\varphi \notin g^L(\Theta)$. Then $\varphi \in ({\sim}\Theta]_L$ and so ${\sim}\varphi \in \Theta$.

(2) Let $\Theta \in \mathcal{F}_{\mathrm{p}}(L)$. Suppose $\varphi \in \Theta$ and $\varphi \in ({\sim}\overline{({\sim}\Theta]_L}]$. Then there exists $\varnothing \neq \Delta \in \mathcal{P}_{<\omega}({\sim}\overline{({\sim}\Theta]_L})$ with $\varphi \vdash_L \bigvee \Delta$. Then $\bigvee \Delta \in \Theta$. Then $\psi \in \Theta$ for some $\psi \in \Delta$. Let $\psi = {\sim}\chi$ and $\chi \notin ({\sim}\Theta]_L$. Then ${\sim}\chi \in \Theta$ and so $\chi \in ({\sim}\Theta]$ which yields a contradiction. Hence $\Theta \subseteq g^L(g^L(\Theta))$. Suppose $\varphi \in g^L(g^L(\Theta))$ and $\varphi \notin \Theta$. If ${\sim}\varphi \in ({\sim}\Theta]_L$, then $\varphi \in \Theta$. Hence ${\sim}\varphi \in \overline{({\sim}\Theta]_L}$ and so $\varphi \in ({\sim}\overline{({\sim}\Theta]_L}]$ which contradicts $\varphi \in g^L(g^L(\Theta))$. Then $g^L(g^L(\Theta)) \subseteq \Theta$. Hence $g^L(g^L(\Theta)) = \Theta$.

(3) Assume $\Theta R^L \Sigma$. Suppose $\Box\varphi \in g^L(\Theta)$. By (1), ${\sim}\Box\varphi \notin \Theta$. Then $\Diamond{\sim}\varphi \notin \Theta$ and so ${\sim}\varphi \notin \Sigma$. By (1), $\varphi \in g^L(\Sigma)$. Suppose $\Diamond\varphi \notin g^L(\Theta)$. By (1), ${\sim}\Diamond\varphi \in \Theta$. Then $\Box{\sim}\varphi \in \Theta$ and so ${\sim}\varphi \in \Sigma$. By (1), $\varphi \notin g^L(\Sigma)$. □

Lemma 8. *Let $\Theta \in \mathcal{F}_{\mathrm{p}}(L)$. Then (1) $\Box\varphi \in \Theta$ if and only if for all $\Sigma \in \mathcal{F}_{\mathrm{p}}(L)$, $\Theta R^L \Sigma$ implies $\varphi \in \Sigma$; and (2) $\Diamond\varphi \in \Theta$ if and only if there exists $\Sigma \in \mathcal{F}_{\mathrm{p}}(L)$ with $g^L(\Theta) R^L \Sigma$ and $\varphi \in g^L(\Sigma)$.*

Proof. For (1), assume $\Box\varphi \in \Theta$ and $\Theta R^L \Sigma$. Then $\varphi \in \Sigma$. Assume $\Box\varphi \notin \Theta$. Then $\varphi \notin \Box^{-1}(\Theta)$. Clearly $\Box^{-1}(\Theta)$ is an L-filter. By Lemma 6 (5), there exists $\Sigma \in \mathcal{F}_{\mathrm{p}}(L)$ with $\Box^{-1}(\Theta) \subseteq \Sigma$ and $\overline{\Box^{-1}(\Theta)} \subseteq \overline{\Sigma}$. Then $\varphi \notin \Sigma$. If $\Box\psi \in \Theta$, then $\psi \in \Sigma$. Suppose $\Diamond\psi \in \overline{\Theta}$ and $\psi \notin \overline{\Sigma}$. Then $\psi \notin \overline{\Box^{-1}(\Theta)}$ and so $\psi \in \Box^{-1}(\Theta)$. Then $\Box\psi \in \Theta$. By $\Box\psi \vdash_L \Diamond\psi$, we have $\Diamond\psi \in \Theta$ which contradicts $\Diamond\psi \in \overline{\Theta}$. Hence $\Theta R^L \Sigma$. For (2), assume $g^L(\Theta) R^L(\Sigma)$ and $\varphi \in g^L(\Sigma)$. Then $\varphi \notin ({\sim}\Sigma]_L$. Then ${\sim}\varphi \notin \Sigma$. By Lemma 7 (1), $\Diamond{\sim}\varphi \in g^L(\Theta)$. Then $\Box\varphi \in \Theta$. By $\Box\varphi \vdash_L \Diamond\varphi$, we have $\Diamond\varphi \in \Theta$. Assume $\Diamond\varphi \in \Theta$. By Lemma 7 (1), $\sim\Diamond\varphi \notin g^L(\Theta)$ and so $\Box\sim\varphi \notin g^L(\Theta)$. By (1), there exists $\Sigma \in \mathcal{F}_{\mathrm{p}}(L)$ with $g^L(\Theta)R^L\Sigma$ and $\sim\varphi \notin \Sigma$. Then $\varphi \in (\sim\Sigma]_L$ and so $\varphi \in g^L(\Theta)$. □

Lemma 9. *For every formula φ, $\mathfrak{M}^L, \Theta \models \varphi$ if and only if $\varphi \in \Theta$.*

Proof. The proof proceeds by induction on the complexity of φ. Atomic cases are trivial. The cases $\varphi = \psi \odot \chi$ with $\odot \in \{\wedge, \vee\}$ are shown by induction hypothesis. Let $\varphi := \sim\psi$. Then $\Theta \models \sim\psi$ if and only if $g^L(\Theta) \not\models \psi$. By Lemma 7 (1), $g^L(\Theta) \not\models \psi$ if and only if $\sim\psi \in \Theta$. Let $\varphi := \Box\psi$. Assume $\Box\psi \in \Theta$ and $\Theta R^L \Sigma$. Then $\psi \in \Sigma$. By induction hypothesis, $\Sigma \models \psi$. Hence $\Theta \models \Box\psi$. Assume $\Box\psi \notin \Theta$. By Lemma 8, there exists $\Sigma \in \mathcal{F}_{\mathrm{p}}(L)$ with $\Theta R^L \Sigma$ and $\psi \notin \Sigma$. By induction hypothesis, $\Sigma \not\models \psi$. Hence $\Theta \not\models \Box\psi$. □

Theorem 3 (Completeness). wS4 *is Kripke-complete.*

Proof. Assume $\varphi \not\vdash_{\mathsf{wS4}} \psi$. By Lemma 6 (2), there exists $\Theta \in \mathcal{F}_{\mathrm{p}}(\mathsf{wS4})$ such that $\varphi \in \Theta$ and $\psi \notin \Theta$. By Lemma 9, $\mathfrak{M}^{\mathsf{wS4}}, \Theta \models \varphi$ and $\mathfrak{M}^{\mathsf{wS4}}, \Theta \not\models \psi$. Clearly $\mathfrak{F}^{\mathsf{wS4}} \models \mathsf{wS4}$. Hence wS4 is Kripke-complete. □

A ntqBl L is called *canonical* if $\mathfrak{F}^L \models L$. Every canonical ntqBl is obviously Kripke-complete. A celebrated result in classical modal logic is that all Sahlqvist formulas are first-order definable and all Sahlqvist logics are canonical (cf. e.g. [3]). One could generalize this result to ntqBls. Here we give some examples.

Lemma 10. *For every involutive frame $\mathfrak{F} = (W, g, R)$, the following hold:*

(1) *$p \models_{\mathfrak{F}} \Box\Diamond p$ if and only if $\forall w, u \in W(wRu \Rightarrow uRw)$.*
(2) *$\Diamond p \models_{\mathfrak{F}} \Box\Diamond p$ if and only if $\forall w, u \in W(wRu \ \&\ wRv \Rightarrow uRv)$.*
(3) *$\top \models_{\mathfrak{F}} \Box p \vee \neg\Box p$ if and only if $\forall w, u \in W(wRu \Rightarrow g(w)Ru)$.*

Proof.(1) Assume $\forall w, u \in W(wRu \Rightarrow uRw)$. Let V be a valuation in $\mathfrak{F}$ and $w \models p$. Suppose wRu. Then uRw and so $g(u)Rg(w)$ by (F3). By $w \models p$ and $w = g(g(w))$, we have $g(g(w)) \models p$. Then $u \models \Diamond p$. Hence $w \models \Box\Diamond p$. Assume $p \models_{\mathfrak{F}} \Box\Diamond p$ and wRu. Let U be a valuation in $\mathfrak{F}$ such that $U(p) = \{w\}$. Then $w \models p$ and so $w \models \Box\Diamond p$. Then $u \models \Diamond p$. There exists $v \in R(g(u))$ with $g(v) \models p$. Then $g(v) = w$ and so $g(u)Rv$. By (F3), $g(g(u))Rg(v)$ and so uRw.

(2) Assume $\forall w, u \in W(wRu \ \&\ wRv \Rightarrow uRv)$. Let V be a valuation in $\mathfrak{F}$ and $w \models \Diamond p$. Suppose wRu. There exists $v \in W$ with $g(w)Rv$ and $g(v) \models p$. By wRu and (F3), we have $g(w)Rg(u)$. By the assumption, $g(u)Rv$. Then $u \models \Diamond p$. Hence $w \models \Box\Diamond p$. Assume $\Diamond p \models_{\mathfrak{F}} \Box\Diamond p$ and $u, v \in R(w)$. By (F3), $g(w)Rg(v)$. Let U be a valuation in $\mathfrak{F}$ such that $U(p) = \{v\}$. Then $v = g(g(v))$ and so $w \models \Diamond p$. Then $w \models \Box\Diamond p$. By wRu, we have $u \models \Diamond p$. Then there exists $t \in W$ with $g(u)Rt$ and $g(t) \models p$. Then $g(t) = v$ and so $t = g(v)$. Then $g(u)Rg(v)$. By (F3), uRv.

(3) Assume $\forall w, u \in W(wRu \Rightarrow g(w)Ru)$. Let V be a valuation in $\mathfrak{F}$ and $w \not\models \Box p$. Then there exists $u \in W$ with wRu and $u \not\models p$. By the assumption, $g(w)Ru$. By $u \not\models p$, we have $g(w) \not\models \Box p$. Then $w \models \sim\Box p$. Assume $\top \models_{\mathfrak{F}} \Box p \vee \neg\Box p$ and wRu. Let U be a valuation in $\mathfrak{F}$ such that $U(p) = W \setminus \{u\}$. Then $u \not\models p$ and so $w \not\models \Box p$. By the assumption, $w \models \sim\Box p$. Then $g(w) \not\models \Box p$. Then there exists $v \in W$ with $g(w)Rv$ and $v \not\models p$. Then $u = v$ and so $g(w)Ru$. □

Let $\mathsf{wS5} = \mathsf{wS4} \oplus p \vdash \Box\Diamond p = \mathsf{wS4} \oplus \Diamond p \vdash \Box\Diamond p$ and $\mathsf{wS4C} = \mathsf{wS4} \oplus \top \vdash \Box p \vee \sim\Box p$. Here note that algebras for wS4C are very close to the variety of classical modal De Morgan algebras given in [6] where the involutive function g in a De Morgan frame is required to be antitone, i.e., $x \leq y$ implies $g(y) \leq g(x)$.

Theorem 4. *wS5 and wS4C are canonical and hence Kripke-complete.*

Proof. Let $\mathfrak{F} = (W, g, R)$ be the canonical involutive frame and $\mathfrak{M} = (W, g, R, V)$ be the canonical involutive model for $\mathfrak{F}^L$ with $L \in \{\mathsf{wS5}, \mathsf{wS4C}\}$.

(1) $p \models_{\mathfrak{F}^{\mathsf{wS5}}} \Box\Diamond p$. Assume $\Theta R_{\mathsf{wS5}} \Sigma$. By Lemma 7 (3), $g^{\mathsf{wS5}}(\Theta) R^{\mathsf{wS5}} g^{\mathsf{wS5}}(\Sigma)$. Suppose $\varphi \notin \Theta$. Clearly $\Diamond\Box\varphi \vdash_{\mathsf{wS5}} \varphi$. Then $\Diamond\Box\varphi \notin \Theta$. By Lemma 8 (2), $\Box\varphi \notin \Sigma$. Suppose $\Diamond\varphi \notin \Theta$. By Lemma 8 (2), $\varphi \notin \Sigma$. Hence $\Sigma R_{\mathsf{wS5}} \Theta$.
(2) $\top \models_{\mathfrak{F}^{\mathsf{wS4C}}} \Box p \vee \sim\Box p$. Assume $\Theta R_{\mathsf{wS4C}} \Sigma$. Suppose $\Box\varphi \in g^{\mathsf{wS4C}}(\Theta)$. By Lemma 7 (1), $\sim\Box\varphi \notin \Theta$. Clearly $\top \vdash_{\mathfrak{F}^{\mathsf{wS4C}}} \Box\varphi \vee \sim\Box\varphi$. Then $\Box\varphi \in \Theta$. Hence $\varphi \in \Sigma$. Suppose $\Diamond\varphi \notin g^{\mathsf{wS4C}}(\Theta)$. By the assumption and Lemma 7 (3), we have $g^{\mathsf{wS4C}}(\Theta) R^{\mathsf{wS4C}} g^{\mathsf{wS4C}}(\Sigma)$. By Lemma 8 (2), $\varphi \notin \Sigma$. Hence $g^{\mathsf{wS4C}}(\Theta) R_{\mathsf{wS4C}} \Sigma$.

□

5 Finite Model Property

In this section, we show some results on the finite model property of tqBls. We extend the filtration method in classical modal logic to tqBls. Recall some results from e.g. [3,9]. A tqBl L has the *finite model property* (FMP) if for every consequence $s \notin L$ there exists a finite involutive model $\mathfrak{M} \models L$ such that $\mathfrak{M} \not\models s$.

Let $\Sigma \neq \varnothing$ be a set of formulas. We say that Σ is *subformula-closed* if every subformula of a member in Σ belongs to Σ. Let Σ^* be the $\sim$-closure of Σ, i.e., if $\varphi \in \Sigma^*$, then $\sim\varphi \in \Sigma^*$. For any involutive model $\mathfrak{M} = (W, g, R, V)$, the binary relation $\leftrightsquigarrow_{\Sigma^*}$ on W is defined by setting

$$w \leftrightsquigarrow_{\Sigma^*} u \text{ if and only if } \forall\varphi \in \Sigma^*(\mathfrak{M}, w \models \varphi \Leftrightarrow \mathfrak{M}, u \models \varphi).$$

Then $\leftrightsquigarrow_{\Sigma^*}$ is an equivalence relation. Let $[w]_{\Sigma^*} = \{u \in W : w \leftrightsquigarrow_{\Sigma^*} u\}$ and $W_{\Sigma^*} = \{[w]_{\Sigma^*} : w \in W\}$. We write $[w]$ for $[w]_{\Sigma^*}$ if no confusion arises. A set of formulas Σ is *finitely based* in $\mathfrak{M}$, if there exists a finite subset $\Theta \subseteq \Sigma$ such that for all $\varphi \in \Sigma$ there exists $\psi \in \Theta$ such that $\forall\varphi \in \Sigma(\mathfrak{M}, w \models \varphi \Leftrightarrow \mathfrak{M}, u \models \varphi)$. Such a finite subset Θ is called a *finite base* for Σ. If Θ is finite, then $|W_\Sigma| \leq 2^{|\Theta|}$. Note that, if Σ is finitely based, then Σ^* is also finitely based.

Definition 7. Let $\mathfrak{M} = (W, g, R, V)$ be an involutive model and $\Sigma \neq \varnothing$ a subformula-closed set of formulas. The *filtration* of $\mathfrak{M}$ through Σ^* is defined as the involutive model $\mathfrak{M}_{\Sigma^*} = (W_{\Sigma^*}, g_{\Sigma^*}, R_{\Sigma^*}, V_{\Sigma^*})$ where (C1) $g_{\Sigma^*}([w]) = [g(w)]$; (C2) $[w]R_{\Sigma^*}[u]$ if and only if for any $\Box\varphi \in \Sigma^*$, (i) if $\mathfrak{M}, w \models \Box\varphi$, then $\mathfrak{M}, u \models \Box\varphi$ and (ii) if $\mathfrak{M}, u \models {\sim}\Box\varphi$, then $\mathfrak{M}, w \models {\sim}\Box\varphi$; and (C3) $V_{\Sigma^*}(p) = \{[w] : w \in V(p)\}$ for every $p \in \mathbb{P}$. Let $\mathfrak{F}_{\Sigma^*} = (W_{\Sigma^*}, g_{\Sigma^*}, R_{\Sigma^*})$.

Lemma 11. *Let $\mathfrak{M} = (W, g, R, V)$ be involutive and $\Sigma \neq \varnothing$ subformula-closed. Then (1) g_{Σ^*} is an involution on W_{Σ^*}; (2) if wRu, then $[w]R_{\Sigma^*}[u]$; (3) if $[w]R_{\Sigma^*}[u]$, then $[g(w)]R_{\Sigma^*}[g(u)]$; (4) R_{Σ^*} is reflexive and transitive.*

Proof. For (1), assume $u \in [w]$. Then $w \leftrightsquigarrow_{\Sigma^*} u$. Let $\varphi \in \Sigma^*$. Then $g(w) \models \varphi$ if and only if $w \not\models {\sim}\varphi$ and $g(u) \models \varphi$ if and only if $u \not\models {\sim}\varphi$. Hence $g(w) \leftrightsquigarrow_{\Sigma^*} g(u)$. Then g_{Σ^*} is well-defined. Moreover, $g_{\Sigma^*}(g_{\Sigma^*}([w])) = [g(g(w))] = [w]$. Then g_{Σ^*} is an involution. For (2), assume wRu. Let $\Box\varphi \in \Sigma^*$ and $w \models \Box\varphi$. Then $w \models \Box\Box\varphi$ and so $u \models \Box\varphi$. Hence $[w]R_{\Sigma^*}[u]$. For (3), assume $[w]R_{\Sigma^*}[u]$. Let $\Box\varphi \in \Sigma^*$. Suppose $g(u) \not\models \Box\varphi$. Then $u \models {\sim}\Box\varphi$. Note that ${\sim}\Box\varphi \in \Sigma^*$. Then $w \models {\sim}\Box\varphi$. Hence $g(w) \not\models \Box\varphi$. Suppose $g(u) \models {\sim}\Box\varphi$. Then $u \not\models \Box\varphi$ and so $w \not\models \Box\varphi$. Then $g(w) \models {\sim}\Box\varphi$. Hence $[g(w)]R_{\Sigma^*}[g(u)]$. For (4), R_{Σ^*} is reflexive by (2). The transitivity of R_{Σ^*} follows from the definition. □

Lemma 12. *Let $\mathfrak{M} = (W, g, R, V)$ be involutive and $\Sigma \neq \varnothing$ subformula-closed. For every $\varphi \in \Sigma^*$ and $w \in W$, $\mathfrak{M}, w \models \varphi$ if and only if $\mathfrak{M}_{\Sigma^*}, [w] \models \varphi$.*

Proof. The proof proceeds by induction on the complexity of φ. Atomic cases are trivial. The cases of $\wedge$ and $\vee$ are shown by induction hypothesis. Suppose $\varphi = {\sim}\psi$. Then $w \models {\sim}\psi$ if and only if $g(w) \not\models \psi$ if and only if $[g(w)] \not\models \psi$ if and only if $[w] \models {\sim}\psi$. Suppose $\varphi = \Box\psi$. Assume $w \models \Box\psi$ and $[w]R_{\Sigma^*}[u]$. Then $u \models \Box\psi$ and so $u \models \psi$. By induction hypothesis, $[u] \models \psi$. Hence $[w] \models \Box\varphi$. Assume $[w] \models \Box\psi$ and $wR_{\Sigma^*}u$. By Lemma 11 (2), $[w]R_{\Sigma^*}[u]$. Then $[u] \models \psi$. By induction hypothesis, $u \models \psi$. Hence $w \models \Box\varphi$. □

Theorem 5. wS4 *has the FMP.*

Proof. Assume $\varphi \not\vdash_{\mathsf{wS4}} \psi$. Let $\mathfrak{M} = (W, g, R, V)$ be the canonical involutive model for wS4. Then $\mathfrak{M}, \Theta \models \varphi$ and $\mathfrak{M}, \Theta \not\models \psi$ for some $\Theta \in \mathcal{F}_{\mathrm{p}}(wS4)$. Let $\Sigma = sub(\varphi) \cup sub(\psi)$. Clearly $\Sigma \cup {\sim}\Sigma$ is a finite base for Σ^* in $\mathfrak{M}$. Then W_{Σ^*} is finite. By Lemma 12, $\mathfrak{M}_{\Sigma^*}, [\Theta] \models \varphi$ and $\mathfrak{M}_{\Sigma^*}, [\Theta] \not\models \psi$. By Lemma 11, $\mathfrak{M}_{\Sigma^*}$ is an involuntive model for wS4. □

Lemma 13. *Let $\mathfrak{F} = (W, g, R)$ and $\mathfrak{M} = (\mathfrak{F}, V)$ be involutive model and $\Sigma \neq \varnothing$ subformula-closed. If $\mathfrak{F} \models$ wS4C, then $\mathfrak{F}_{\Sigma^*} \models$ wS4C.*

Proof. Assume $\mathfrak{F} \models$ wS4C. Suppose $[w]R_{\Sigma^*}[u]$. Assume $\Box\varphi \in \Sigma^*$ and $g(w) \models \Box\varphi$. Then $w \not\models {\sim}\Box\varphi$ and so $u \not\models {\sim}\Box\varphi$. By $u \models \Box\varphi \vee {\sim}\Box\varphi$, we have $u \models \Box\varphi$. Assume $u \models {\sim}\Box\varphi$. Then $g(u) \not\models \Box\varphi$. By $[w]R_{\Sigma^*}[u]$ and Lemma 11 (3), $[g(w)]R_{\Sigma^*}[g(u)]$. Then $g(w) \not\models \Box\varphi$. By $g(w) \models \Box\varphi \vee {\sim}\Box\varphi$, we have $g(w) \models {\sim}\Box\varphi$. Hence $g(w)R_{\Sigma^*}[u]$. By Lemma 10, $\mathfrak{F}_{\Sigma^*} \models$ wS4C.

Theorem 6. wS4C *has the FMP.*

Proof. Assume $\varphi \nvdash_{\mathsf{wS4C}} \psi$. Let $\mathfrak{M} = (W, g, R, V)$ be the canonical involutive model for wS4C. Let $\mathfrak{F} = (W, g, R)$. Then $\mathfrak{F} \models \mathsf{wS4C}$ by Theorem 4. Let $\Sigma = sub(\varphi) \cup sub(\psi)$. By the same proof of Theorem 5, $\varphi \vdash \psi$ is not true in the finite involutive model $\mathfrak{M}_{\Sigma^*}$. By Lemma 13, $\mathfrak{F}_{\Sigma^*} \models \mathsf{wS4C}$. □

Finally, we show the FMP of wS5. In this case the definition of filtration is slightly changed. Let $\mathfrak{M} = (W, g, R, V)$ be involutive and $\Sigma \neq \varnothing$ subformula-closed. The wS5-*filtration* of $\mathfrak{M}$ through Σ^* is defined as the involutive model $\mathfrak{M}^{\mathsf{wS5}}_{\Sigma^*} = (W_{\Sigma^*}, g_{\Sigma^*}, R^{\mathsf{wS5}}_{\Sigma^*}, V_{\Sigma^*})$ where (C1) and (C3) in Definition 7 hold, and (C2) is replaced by the condition (C2$^{\mathsf{wS5}}$): $[w]R^{\mathsf{wS5}}_{\Sigma^*}[u]$ if and only if for any $\Box\varphi \in \Sigma^*$, (i) $\mathfrak{M}, w \models \Box\varphi$ if and only if $\mathfrak{M}, u \models \Box\varphi$, and (i) $\mathfrak{M}, w \models {\sim}\Box\varphi$ if and only if $\mathfrak{M}, u \models {\sim}\Box\varphi$.

Lemma 14. *Let $\mathfrak{M} = (W, g, R, V)$ be an involutive model based on a frame $\mathfrak{F} = (W, g, R)$ such that $\mathfrak{F} \models \mathsf{wS5}$. Let $\Sigma \neq \varnothing$ be a subformula-closed set of formulas. Then* (1) *if wRu, then $[w]R^{\mathsf{wS5}}_{\Sigma^*}[u]$;* (2) *$[w]R^{\mathsf{wS5}}_{\Sigma^*}[u]$ if and only if $[g(w)]R^{\mathsf{wS5}}_{\Sigma^*}[g(u)]$;* (3) *$R^{\mathsf{wS5}}_{\Sigma^*}$ is an equivalence relation.*

Proof. For (1), assume wRu. Let $\Box\varphi \in \Sigma^*$. Since R is an equivalence relation, we have (i) $w \models \Box\varphi$ if and only if $u \models \Box\varphi$ and (ii) $w \models {\sim}\Box\varphi$ if and only if $u \models {\sim}\Box\varphi$. Hence $[w]R_{\Sigma^*}[u]$. For (2), assume $[w]R^{\mathsf{wS5}}_{\Sigma^*}[u]$. Let $\Box\varphi \in \Sigma^*$. Suppose $g(w) \models \Box\varphi$. Then $w \models {\sim}\Box\varphi$ and so $u \models {\sim}\Box\varphi$. Then $g(u) \models \Box\varphi$. Similarly, if $g(u) \models \Box\varphi$, then $g(w) \models \Box\varphi$. Suppose $g(w) \models {\sim}\Box\varphi$. Then $w \not\models \Box\varphi$ and so $u \not\models \Box\varphi$. Then $g(u) \models {\sim}\Box\varphi$. Similarly, if $g(u) \models {\sim}\Box\varphi$, then $g(w) \models {\sim}\Box\varphi$. Hence $[g(w)]R^{\mathsf{wS5}}_{\Sigma^*}[g(u)]$. Conversely, by a similar proof, if $[g(w)]R^{\mathsf{wS5}}_{\Sigma^*}[g(u)]$, then $[w]R^{\mathsf{wS5}}_{\Sigma^*}[u]$. Note that (3) follows from the definition of $R^{\mathsf{wS5}}_{\Sigma^*}$. □

Lemma 15. *Let $\mathfrak{M} = (W, g, R, V)$ be an involutive model based on a frame $\mathfrak{F} = (W, g, R)$ such that $\mathfrak{F} \models \mathsf{wS5}$. Let $\Sigma \neq \varnothing$ be a subformula-closed set of formulas. For every $\varphi \in \Sigma^*$ and $w \in W$, $\mathfrak{M}, w \models \varphi$ if and only if $\mathfrak{M}^{\mathsf{wS5}}_{\Sigma^*}, [w] \models \varphi$.*

Proof. The proof is similar to Lemma 12. Note that the case $\varphi = \Box\psi$ is trivial by Lemma 14. Details are omitted. □

Theorem 7. wS5 *has the FMP.*

Proof. Assume $\varphi \nvdash_{\mathsf{wS5}} \psi$. Let $\mathfrak{M} = (W, g, R, V)$ be the canonical involutive model for wS5. By the same proof of Theorem 5, using Lemma 15, $\varphi \vdash \psi$ is not true in the finite involutive model $\mathfrak{M}^{\mathsf{wS5}}_{\Sigma^*}$. By Lemma 14, $\mathfrak{F}^{\mathsf{wS5}}_{\Sigma^*} \models \mathsf{wS5}$. □

It follows from the FMP of wS4, wS4C and wS5 that these ntqBls are decidable. A ntqBl L is *finitely approximable* if for any consequence $s \notin L$, there exists a finite involutive frame $\mathfrak{F} \models L$ such that $\mathfrak{F} \not\models s$. One could prove that the FMP and the finite approximability of ntqBls coincide. Thus many results on the FMP of ntqBls could be derived.

6 Concluding Remarks

This paper contributes a relational semantics for normal topological quasi-Boolean logics by using involutive frames. A discrete duality for these algebras and frames is developed. The Kripke-completeness and finite model property are obtained for some ntqBls. There are certainly many problems left open for these modal logics. Duality theory, completeness theory and correspondence theory are interesting for exploration in future work. Moreover, Blok's work on interior algebras (cf. [4]) can be considered for normal topological quasi-Boolean algebras.

References

1. Banerjee, M., Chakraborty, M.K.: Rough sets through algebraic logic. Fundamenta Informatica **28**(3–4), 211–221 (1996)
2. Belnap, N.: A useful four-valued logic. In: Dunn, J.M., Epstein, G. (eds.) Modern Uses of Multiple-Valued Logic, pp. 5–37. D. Reidel Publishing Company, Dordrecht (1977)
3. Blackburn, P., de Rijke, M., Venema, Y.: Modal Logic. Cambridge University Press, Cambridge (2001)
4. Blok, W.J.: Varieties of interior algebras. Ph.D. dissertation. University of Amsterdam (1976)
5. Białynicki-Birula, A., Rasiowa, H.: On the representation of quasi-Boolean algebras. Bulletin de l'Académie Polonaise des Sciences, Classe III, Bd. **5**, 259–261 (1957)
6. Celani, S.: Classical modal De Morgan algebras. Stud. Logica **98**, 251–266 (2011)
7. Celani, S.A., Jansana, R.: Priestley duality, a Sahlqvist theorem and a Goldblatt-Thomason theorem for positive modal logic. Log. J. IGPL **7**(6), 683–715 (1999)
8. Celani, S.A.: Notes on the representation of distributive modal algebras. Miskolc Math. Notes **9**(2), 81–89 (2008)
9. Chagrov, A., Zakharyaschev, M.: Modal Logic. Clarendon Press, Oxford (1997)
10. Davey, B.A., Priestley, H.A.: Introduction to Lattices and Order, 2nd edn. Cambridge University Press, Cambridge (2002)
11. Dunn, J.M.: The algebra of intensional logics. Ph.D. thesis. University of Pittsburg (1966)
12. Dunn, M.: Positive modal logic. Stud. Logica **55**, 301–317 (1995)
13. Font, J.M., Rius, M.: An abstract algebraic logic approach to tetravalent modal logics. J. Symbolic Log. **65**, 481–518 (2000)
14. Font, J.M.: Belnap's four-valued logic and De Morgan lattices. Log. J. IGPL **5**(3), 413–440 (1997)
15. Lin, Y., Ma, M.: Belnap-Dunn modal logic with value operators. Stud. Logica **109**, 759–789 (2021)
16. Lin, Z., Ma, M.: Residuated algebraic structures in the vicinity of pre-rough algebras and decidability. Fundamenta Informatica **179**, 239–174 (2021)
17. Loureiro, I.: Prime spectrum of a tetravalent modal algebra. Notre Dame J. Formal Log. **24**, 389–394 (1983)
18. Odintsov, S.P., Speranski, S.O.: Belnap-Dunn modal logics: truth constants vs. truth values. Rev. Symbolic Log. **13**(2), 416–435 (2020)

19. Odintsov, S.P., Wansing, H.: Modal logics with Belnapian truth values. J. Appl. Non-Classical Log. **20**(3), 279–301 (2010)
20. Rasiowa, H.: An Algebraic Approach to Non-Classical Logics. North-Holland Publishing Company, Amsterdam (1974)
21. Saha, A., Sen, J., Chakraborty, M.K.: Algebraic structures in the vicinity of pre-rough algebra and their logics. Inf. Sci. **282**, 296–320 (2014)

Author Index

M. Banerjee and A. V. Sreejith (Eds.): ICLA 2023, LNCS 13963, p. 223, 2023.
https://doi.org/10.1007/978-3-031-26689-8

Printed in the United States
by Baker & Taylor Publisher Services